T0075331

MATLAB® With Applications in Mechanics and Tribology

Leonid Burstein
Independent Researcher, Israel

A volume in the Advances in Systems Analysis, Software Engineering, and High Performance Computing (ASASEHPC) Book Series

Published in the United States of America by
IGI Global
Engineering Science Reference (an imprint of IGI Global)
701 E. Chocolate Avenue
Hershey PA, USA 17033
Tel: 717-533-8845
Fax: 717-533-8661
E-mail: cust@igi-global.com
Web site: http://www.igi-global.com

Library of Congress Cataloging-in-Publication Data

Names: Burstein, Leonid, author.
Title: MATLAB with applications in mechanics and tribology / by Leonid Burstein.
Description: Hershey, PA : Engineering Science Reference, an imprint of IGI Global, [2021] | Includes bibliographical references and index. | Summary: "This book provides basics of programming tools by examples taken from such widespread areas as mechanics and tribology (surfaces, friction, lubrication, and wear)"-- Provided by publisher.
Identifiers: LCCN 2020042248 (print) | LCCN 2020042249 (ebook) | ISBN 9781799870784 (hardcover) | ISBN 9781799870791 (paperback) | ISBN 9781799870807 (ebook)
Subjects: LCSH: MATLAB. | Mechanics--Mathematics. | Tribology--Mathematics.
Classification: LCC TA345.5.M42 B87 2021 (print) | LCC TA345.5.M42 (ebook) | DDC 621.0285/53--dc23
LC record available at https://lccn.loc.gov/2020042248
LC ebook record available at https://lccn.loc.gov/2020042249

This book is published in the IGI Global book series Advances in Systems Analysis, Software Engineering, and High Performance Computing (ASASEHPC) (ISSN: 2327-3453; eISSN: 2327-3461)

British Cataloguing in Publication Data
A Cataloguing in Publication record for this book is available from the British Library.

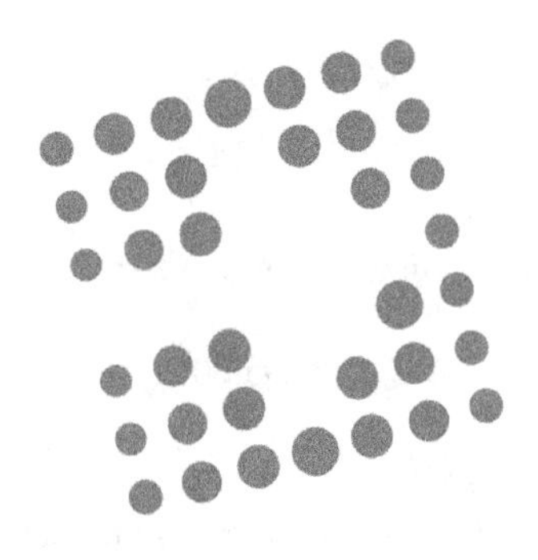

Advances in Systems Analysis, Software Engineering, and High Performance Computing (ASASEHPC) Book Series

Vijayan Sugumaran
Oakland University, USA

ISSN:2327-3453
EISSN:2327-3461

Mission

The theory and practice of computing applications and distributed systems has emerged as one of the key areas of research driving innovations in business, engineering, and science. The fields of software engineering, systems analysis, and high performance computing offer a wide range of applications and solutions in solving computational problems for any modern organization.

The **Advances in Systems Analysis, Software Engineering, and High Performance Computing (ASASEHPC) Book Series** brings together research in the areas of distributed computing, systems and software engineering, high performance computing, and service science. This collection of publications is useful for academics, researchers, and practitioners seeking the latest practices and knowledge in this field.

Coverage

- Software Engineering
- Enterprise Information Systems
- Storage Systems
- Network Management
- Parallel Architectures
- Engineering Environments
- Virtual Data Systems
- Human-Computer Interaction
- Computer Graphics
- Distributed Cloud Computing

IGI Global is currently accepting manuscripts for publication within this series. To submit a proposal for a volume in this series, please contact our Acquisition Editors at acquisitions@igi-global.com or visit: https://www.igi-global.com/publish/.

The Advances in Systems Analysis, Software Engineering, and High Performance Computing (ASASEHPC) Book Series (ISSN 2327-3453) is published by IGI Global, 701 E. Chocolate Avenue, Hershey, PA 17033-1240, USA, www.igi-global.com. This series is composed of titles available for purchase individually; each title is edited to be contextually exclusive from any other title within the series. For pricing and ordering information please visit https://www.igi-global.com/book-series/advances-systems-analysis-software-engineering/73689. Postmaster: Send all address changes to above address.

Titles in this Series

For a list of additional titles in this series, please visit: www.igi-global.com/book-series

Handbook of Research on Software Quality Innovation in Interactive Systems
Francisco Vicente Cipolla-Ficarra (Latin Association of Human-Computer Interaction, Spain & International Association of Interactive Communication, Italy)
Engineering Science Reference • ©2021 • 501pp • H/C (ISBN: 9781799870104) • US $295.00

Handbook of Research on Methodologies and Applications of Supercomputing
Veljko Milutinović (Indiana University, Bloomington, USA) and Miloš Kotlar (University of Belgrade, Serbia)
Engineering Science Reference • ©2021 • 393pp • H/C (ISBN: 9781799871569) • US $345.00

Advancements in Fuzzy Reliability Theory
Akshay Kumar (Graphic Era Hill University, India) Mangey Ram (Department of Mathematics, Graphic Era (Deemed to be University), India) and Om Prakash Yadav (North Dakota State University, USA)
Engineering Science Reference • ©2021 • 322pp • H/C (ISBN: 9781799875642) • US $245.00

Impacts and Challenges of Cloud Business Intelligence
Shadi Aljawarneh (Jordan University of Science and Technology, Jordan) and Manisha Malhotra (Chandigarh University, India)
Business Science Reference • ©2021 • 263pp • H/C (ISBN: 9781799850403) • US $195.00

Handbook of Research on Modeling, Analysis, and Control of Complex Systems
Ahmad Taher Azar (Faculty of Computers and Artificial Intelligence, Benha University, Benha, Egypt & College of Computer and Information Sciences, Prince Sultan University, Riyadh, Saudi Arabia) and Nashwa Ahmad Kamal (Faculty of Engineering, Cairo University, Giza, Egypt)
Engineering Science Reference • ©2021 • 685pp • H/C (ISBN: 9781799857884) • US $295.00

Artificial Intelligence Paradigms for Smart Cyber-Physical Systems
Ashish Kumar Luhach (The PNG University of Technology, Papua New Guinea) and Atilla Elçi (Hasan Kalyoncu University, Turkey)
Engineering Science Reference • ©2021 • 392pp • H/C (ISBN: 9781799851011) • US $225.00

Advancements in Model-Driven Architecture in Software Engineering
Yassine Rhazali (Moulay Ismail University of Meknes, Morocco)
Engineering Science Reference • ©2021 • 287pp • H/C (ISBN: 9781799836612) • US $215.00

701 East Chocolate Avenue, Hershey, PA 17033, USA
Tel: 717-533-8845 x100 • Fax: 717-533-8661
E-Mail: cust@igi-global.com • www.igi-global.com

To everyone I love and
those who love me.
In good memory.

Table of Contents

Preface

Mechanics, historically, was the first discipline of the exact sciences. Respectively, mechanical engineering is as an oldest practical discipline applying the scientific knowledge and mathematical support to the design, development, manufacturing, and maintenance of the mechanical parts and systems. Many other classical and modern scientific disciplines were worked out inside mechanics and became independent sciences. One of these is tribology - a science and technology, which deals with the friction, lubrication, and wear of the rubbing surfaces and mechanical parts. Naturally, in mechanics and in tribology (M&T), like other technical and scientific fields, computer calculations, simulation and modeling are widely used. Scientists, engineers, and students in mechanics and tribology (M&T) use for this some special programs. Among them, MATLAB® is very practical and popular software used in technical computing because of its efficiency and simplicity.

This book presents a primer on technical programming in MATLAB® with examples from the field of mechanics and tribology sciences and addresses a wide M&T audience – scientists and practicing engineers as well as undergraduate and graduate students. First of all, the material of the book is intended for young scientists and college students performing M&T calculations, nevertheless the material is also aimed at non-programmer M&T specialists looking for efficient software and desired to use MATLAB® as the software becomes more popular. It presents MATLAB® fundamentals and provides a variety of application examples and problems taken from statics, kinematics, dynamics, thermodynamics, fluid and gas dynamics, as well as M&T phenomenon simulations in order to facilitate studying of the software language. I hope that many non-programmer engineers, scientists, and students from the M&T fields will find the software a comfortable tool and use it for solving their specific problems.

The book accumulates many years of experience in teaching introductory and advanced courses in the field of MATLAB® and its applications in mechanics, material properties, and tribology for students, engineers and scientists specializing in these areas.

I am grateful to The MathWorks Inc. which kindly granted permission to use certain materials and I also thank Elsevier Inc. for permission to reuse some portions of text in tables, figures, or screenshots from my previous book “ A MATLAB® Primer for Technical Programming in Material Science and Engineering” (Woodhead Publishing Series in Technology and Engineering). I would like to appreciate Stephen Rifkind, an ITA Recognized Translator, who edited the book chapters.

I hope the primer will prove useful to engineers, scientists, and students in M&T areas and enable them to work with the available fine software.

The author will gratefully accept any reports of bugs or errata, as well as comments and suggestions for improving the book.

Leonid Burstein
Independent Researcher, Israel
December 2019 - May 2020

Chapter 1
Introduction

ABSTRACT

The chapter serves as the introduction to a MATLAB® primer oriented for scientists, specialists, and students of mechanical and tribological (M&T) sciences. The objectives of the book and the principal audience are outlined and accompanied by a brief description of the topics and structure of the book chapters and its appendices. The history of MATLAB® is briefly described together with the advantages of the software. In addition, the order of studying the material of the book is discussed.

Mechanics was the first discipline of the exact sciences. Thus, mechanical engineering is the oldest practical discipline applying knowledge of mechanics and mathematical support to the design, development, manufacturing, and maintenance of the mechanical parts and systems. Many other classical and modern scientific and applied disciplines were developed inside mechanics and became independent sciences, including celestial mechanics, thermodynamics, fluid and gas mechanics, materials science, and construction mechanics. One of the relatively new areas arising from mechanics is tribology, which is a science and technology dealing with the friction, lubrication, and wear of rubbing surfaces and mechanical parts. Naturally, in mechanics and in tribology (M&T), like in other technical and scientific fields, computer calculations, simulation, and modeling are widely used. For these purposes, a wide variety of software is available. Among them, MATLAB® software is one of the most common and popular programming tools used for various scientific and technical applications. Modern mechanical and tribology science specialists work widely with computers and some special programs, but require a universal tool for solving, simulating, and modeling specific problems. Numerous books on MATLAB® have been written for engineers in general or for those specializing in some narrowly-specific area, but none have been designed contemporaneously, especially for tribology and mechanics, two particularly broad, interdisciplinary topical areas. The present edition aims to fill this gap.

DOI: 10.4018/978-1-7998-7078-4.ch001

1.1. HISTORY AND BENEFITS OF THE SOFTWARE

The foundations and language of MATLAB® were established in the 1970s by mathematician Cleve Moler. Later, the language was rewritten in C and improved by specialists that had joined the founder. Initially, the language was intended for students to adapt LINPACK and EISPACK mathematical packages but then was applied to control engineering. In a fairly short period, researchers and engineers recognized MATLAB® as an effective and convenient tool for solving not only mathematical, but also many technological problems. Since the mid-80s, commercial versions of MATLAB® have appeared on the common software market. Following this, graphics and special engineering-oriented means were incorporated into MATLAB®, giving it its modern form. Today's MATLAB® is a unique assembly of implemented classical and modern numerical methods and specialized toolboxes for engineering calculations, designed for specialists in various fields. MATLAB® competes with other programming tools and occupies a valued place among them as the software for technical computing. Without going deep into detail, the following factors and their interactions explain a sustainable preference for MATLAB®:

- Substantial universality and the ability to solve both simple and complex problems using user-friendly programming tools;
- Excellent applicability in various fields of technology and science provided by a significant number of problem-oriented tools - toolboxes;
- Diversity and a convenience means of visualization for solving general and specific scientific and engineering problems, e. g. M&T problems;
- Accessibility and promptness in obtaining help using the well-organized, extensive documentation.

To these characteristics, we should add the latest innovation - Live Editor, which presents the ability to display images and tables in one window with codes immediately visible after entering them.

1.2. THE GOALS OF THE BOOK AND ITS AUDIENCE

The aim of this book is to provide researchers, engineers, teachers, students, and M&T professionals with a manual that will teach them MATLAB® and demonstrate the content by developing programs that solve their professional problems and present the results in descriptive, graphic, and tabular forms.

It is assumed that the reader has no programming experience and will be using the software for the first time. In order to make the primary programming steps and use of commands clear to the target audience, they are demonstrated by problems taken from different areas of materials science whenever possible. It is expected that the reader has no programming experience and will use the software for the first time. To make the basic steps of programming and use of commands understandable for the target audience, they are demonstrated by problems, when possible, from different areas of tribology and mechanics. As basic programming knowledge increases, M&T problems are solved with special tools, including the Partial Differential Equation Toolbox™, Symbolic Math Toolbox™, and Statistics and Machine Learning Toolbox™. MATLAB® is positioned as a tool for technical analysis and calculations that distinguish it from the many other available programs. Importantly, it is updated and improved in parallel with the development of modern technologies. The technical analysis and calculations conducted with MATLAB® have led it to be recognized as a convenient and effective tool for modern science and technology. Thus, mastering its latest versions and practical solutions with their help is increasingly essential for the creation of new products in mechanics, electronics, chemistry, life sciences, and modern industry as a whole. Modern mechanical and tribology sciences specialists widely used computers and some special programs, but needs some universal tool for solving, simulating and modeling specific problems from their area.

The existing books on the various aspects of MATLAB® can be divided contingently into two categories: (a) books on MATLAB® programming itself and (b) books on advanced technology, science, or mathematics with introductory section/s on MATLAB. The first category assumes that the reader is already familiar with math methods and focuses on advanced programming technique. The second category is generally devoted to special engineering or scientific subjects that require some advanced level and do not actually teach the programming language itself. This book is different in that it is adopted for the reader with a modest mathematical background and introduces the programming or technical concepts using some simplifications in the traditional approach. MATLAB® is then used as a tool for subsequent computer solutions, applying it to solve problems and case studies in M&T. Another distinction of the book is its relatively moderate size combined with a variety of examples from broad range of modern and classical mechanics and tribology sciences, what help solidify the understanding of the presented material and show M&T specialists the possibility of using software in their specific fields. Thus, the book can serve as a guide for two categories of users:

- Researchers just wanting to learn and apply MATLAB® in their industry, not necessarily in M&T;
- M&T engineers and scientists that would like to see how to use MATLAB® in their specific areas.

Considering the above, the principal audiences of the book include:

- Scientists, engineers, and specialists that seek to solve M&T problems and search for similar problems solvable by computer;
- Non-programmer professionals and the academic community dealing with modeling and simulation machinery and processes in areas of technique and technology from mechanics or electricity to the live sciences;
- Students, engineers, managers, and teachers from the academic and scientific communities in the field of M&T;
- Instructors and their audiences in M&T study courses where MATLAB® is used as a supplemental but necessary tool;
- Staff, students and non-programmers as well as self-taught readers for quick mastering of the programs for their needs.
- Freshmen and participants in advanced M&T courses, seminars, or workshops where MATLAB® is taught.

Both M&T and non-M&T specialists using a computer for modeling calculations and solving actual engineering problems can also use the book as a reference.

1.3. ABOUT TOPICS IN BOOK CHAPTERS

The topics were selected based on approximately 25 years of research and 18 years of teaching experience in the fields of MATLAB®, tribology, mechanics, and properties of materials and substances. They were presented so that the beginner can progress gradually using only previously acquired material as a prerequisite for each new chapter.

The most important, basic MATLAB® features are introduced in the second chapter. At the beginning of the chapter, the desktop of the software, its toolbars, and main windows are described. Here the elementary functions, input and output commands, numbers and strings, vectors, matrices/arrays, their manipulations, and flow control commands as well as relational and logical operators are discussed. Each command is presented in its most applicable form with practical examples. The commands are applied to elementary problems in the field of mechanics and tribology (M&T), in particular to problems involving the stress intensity factor, stiffness of a threaded bolt and adhesive force in contact between two spheres. Commands of this chapter enables the reader to write, execute, and display the simple calculations directly in the Command Window.

The third chapter introduces the user-defined function and presents the Editor window for writing program scripts and user-defined functions as well as the Live Editor window for writing live scripts and functions. All commands, regular and live scripts, and functions are explained by examples from M&T fields. The application examples of the chapter include stress unit convertors, computing of the stress factor of a shaft with a transverse hole, gear warm K-parameter calculations, installation, and operation stresses on the piston ring.

The visualization means for generating two- and three-dimensional graphs are described in the fourth chapter. It describes formatting commands for inserting labels, headings, texts, and symbols into a plot, as well as color, marker, and line qualifiers. Graphs with more than one curve and graphs with two Y axes are discussed. The possibilities of creating multiple plots on one page are shown. At the end of the chapter, M&T applications are provided, illustrating how to generate 2D and 3D graphs for engine piston velocity, power screw efficiency and engine oil viscosity, among other M&T problems. Understanding of the material of the second, third, and fourth chapters allows the reader to create rather complex programs for technical calculations and their graphical representation.

The fifth chapter presents more advanced topics and includes the numerical methods for solving problems arising in science and engineering in general and M&T fields in particular. The topics of this chapter include interpolation and extrapolation, solving nonlinear equations with one or more unknowns, finding minimum and maximum, integration, and differentiation. The applications presented here illustrate how to interpolate the friction coefficient data, calculate elongation of a scale with two springs, determine the maxima and minima of the pressure-angle function, and solve other M&T problems.

The three next chapters, chapters 6-8, are devoted to numerical solutions of ordinary and partial differential equations (ODE and PDE). The sixth chapter introduces solvers for resolving initial and boundary value problems (IVP and BVP) of the ODE. It begins with a description of the ODE-solver commands applied to the IVP problem and then presents the commands to the BVP problem. The discussed commands are applied to M&T problems such as the spring-mass system and particle falling, single clamped beam, and hydrodynamic lubrication of a sliding surface covered with semicircular pores. The seventh chapter describes the pdepe command, which is used to solve transient and spatially one-dimensional partial differential equations (PDEs). It is shown how various PDEs with different boundary conditions can be represented in standard forms. Applications illustrate how to solve heat transfer PDE with temperature dependent material properties, startup velocities of the fluid flow in a pipe, Burger's PDE, and coupled FitzHugh-Nagumo PDEs. The eighth chapter describes the PDE Modeler tool (provided by the Partial Differential Equation Toolbox), which is used to solve transient and spatially two-dimensional partial differential equations. It describes the standard forms of PDEs and the initial and boundary conditions used by the Modeler. Applications demonstrate the use of the PDE Modeler to solve the Reynolds equation of the hydrodynamic lubrication, implement the Mechanical Stress option of the Modeler to solve PDE for a plate with an elliptical hole, resolve the transient heat equation with temperature-dependent material properties, and study vibration of a rectangular membrane.

The ninth chapter is devoted to matching M&T-field data with polynomial and non-polynomial expressions. The Basic Fitting tool and examples of its use are described. Single and multivariate fitting through optimization is discussed. Application examples demonstrate the curve fitting for the following data: fuel efficiency-velocity, yield strength-grain diameter, and friction coefficient - time as well as data for machine diagnostics.

The tenth chapter represents an independent class of calculations - symbolic calculations (provided by the Symbolic Math Toolbox™), in which the variables and commands operate on mathematical expressions containing symbolic variables. In the chapter are presented examples of the symbolic calculations for some M&T problems that were solved numerically in previous chapters. They include lengthening a two-spring scale, shear stress in a lubrication film, a centroid of a certain plate, and two-way solutions of the ODE describing the second order dynamical system through the traditional method and Laplace transform.

The final, eleventh chapter is intended for readers having at least basic skill in statistics and explains the specialized commands of the Statistics and Machine Learning Toolbox™. Descriptive statistics, the Data Statistics tool, specialized statistical graphs, probability distributions, and hypothesis tests are discussed. The solutions of various applied problems are given here. In particular, surface roughness indices are calculated by the measured data using the descriptive statistics command; the histogram generated by a runout data are matched with the theoretical distribution ; capability plot generation is shown through the data for the piston ring gaps; and friction torques for two different oil additives are compared using a hypothesis test.

Each chapter of the book has an appendix that presents the examples, problems, and applications discussed in that chapter. The appendix presents a summary collection of the more than 250 variables, special characters, operators, and commands discussed in book.

The index contains over 700 alphabetically arranged names, terms, and commands that were explained, implemented, or at least mentioned throughout the book.

1.4. CHAPTER STRUCTURE AND EDITORS FOR PROGRAM DEVELOPMENT

The chapter topics are represented gradually to ensure gradual assimilation of the concepts. Each chapter begins with a preface describing the chapter content. The new material, basic forms of the commands, and their applications are then presented. The commands are usually given in one or the two simplest forms with possible useful extensions. Each subject, if possible, is fully considered in one subsection to allow the readers to attain the knowledge in a focused manner. The tables, if available, list additionally available commands, specifiers, modifiers, and equations that correspond to the topic and examples included in the chapter.

At the end or sometimes in the middle of the chapter, application problems associated with the M&T areas are solved with the commands previously presented in the book. The completed solutions are most accessible to understanding but not necessarily the shortest or most original. The forms of the programs, especially the result representations, vary to show the different possible solutions and thereby provide the skills to manage the solution and visualize the result. Readers are invited to run not only the provided solutions, but also try their own solutions and compare the results with those in the book.

Note that the numerical values and contexts used in the various M&T problems are not factual reference data and serve for demonstration purposes only.

To create program codes and save them in files, MATLAB provides two editors - the old, commonly used *Editor* and the new, less known *Live Editor*, which shows the results along with the written codes. Both editors are described in the book and used interchangeably in order to allow the reader to master writing and running programs with each of the editors and choose the most suitable one in the future.

1.5. MATLAB® AND TOOLBOXES USED IN THE BOOK

Two new versions of MATLAB® appear annually in six-month intervals. Each new version is updated and expanded but also provides the code work developed in the previous versions. Each toolbar also has its own version but the new version supports the previous one. Therefore, the basic commands described in this book will work in any new versions. The version used in this book is R2019b (9.7.0.1190202). For solving partial differential equations, version 3.3 (R2019b) of the Partial Differential Equation Toolbox™ is used. For symbolic calculations, version 3.8 (R2019b) of the Symbolic Math Toolbox™ has been adopted. For statistical calculations, version 11.6 (R2019b) of the Statistics and Machine Learning Toolbox™ is used.

Readers are expected to install MATLAB® along with the above-mentioned toolboxes on their computers to be able to perform all the basic operations presented in the book.

1.6. ORDER OF READING

The book is oriented to a newcomer in MATLAB® and computer calculations. Therefore, the topics are arranged sequentially in a convenient form. This and the four next chapters present the basics of MATLAB. After mastering this basis, M&T specialists can go to any subsequent chapter according to their interests - ODE, PDE, fitting, symbolic, or statistical calculations. Readers are not obligated by the chapter order. For example, the Editor (Subsection 3.1.) and Live Editor (Subsection 3.3) can be learned directly after the output commands (Subsection 2.1.7) to permit the reader to create a simple program for the early studied stages. Those for whom the regular mathematical manipulations are important can study symbolic calculations (Chapter 10) immediately after the five first chapters. Also, the graph generation (Chapter 5) can be studied directly after Chapter 2 to visualize results from the first steps. Likewise, the material regarding creation of script files (Subsection 3.2.) can be read after the input and output commands (Subsection 2.2.). Another way to master and apply programming to solve the M&T problem is to use the aforementioned Life Editor, which can be studied immediately after being familiar with variables and interactive calculations (prior to Subsection 2.1.5). This editor can be used throughout the study with examples of scripts / functions written in regular Editor being easily converted to a live file (explained in Subsection 4.4.3).

As a whole, I hope that the book will help users learn this software and use it effectively to solve M&T problems.

1.7. CONCLUSION

Summarizing discussed above, here we present the MATLAB manual oriented to researchers, professionals, students, and non-programmers from the M&T field that contains description of basic and many supplementary commands, as well as generated programs that solve various practical problems in the field of study. The topics and order of the book's materials are aimed at solving the presented goals.

So, let's get started now.

Chapter 2
Basics of Software Used to Solve M&T Problems

ABSTRACT

At the beginning of the chapter, the desktop of the software, its toolbars, and main windows are introduced. Examples of interactive calculations with MATLAB® language are presented. Elementary functions, input and output commands, numbers and strings, vectors, matrices and arrays, flow control commands, relational and logical operators are discussed. Each command is presented in its most applicable form and with practical examples. At the end of each subsection, the commands studied are applied to elementary problems in the field of mechanical and tribological (M&T) sciences and technology, in particular to such as stress intensity factor, stiffness of a threaded bolt, adhesive force in contact between two spheres, and many others.

INTRODUCTION

Fifty years have passed since the creation of a special computer tool and its language called MATLAB®. The term is formed from the first three letters of two words, *matrix* and *laboratory*, and highlights the basis of this language: a matrix. This approach unifies and speeds up computational and graphical processing. Afterwards, and over a fairly short period, MATLAB® was adapted and has been intensively used in scientific and engineering calculations. Today, its various applications are a vital tool of scientific and technical specialists. Over time, MATLAB® has undergone significant changes, both in the interface and the number of commands. Language and tools have become more diverse, efficient and complex. Therefore, in order to effectively solve current technical problems using computational tool, it is necessary to study the most current tool.

This chapter describes the tool desktop, its toolstrips and available windows; introduces a startup procedure; provides commands for simple arithmetic, algebraic, array and matrix operations; and finally performs the main loops and relational and logical operators.

DOI: 10.4018/978-1-7998-7078-4.ch002

2.1. RUNNING THE MATLAB® SOFTWARE

The user must install MATLAB® before working with him. Therefore, it is assumed that this software was previously installed on your computer. MATLAB®, like other computer software, is controlled by a special set of programs called the operating system (OS), which may differ according to the computing platforms. Therefore, we additionally assume here, that MATLAB® is controlled by the OS Windows.

To open MATLAB®, click the icon showing a red L-shaped membrane. This icon should previously be placed in the Windows Start Menu, the Desktop and on Task bar. Another way to start MATLAB® is to type the word MATLAB® in the Windows Search box. You can also start the program by finding and clicking the matlab.exe file in the MATLAB® bin directory. The path to this file might look like this C:\Program Files\MATLAB\R2019b\bin\matlab.exe. After that, a startup picture appears with the MATLAB® logo and some product information, including the initial version, license number and the name of the development company. This image disappears after a short time after which the MATLAB® desktop opens.

2.1.1. Desktop Layout and Its Toolstrip

The desktop includes the toolstrip and three panels, which can be separated from the desktop and appear as windows: Command, Current Folder and Workspace – see Figure 1.

Figure 1. MATLAB® desktop

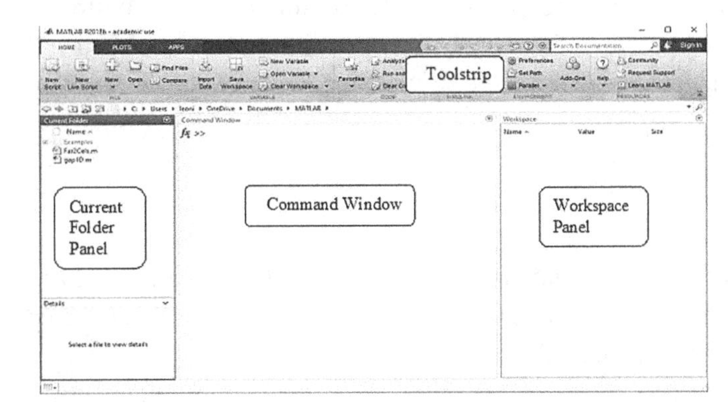

Toolstrip

The top strips of the Desktop, called toolstrip, contains the main target operations and functions of MATLAB®, presented in three global tabs labeled Home, Plots and Apps. Each of tabs is divided into sections with a number of interconnected controls (file, variable, code, etc.) that include buttons, drop-down menus and other interface elements (Figure 2). The sections contain buttons for performing various operations (such as "Open", "New Variable" and"Settings"). The Home is the tab that is used most intensively and includes general-purpose operations such as creating new files, importing data and managing the workspace, as well as desktop layout options. The Plots tab contains buttons for generating the various charts. The Apps tab contains a gallery of applications belonging to the MATLAB® toolboxes.

Figure 2. The toolstrip of the MATLAB® desktop

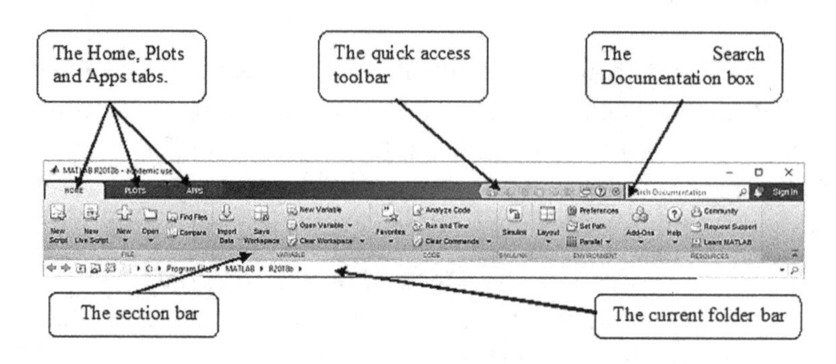

To the right of the top of the strip is a quick access bar. It contains some commonly used options such as Cut, Copy and Paste. Next to the quick access panel is the Search Documentation box allowing to search for documentation regarding commands, windows, applications, and other topics of interest. The current folder toolbar is placed on the bottom of the toolstrip. This bar contains file/directory management buttons and shows the current working directory.

Command Window

This window is the central component of the Desktop. Commands are entered here and the result of their execution is displayed. For convenience, the window can be separated from the desktop by selecting the Undock line of the Show Command Window Actions button. This button is placed to the right of the Command Window title bar. Analogously, it is possible to separate each of the Desktop windows. To assemble separated windows, choose the Dock line or select the Default line in the Layout of the Desktop Home tab.

Workspace Window

In this window, the variable icons and other objects currently located in the MATLAB® workspace appear. It also displays the class of each variable and its values. The latter can be edited here. Data values are automatically updated with calculations.

Current Folder Window

After launching MATLAB®, in this window we can view the start directory, called the startup directory, as well as the files and folders located there. The window has the Details panel located at its bottom where information about the selected file appears. This browser window displays the full path and contents of the current folder. You can change the directory by entering the appropriate pass in the "Current folder" bar or using its folder management buttons.

The above windows are the most commonly used. There are also other windows, namely Command History, Help, Editor, Live Script Editor and Figure, which by default are not displayed when the desktop is opened. They will be described later in the chapters where they are used.

2.1.2. Interactive Calculations and Elementary Math Functions

MATLAB® commands can be entered and executed interactively - written directly in the command window, or programmatically - using a program containing commands previously recorded in the editor and saved in *m*- or *mls*-files. The interactive way is briefly presented here while the programmatic will be explained later.

To realize a command, it must be typed and entered in the Command Window in place where a blinking cursor appears, immediately after the >> sign which is called the "command prompt "or simply "prompt". Figure 3 presents the undocked Command Window with some variables, numbers, and elementary calculating commands entered and executed in the Command Window.

Figure 3. Some commands entered and executed interactively in the command window

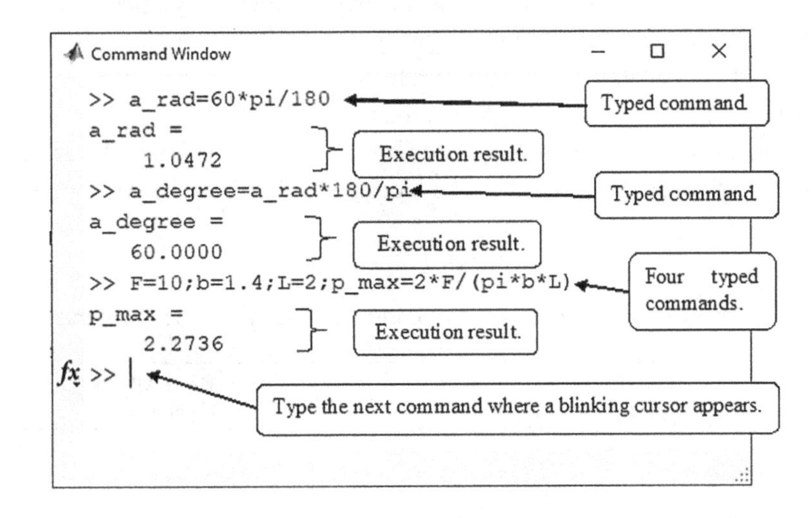

The button *fx*, appearing to the left of the prompt of the current command line is called the Function browser. It helps in finding the necessary commands, their syntax and a description of their use.

Manipulating with commands requires mastery of the following rules:

- The command should be typed next to the last command prompt >>;
- A command located on the preceding line cannot be changed; to correct/repeat a command press the up-arrow key ↑ and then introduce the desirable change in shown command;
- A long command can be continued on the next line after typing an ellipsis ... (three periods); the total length of a single command may not exceed 4096 characters;
- To write two or more commands in the same line, divide them with semicolons (;) or commas (,);
- To execute the command/s after typing it/them on one command line, press the key Enter after the last commands;
- A semicolon entered after the command withholds displaying of the answer;
- The % symbol (percent symbol) designates the comments that must be written after it in the line; the comments are not executed after entering; they appear in green;
- To clear the Command Window, enter the clc command from the command line;

- The result of entering a command is a variable named ans if the result has not been previously named by another name;
- The equal sign (=) is called the assignment operator and is used to set a value for a variable;
- A new entered/calculated value cancels its predecessor in the variable, e.g. if the value of a was assigned to be 5.1916, after entering the new command, >> a=12.7328, the previous value of a is removed and a receives the 12.7328 value.

When calculations are performed, the following symbols are used for arithmetic operations: + (addition), - (subtraction), * (multiplication), / (right division), \ (left division, mostly used for matrices), and ^ (exponentiation).

In arithmetic, algebraic or other mathematical calculations, a wide range of elementary, trigonometric and special mathematical functions are used. The MATLAB® commands in general and, in particular, calculations with these functions must be written as the name with the argument in parentheses, e.g. cos x is written as cos(x). In trigonometric functions, the argument x is given in radians. If the corresponding command is written with the ending d, e.g. cosd(x), the argument x should be given in degrees. Inverse trigonometric functions with the ending d produce the result in degrees.

A short list of commands for calculations with math functions is given in Table 1. Hereinafter, the commands/operators/variables that entered in the Command Window are written after the command line prompt (>>).

Table 1. Some of the elementary, trigonometric and specific mathematical functions.

Description	Math Notation	MATLAB® Command	Example (Input and Output)
modulus (absolute value)	$\|x\|$	abs(x)	>> abs(-12.7328) ans = 12.7328
exponential	e^x	exp(x)	>> exp(1.98) ans = 7.2427
natural (base e) logarithm	$ln\ x$	log(x)	>> log(10.0000) ans = 2.3026
Napierian (base 10) logarithm	$log_{10}\,x$	log10(x)	log10(10.0000) ans = 1
base a logarithm: $\log_a x = \frac{\log_{10} x}{\log_{10} a}$	$log_a\,x$	log10(x)/log10(a)	>> log10(10.0000)/ log10(2.7) ans = 2.3182
square root	$\sqrt{x}$	sqrt(x)	>> sqrt(1.9999/2.9999) ans = 0.8165
the number π (circumference-to-diameter ratio)	π	pi	>> 3/2*pi ans = 4.7124

continued on following page

Table 1. Continued

Description	Math Notation	MATLAB® Command	Example (Input and Output)
factorial; product of the integers from 1 to *n*	$n!$	factorial(n)	>> factorial(4) ans = 24
round towards minus infinity	$\lfloor x \rfloor$	floor(x)	>> floor(-11.21) ans = -12
round towards infinity	$\lceil x \rceil$	ceil(x)	>> ceil(11.21) ans = 12
round toward zero	$fix(x)$	fix(x)	>> fix(-1.89) ans = -1
round to the nearest decimal or integer	$[x]$	round(x,n)	>> round(12.7252,2) ans = 12.7300
sine	$sin\ x$	sin(x)	>> sin(pi/2) ans = 1
hyperbolic sine	$sinh\ x$	sinh(x)	>> sinh(pi/2) ans = 2.3013
sine with x in degrees	$sind\ x$	sind(x)	>> sind(90) ans = 1
cosine	$cos\ x$	cos(x)	>> cos(pi/3) ans = 0.5000
hyperbolic cosine	$cosh\ x$	cosh(x)	>> cosh(pi/3) ans = 1.6003
cosine with x in degrees	$cosd\ x$	cosd(x)	>> cosd(60) ans = 0.5000
tangent	$tan\ x$	tan(x)	>> tan(pi/3) ans = 1.7321
hyperbolic tangent	$tanh\ x$	tanh(x)	>> tanh(pi/3) ans = 0.7807
tangent with x in degrees	$tand\ x$	tand(x)	>> tand(60) ans = 1.7321
secant	$sec\ x$	sec(x)	>> sec(pi/3) ans = 2.0000
Secant with x in degrees	$secd\ x$	secd(x)	>> secd(60) ans = 2.0000

continued on following page

Table 1. Continued

Description	Math Notation	MATLAB® Command	Example (Input and Output)
cotangent	*cot x*	cot(x)	>> cot(pi/3) ans = 0.5774
hyperbolic tangent	*coth x*	coth(X)	>> coth(pi/3) ans = 1.2809
cotangent with *x* in degrees	*cotd x*	cotd(x)	>> cotd(60) ans = 0.5774
inverse sine	*arcsin x*	asin(x)	>> asin(1.0000) ans = 1.5708
inverse sine with *x* between -1 and 1; result in degrees between -90° and 90°	*arcsind x*	asind(x)	>> asind(1.0000) ans = 90
inverse cosine	*arccos x*	acos(x)	>> acos(1.0000) ans = 0
inverse cosine with *x* between -1 and 1; result in degrees between 0° and 180°	*arccosd x*	acosd(x)	>> acosd(1.0000) ans = 0
inverse tangent	*arctan x*	atan(x)	>> atan(1) ans = 0.7854
inverse tangent; result in degrees between -90° and 90° (asymptotically)	*arctand x*	atand(x)	>> atand(1) ans = 45
inverse cotangent	*arccot x*	acot(x)	>> acot(1.0000) ans = 0.7854
inverse cotangent; result in degrees between -90° and 90°	*arccotd x*	acotd(x)	>> acotd(1.0000) ans = 45
inverse secant	*asec x*	asec(x)	>> asec(2) ans = 1.0472
invers secant; results in degrees between 0° and 180°	*asecd x*	asecd(x)	>> >> asecd(2) ans = 60.0000
Bessel function of first kind	$Jv\left(z\right)=\left(\frac{z}{2}\right)^{v}\sum_{(k=0)}^{\infty}\frac{\left(\frac{z^2}{4}\right)^{k}}{k!\Gamma\left(v+k+1\right)}$	besselj(nu,z) nu – real constant	>> besselj(1,19) ans = -0.1057
gamma function	$\Gamma(x)=\int_0^{\infty}e^{-t}t^{x-1}dt$	gamma(x)	>> gamma(6) ans = 120
error function	$erf\ x=\frac{2}{\sqrt{\pi}}\int_0^{x}e^{-t^2}dt$	erf(x)	>> erf(2.9/sqrt(1.98)) ans = 0.9964

When the calculations are executed, operations are performed in the following order:

- operations in parentheses (starting with the innermost);
- Exponentiation;
- Multiplication and division;
- Addition and subtraction.

If the calculated expression contains operations of the same priority, they run sequentially from left to right.

Examples of arithmetic expressions written in the Command Window with explanations about the order of operations are provided below. Each command line is provided with explanations written as MATLAB® comments, which can be omitted when typing the operations in the Command Window (explanatory comments are used from hereinafter in this chapter).

```
>>6.1+3/5/2   %first 3/5 is executed, the result is divided by 2, and then 6.1+
ans =
  6.4000
>>2/5*4                                      %first 2/5 is executed and then *4
ans =
  1.6000
>>2/(5*4)                                    %first 5*4 is executed and then *2/
ans =
  0.1000
>>ans =
  4.2883
>>2.98^4/3                                   %first 2.98^4 is executed and then /3
ans =
  26.2872
>>2.98^(4/3)                                 %first 4/3 is executed and then 2.98^
ans =
  4.2883
>>2.98^4/3,2.98^(4/3)        %two above expressions are written in the same line
ans =
  26.2872
ans =
  4.2883
>>(7+2)/5                                    %first 7+2 is executed and then /5
ans =
  1.8000
>>8+3\6                     %left division: divide 6 by 3 (not 3 by 6) and 8+ next
ans =
  10
>>3.3318*10^23  %result is displayed in scientific notations (Subsection 2.1.5)
ans =
```

```
  3.3318e+23
>>2^.1-1.72^(1/4)+log(15*.1005)/asin(pi/8)...      %write ellipsis, then Enter,
-sqrt(8.3)                              % and continue the expression on the next line
ans =
-1.9373
```

The calculated numbers are shown on the display in the default format – a fixed point followed by four decimal digits, with the last decimal digits rounded (information on the output formats is provided in Subsection 2.1.5).

2.1.3. Online Help and Toolboxes

Information about the commands, examples of their use, descriptions of the toolbox capabilities, etc. can be obtained using the commands or the Help window. To access the command functionality description, type and enter help and the command name after a space. For example, entering

```
>> help format
```

provides explanations about available output formats and gives examples of its usage strictly in the Command Window.

When a command is searched concerning a topic of interest, the lookfor command may be used. For example, to obtain list MATLAB® command/s on the subject of the pressure you can type and enter

```
>>lookfor pressure
```

after a rather long search, a list of commands appears on the screen as shown below (incomplete)

```
atmospalt     -Calculate pressure altitude based on ambient pressure.
convpres      -Convert from pressure units to desired pressure units.
dpressure     -Compute dynamic pressure using velocity and density.
rrdelta       -Compute relative pressure ratio.
...
```

For further information, click on the selected command in the list (for example on convpres) or use the help command (e.g. >>help convpres). To interrupt the search process, press the two abort keys together - ctrl and c . In this way, you can interrupt any other running process, such as a program or a single command.

Note:

- On different computers, the lookfor command may produce a different list of information as this is determined by the toolbox set installed with MATLAB®. For example, for the requested convpres, Aerospace toolbox ™ must be installed on your computer. For detailed information about the defined command, the doc command can be used:

```
>>doc convpress
```

In this case, the Help window will open (Figure 4).

Figure 4. Help window containing the contents and results pains with information about the convpres *command, which converts pressure units*

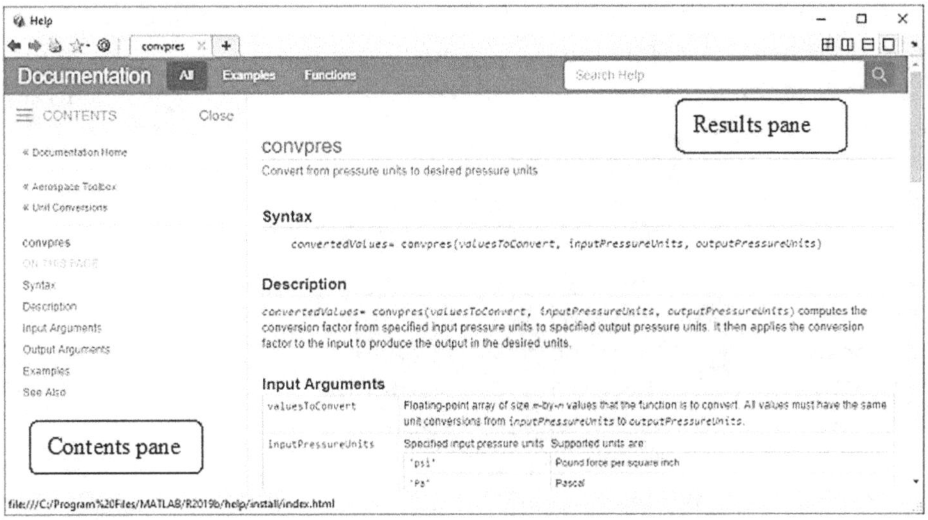

The Help window, opened with above command, comprises the menu strip, the Search Help field of the lower menu line, the narrow Contents pane (to the left) with Contents button and the broad pane (to the right) with documentation chapters containing the searched topic as well as the search results with the specific information on the subject. Information of interest can also be obtained by typing the word/s into the Search Help field at the top of the window. In this case, the Contents pane represents the information by product, category or type, while the Results pane presents list of available informational units with a preview of each of defined information, including the toolbox name for which the information is relevant. For example, the Documentation > Aerospace Toolbox > Unit Conversions line designates the location of the page with explanations on the requested convpres command.

The Help window can also be opened without the doc command in various ways, for example, by selecting the Documentation line Documentation in the popup menu of the Help button on the toolstrip located in the Resources group of the desktop Home tab.

As mentioned, MATLAB® help provides information pertained to the toolboxes,which may include mathematical functions/operations (such as **sin**, **cos**, **sqrt**, **exp**, **log**, etc.) used in a wide range of sciences and specific mathematical functions/actions used in the narrow scientific/technical areas that require special commands to solve their problems. For these purposes, basic and several problem-oriented tools have been developed – the toolboxes. For example, while most of the commands discussed earlier in the chapter are collected in the MATLAB® toolbox, commands for aerospace problems are in the Aerospace toolbox; commands related to statistics in the Statistics toolbox; and commands for neural networks in the Neural Network toolbox, to name just a few.

To verify which toolboxes are available on your computer, the ver command is used. After entering this command, the header displays with the product information and lists the toolbox names, versions and releases as follows:

```
>> ver
-----------------------------------------------------------------
MATLAB Version: 9.7.0.1190202 (R2019b)
MATLAB License Number: XXXXXXXX
Operating System: unknown Version 10.0 (Build 18362)
Java Version: Java 1.8.0_202-b08 with Oracle Corporation Java HotSpot(TM) 64-
Bit Server VM mixed mode
-----------------------------------------------------------------
MATLAB                     Version 9.7    (R2019b)
Simulink                   Version 10.0   (R2019b)
5G Toolbox                 Version 1.2    (R2019b)
AUTOSAR Blockset           Version 2.1    (R2019b)
Aerospace Blockset         Version 4.2    (R2019b)
Aerospace Toolbox          Version 3.2    (R2019b)
Antenna Toolbox            Version 4.1    (R2019b)
Audio Toolbox              Version 2.1    (R2019b)
Automated Driving Toolbox  Version 3.0    (R2019b) ...
```

Here is a shortened list of toolboxes. In reality. it can be quite long, depending on the installed toolboxes. The information about available toolboxes can also be obtained from by selecting the Manage Add-Ons Manage Add-Ons line in the Add-Ons button Add-Ons in the Environment section of the MATLAB® Home tab.

2.1.4. Variables and Commands for Managing Them

In computer programming, a variable is a symbol or several symbols, namely letter/s and/or number/s, to which some value is assigned. A variable can be assigned a single number (scalar) or a table of numbers (array). To store both variable names and their values, MATLAB® allocates space in computer memory.

The variable name consists of letters, digits and underscores and can contain up to 63 characters, the first character being a letter. Available command names (e.g. log, sin, cos, sqrt, etc.) are not recommended for use as variable names because they would confuse the system. The assignment of numbers to variables and their use in algebraic calculations is demonstrated in the following screenshot:

```
>>a1=3.149                       %the value 3.149 is assigned to variable a1
a1 =
  3.1490
>>a2=2.135                       %the value 2.135 is assigned to variable a2
a2 =
  2.1350
>>c=sqrt(a1^2+a2^2)              %calculated value is assigned to variable c
```

```
c =
  3.8045
>> d=log(c)                              %calculated value is assigned to variable d
d =
  1.3362
```

Certain variables with assigned values are permanently stored in MATLAB®. Such variables are termed *predefined*. Apart from the previously mentioned pi and ans, other predefined variables include inf (infinity; is produced, for example, by dividing by zero), i or j (square root of -1), NaN (not-a-number, appears when a numerical value is moot, e.g., 0/0) and eps (smallest allowed difference between two numbers, its value being 2.2204e-16).

The following several commands are used to manage variables: clear – to remove all variables from memory, or clear var1 var2 … – to remove the named variables; who – for displaying the names of variables and whos – for displaying variable names, matrix sizes, variable byte sizes and variable storage classes (numeric variables are double precision by default). In the Workspace Window, in addition, each variable with the same information as in the case of whos, and their possible additional parameters are presented with each variables marked by the icon ⊞ (this icon varies for different variable classes). To obtain additional information about a variable in the Workspace window, click the mouse's right button with the cursor placed on the variable header line. The pop-up menu appears with a list of possible additional information (namely Size, Class, Min, Max, etc.).

2.1.5. Output Formats

MATLAB® displays the numeric result of a command execution on the screen in a specific format; by default, the four decimal digits are displayed, e.g. 3.1416, with the last digit rounded. This format termed short. When the real number is lesser than 0.001 or greater than 1000, the number is shown in the shortE format. The latter format uses so-called scientific notations: a number between 1 and 10 multiplied by a power of 10. For example, the axial moment of inertia of the rectangle 0.4x0.4 m displayed as 2.1333e-3 (m^4) should be read as $2.1333\cdot10^{-3}$, the Plank's constant shown as 6.6261e-34 (in m·kg·s-1) should be read as $6.6261\cdot10^{-34}$, the Young's modulus for Aluminum displayed as 6.9000e-8 (in Pa) should be read as $6.9\cdot10^{-8}$. The number a=1000.1 is displayed the same in short and shortE formats as 1.0001e+03 where e+03 is 10^3 while the whole number should be read as $1.0001\cdot10^3$. Note, that scientific notations can also be used for variable input. For example, the command >> d=5.6e8 enters value $5.6\cdot10^8$ in the d variable.

The format in which the number is displayed can be governed with the format command. The command takes the form

```
format
```

```
format format_type
```

The first command sets the short format type. The format_type parameter in the second command is a word that specifies the type of the displayed numbers. There are more than two described format types. The long or longE format should be used to show 15 decimal digits. For example, when setting the longE format type and inputting the Boltzmann constant $1.38064852 \cdot 10^{-23}$, (in $m^2 \cdot kg \cdot s^{-2} \cdot K^{-1}$), MATLAB® yields the following results:

```
>> format longE
>> B=1.38064852e-23
B =

   1.380648520000000e-23
```

Note:

- Once a certain format type is specified, all subsequent numbers are displayed within it.
- To return to the replaced short format, the format or format short command must be entered.
- The format commands change the length of the numbers on the display but do not make changes in the computer memory nor affect the type of the numbers entered.
- The format_type parameter can be written with a space between short and E and long and E while the capital E can be written as conventional e.
- Short or long formats written with the letter G provide the best representation on the display of a 5-digit or 15-digit number, respectively. For example, in the longG format, the Boltzmann constant appears without trailing zeros (compare with the previously outputted B):

```
>> format longG
B =
        1.38064852e-23
```

This format is suitable for cases when tables with numbers very different in size must be displayed since the format picks where it is best to use the regular (e.g., short) and scientific (e.g., shortE) notation. More details about the described and other available formats can be obtained using the help format or doc format command.

2.1.6. Output Commands

The result is immediately displayed in the assigned format after entering the command (but not when a semicolon follows the command). For additional control of the output, two commands are used - disp and fprintf.

The disp command displays texts or variable values without outputting the name of the variable and the = (equal) sign. Every new disp command yields its result in a new line. In its general form, the command reads:

```
disp('Text string')
```

or

```
disp(Variable name)
```

The text between the quotes is displayed in blue.

As an example, the following lines input the value of the friction coefficient for steel μ=0.65 and display it without and with the disp command:

```
>> mu=0.65          % input and displaying the mu value without the disp command
mu =
  0.6500
>>disp('Steel friction coefficient'),disp(mu) % first disp displays the string;
second - mu value
Steel friction coefficient
  0.6500
```

The fprintf command is used to display texts and data or save them in a file. Here we present the simplest forms of the command, which are sufficient for displaying the calculating results.

The following form is used for formatted output and displaying text and numbers in one line.

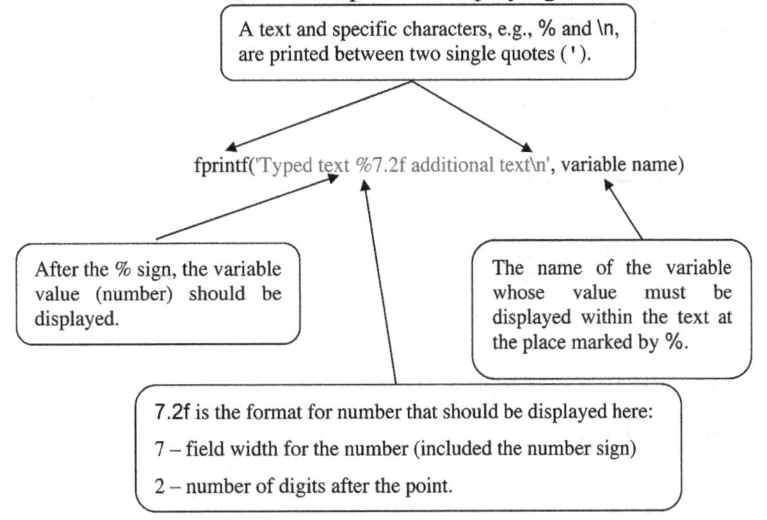

To display text with a new line or divide text on two or more lines, write the \n (slash n) before the word that you want to see on the new line. In addition, for appearance of the >> prompt on the new line, after executing the fprintf command, the \n characters should be also written. The % sign and the character f, termed the conversion characters, are obligatory but the field width number and number of digits after the point (7.2 in the presented example) are optional. For example, if the %f characters were written instead of %7.2f, the number will be displayed by default with 6 digits after the point. The f specifies the fixed-point notation in which the number is displayed and converts a real value to the value with fixed number of digits after the decimal point. Some additional conversion characters that can be used include are: i (or d) – integer notation, e – exponential notation (e.g., 2.309123e+001), and g – the shortest of e or f notations and without trailing zeros. The addition of several %f units (or full formatting elements) permits the inclusion of multiple variable values in the text.

The following example uses the fprintf command to display three lines containing: the title "Coefficient of friction", the word "Aluminum" and Aluminum friction coefficient value μ=1.2 with two digits after the point and the "Steel" word and its friction coefficient value μ=0.65 with two digits after the point:

```
>>Al=1.2;St=0.65;
>>fprintf('Coefficient of friction\n  Aluminum %7.2f\n  Steel   %7.2f\n',Al,St)
Coefficient of friction
   Aluminum  1.20
   Steel     0.65
```

Here, exemplarily, for each of the numbers, seven positions were assigned, which include three positions for decimal digits and decimal point as well as four positions for the integer part of a number, number sign and two previous spaces. The color of the characters in the quotes is blue, the same as in the disp command.

The presented output commands can also be used to display tables when the variables are vectors, matrices or arrays (Section 2.2.5).

2.1.7. Application Examples

The following are some examples of using the elementary MATLAB® commands in M&T calculations.

2.1.7.1. The Stiffness of a Threaded Bolt

The threaded bolt stiffness k can be calculated with following expression (Burr & Cheatem, 1995):

$$k = \frac{\pi E d_0}{\ln\left(\frac{\left(d_1 + h\tan\left(30^o\right) - d_0\right)\left(d_1 + d_0\right)}{\left(d_1 + h\tan\left(30^o\right) + d_0\right)\left(d_1 - d_0\right)}\right)}$$

Where E is the Young's modulus of a material, d_0 and d_1 are the hole and washer face diameter respectively, and h is the bolt height.

Problem: write the MATLAB® commands and calculate the stiffness of the bolt when h=3 cm, d_0=6 mm, d_1=16 mm and E=206· 10^9 N/m^2; write commands that display the result without and with the disp command.

The solution is:

```
>>h=3e-2;                                          %scientific notations, h in m
>>d0=6e-3;                                        %scientific notations, d0 in m
>>d1=16e-3;                                       %scientific notations, d1 in m
>>E=206e9;                                      %scientific notations, E in N/m^2
>>d2=d1+h*tand(30);                                %tand uses the angle in degrees
>>k=pi*E*d0*tand(30)/log((d2-d0)*(d1+d0)/(...
(d2+d0)*(d1-d0)))                    % ... - for continuing expr.on the next line
k =
  5.2831e+09
>>disp('Bolt stiffness, N/m'),disp(k)                        %output with two DISPs
Bolt stiffness, N/m
  5.2831e+09
```

Note:

- As stated above, the angles in trigonometric functions should be given in radians. However, if the appropriate function is written with the final letter d, the angle can be specified in degrees; this feature is activated with the tand command;
- In interactive calculations, two disp must be written on one line to display the text (header string) and number on the two consecutive lines; if we want to write these commands in two separate lines, we input the first disp and get the line with the text, then input the second disp and get the line with the number k:

```
>>disp('Bolt stiffness, N/m')
Bolt stiffness, N/m
>>disp(k)
  5.2831e+09
```

2.1.7.2. The Friction Factor of the Pipe Surface

The friction factor f of the roughened surface of a pipe depends on the roughness height ε, pipe diameter D and Reynolds number $Re=VD/\nu$ where V is the velocity of the fluid flowing in the pipe and ν is the fluid kinematic velocity. The mathematical expression that combines these factors is

$$f=\left\{\left(\frac{64}{\mathrm{Re}}\right)^{8}+9.5\left[\ln\left(\frac{\varepsilon}{3.7D}+\frac{5.74}{\mathrm{Re}^{0.9}}\right)-\left(\frac{250}{\mathrm{Re}}\right)^{6}\right]^{-16}\right\}^{\frac{1}{8}}$$

Problem: write the MATLAB® commands to calculate the pipe friction factor when $\varepsilon=0.21\cdot10^{-3}$ m, $D=0.35$ m, $\nu=1.021\cdot10^{-6}$ m²/s and $V=2.75$ m/s; display the calculated Reynolds number and the friction factor without and with the disp command.

The solution is:

```
>>epsilon=.21e-3;                                    %scientific notations, h in m
>>D=.35;                                                               % D in m
>>nu=1.021e-6;                                    %scientific notations, nu in m^2/s
>>V=2.75;                                                            % V in m/s
>>Re=V*D/nu
Re =
  9.4270e+05
>>f=((64/Re)^8+9.5*(log(epsilon/(3.7*D)+5.74/Re^.9)-…
(250/Re)^6)^-16)^(1/8)                  %… - for continuing expr. on the next line
f =
  0.0180
>>disp('Reynolds Number '),disp(Re)                          %output with two DISPs
Reynolds Number
  9.4270e+05
>>disp('Friction factor'),disp(f)                            %output with two DISPs
Friction factor
  0.0180
```

Note: the leading zero before the decimal point can be omitted; this feature was used above when the numbers were assigned to the epsilon and D variables.

2.1.7.3. Stress Intensity Near a Centrally Located Crack

For a rectangular plate of width $2b$ and height $2h$, the stress intensity factor of the centrally located crack can be calculated by the expression (Rooke & Cartwright, 1976)

$$K = \sigma\sqrt{\pi a}\left[\frac{1-\frac{a}{2b}+0.326\left(\frac{a}{b}\right)^2}{\sqrt{1-\frac{a}{b}}}\right]$$

where a is the crack radius, b – half width of the plate and σ is the uniaxial tension.

Problem: write the MATLAB® commands to calculate stress intensity factor K, kPa·m$^{1/2}$ when $a=0.0381$ m, $b=0.1016$ m and $\sigma=82.7371\cdot10^3$ kPa. Show results with and without the fprintf command. In the latter case, on one line display the words "Stress intensity factor", the resulting K value with one digit after the decimal point, and the units of K.

The solution:

```
>> a=.0381;b=.1016;sigma=82.7371e3;                  % all assignments in one line
>> ab=a/b;
>> K=sigma*sqrt(pi*a)*(1-ab/2+0.326*ab^2)/sqrt(1-ab)
 K =
   3.1078e+04
>> fprintf('Stress intensity factor=%8.1f kPa·m^1/2\n',K)
Stress intensity factor= 31078.4 kPa·m^1/2
```

2.1.7.4. The Minimal Number of Teeth of a Spur Gear

The smallest number of the interference-free teeth on spoon pinion and gear N_P can be calculated with the equation (Single & Mischke, 1989):

$$N_P = \frac{2k}{3\sin^2\varphi}\left(1+\sqrt{1+3\sin^2\varphi}\right)$$

Where φ is the pressure angle, k - is the coefficient that equal to one or to 0.8 for the full-depth and stub teeth respectively.

Problem: write the MATLAB® commands to calculate the number of teeth for $\varphi = 25^\circ$ and k=0.8. Use the radian notation for φ. Output the N_P rounded towards infinity using the disp and fprint commands in that way that each of them displays the words "Teeth number" in the first line and the resulting N_P in the second.

The solution is:

```
>>fi=25;                                            % angle, degree units
>>fi_rad=20*pi/180;                           % converts degrees to radians
>>k=0.8;                                                          % inputs k
>>sfi3=3*sin(fi_rad)^2;               %calculates the 3sin^2 term of the Np
>>Np=ceil(2*k*(1+sqrt(1+sfi3))/sfi3);          % ceil rounds Np towards ∞.
>>disp('Teeth number'),disp(Np)             % requires two disps in one line
Teeth number
  10
>>fprintf('Teeth number\n %5i\n',Np)
Teeth number
  10
```

Note: when outputting with the fprintf command, the single command can be used in the interactive calculations. Nevertheless, when outputting with the disp command, we need to use two of them in one line, one for the text and one for the number; to combine text and number in the one disp command see subsection 2.2.7.2.

2.1.7.5. Adhesive Force in the Contact Between Two Spheres

Adhesive force F_s arising in the contact between two equal-sized spheres of radius R can be calculated based on the JKR (Johnson-Kendall-Roberts) or DMT (Derjaguin-Muller-Toporov) models by the following equations (Bhushan, 2013):

$$F_s = \frac{3}{2}\pi R\gamma \quad \text{- JKR model}$$

$$F_s = 2\pi R\gamma \quad \text{- DMT model}$$

where $R = \frac{R_1 R_2}{R_1 + R_2}$ is the composite radius, mm, and $\gamma = \gamma_1 + \gamma_2$ is the total surface energy of both surfaces of the contacting spheres, mN/m.

Problem: write the MATLAB® commands to calculate the adhesive force by the JKR and DMT equations for two mica spheres of R_1=R_2=25 mm diameter. Assume the free surface energy for mica $\gamma_1 = \gamma_2 =$ 250 mN/m. Output the F_s values with one digit after the decimal point. Use the fprint command so that the string "Adhesive force =XXXX.X mN for …" is displayed in the first line with the values obtained for the JKR model while the appropriate result for DMT model is displayed on the second line (the X denotes the places for the calculated values, and … (ellipsis) denotes JKR or DMT words).

The solution:

```
>>R1=25;                                              % sphere radius, mm
>>R2=R1;
>>gamma1=250;                                       % surface energy, mN/m
>>gamma2=gamma1;
>>R=R1*R2/(R1+R2)                                       % composite radius
>>d_gamma=(gamma1+gamma2)*1e-3;             %1e-3 to convert mN/m m to mN/mm
>>Fs_JKR=3/2*pi*R*d_gamma;
>>Fs_DMT=2*pi*R*d_gamma;
>>fprintf('Fs=%4.1f for JKR\nFs=%4.1f for DMT\n',[Fs_JKR Fs_DMT])

Fs=29.5 for JKR
Fs=39.3 for DMT
```

2.2. GENERATING AND WORKING WITH VECTORS, MATRICES AND ARRAYS

A table containing numerical data is essentially an array or matrix. Heretofore, the variables in scalar form were used containing a single variable each. However, in MATLAB®, each of these variables is a 1x1 matrix. Handling operations with matrices and arrays is more complicated than with scalars: linear algebra operations should be applied for matrices while element-wise operations for arrays.

2.2.1. Vectors, Matrices, and Arrays: Generating and Operating With Them

Generation of Vectors

Sequences of numbers organized in rows or columns are called row or column vectors, respectively. They can also be organized as a series of letters and/or equations.

A vector in MATLAB® is generated by typing the numbers in square brackets. In case of a row vector, the spaces or commas should be typed between the numbers. In the case of a column vector, the semicolons are typed between the numbers. Pressing **Enter** between the numbers also generates a column vector.

An example is a table with data containing the nominal (major) thread diameter and the thread pitch - Table 2.

Table 2. Thread-pitch data

Thread diameter, mm	1.00	5.00	10.00	20.00	30.00	45.00	60.00
Thread pitch, mm	0.25	0.8	1.5	2.5	3.5	4.5	5.5

This table data can be presented for example as two vectors the first table row as row-vector with name diameter and the second as column vector with name pitch.

```
>>diameter=[1 5 10 20 30 45 60]                    %generates row vector data
    diameter =
   1    5   10   20   30   45   60
>>pitch=[0.25;0.8;1.5;2.5;3.5;4.5;5.5]               %generates column vector
pitch =
   0.2500
   0.8000
   1.5000
   2.5000
   3.5000
   4.5000
   5.5000
```

If the Command Window is not wide enough to display all the vector elements in one line, they are displayed in two or more lines with a message line informing which column is presented in the line, for example:

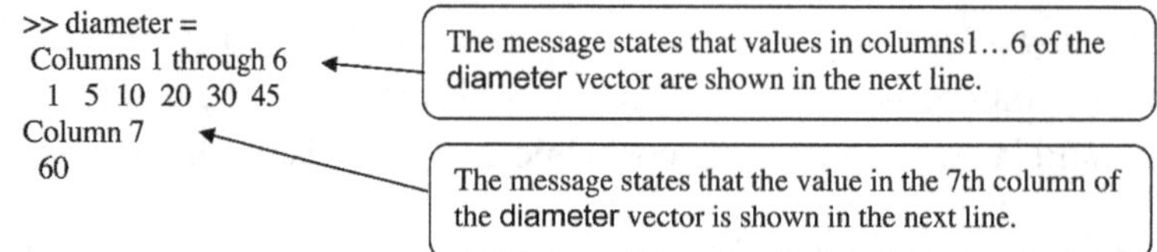

Vectors are also used in vector analysis where the position of a point in three-dimensional coordinates, e.g. *A*-point, is presented by a position vector r_A=2i+7j-2k (where i, j and k are so-called unit vectors and 2, 7 and -2 are the vector projections on the axis). In MATLAB® notations it can be written as row vector A=[2,7,-2].

Frequently adjacent numbers within a vector differ by the same value. For example, in vector v =[1 3 5 7 9], the spacing between the elements is 2. Two commands can be used to create such vectors, namely the ':' (colon) and linspace. The first of them has the form

```
vector_name=a:k:b
```

where a, b and k are, respectively, the first and last term and the step between adjacent terms within the vector with the name vector_name. The last number in generated vector cannot exceed the last term b. The step k can be omitted. In this case, it is assumed by default that k =1. Examples of using this command are:

```
>> p=1.015:0.14:2               % 1st number 1.015, last number <= 2, step 0.14
p =
 Columns 1 through 5
  1.0150   1.1550   1.2950   1.4350   1.5750
 Columns 6 through 8
  1.7150   1.8550   1.9950
>> x=-3:4                      % 1st number -3, last number 4, by default step 1
x =
  -3   -2   -1    0    1    2    3    4
>>y=0.5:3.1/25.5:1              %1st number 0.5, last number <=1,step 3.1/25.5
y =
  0.5000   0.6216   0.7431   0.8647   0.9863
>>z=9.2:-2.2:0.35               %1st number 2.2, last number >=1.3, step -2.2
z =
  9.2000   7.0000   4.8000   2.6000   0.4000
```

The second command that creates a vector with a constant step between adjacent numbers is the linspace command, which has the form

```
vector_name=linspace(a,b,n)
```

where a and b are the first and last numbers, respectively, and n is the amount of numbers. When n is not specified, n is set to 100 by default.

For example:

```
>>x=linspace(1,3,5)                    %5 numbers, 1st number 1, last number 3
x =
  1.0000  1.5000  2.0000  2.5000  3.0000
>>y=linspace(-11,11,4)               %4 numbers, 1st number -11, last number 11
y =
 -11.0000  -3.6667  3.6667  11.0000
>> z=linspace(15.2,1.3,5)
z =
  15.2000  11.7250  8.2500  4.7750  1.3000
>> v=linspace(0,10)                %100 numbers(default), 1st number 0, last 10
v =
 Columns 1 through 5
     0  0.1010  0.2020  0.3030  0.4040
... (shortened, only the five first numbers out of 100 are given)
```

The position of an element in the generated vector is its address, which should be a positive integer and not zero. In the vector diameter (containing the first line numbers of the Table 1):

- The fifth position can be addressed as diameter(5), the element located here is the number 30.
- If the diameter(0) command was entered, an error message appears since the element number cannot be zero.

The last position in a vector may be addressed with the end terminator, e.g., diameter(end) is the last position in the vector diameter and marks the number 60. Another way to address the last element position of the number, namely diameter (7).

Generating Matrices and Arrays

A two-dimensional array or matrix contains the rows and columns with numerical data and resembles a numerical table. The difference between a matrix and the ordinal table arises only when we perform some mathematical operations. When the number of rows and columns is equal, the matrix is called square. Otherwise, it is called rectangular. Like a vector, a matrix is generated by entering elements in square brackets with spaces or commas between them as well as using a semicolon or pressing the ENTER key between lines. Each row should have an equal number of elements. The elements can be specified also with the variable names or mathematical expressions.

As an example, Table 3 shows the height H of the spur gear tooth and its circular thickness S for various numbers of teeth N.

The matrix presentation of the table numbers and some other matrix generation examples are:

Table 3. Gear teeth height and circular thickness at different number of teeth

N	*H*, mm	*S*, mm
10	1.0616	1.5643
20	1.0308	1.5692
30	1.0206	1.5700
40	1.0154	1.5704

```
>>M = [10 1.0616 1.5643;20 1.0308 1.5692;30 1.0206 1.5700;40 1.0154
1.5704]
M =
   10.0000   1.0616   1.5643
   20.0000   1.0308   1.5692
   30.0000   1.0206   1.5700
   40.0000   1.0154   1.5704
>> B=[30 -41.7  % Press Enter before inputting the next line
19 6.23]
B =
   30.0000 -41.7000
   19.0000   6.2300
>>a1=3.14; a2=pi/3;
C=[a1 a2;          %elements are assigned with variables
cos(pi*a1/180) exp(a2)] %elements are defined using expressions
C =
   3.1400   1.0472
   0.9985   2.8497
```

A semicolon is placed before inputting the new line with numbers.

Row and column addressing is used to refer to matrix element/s. For instance, in matrix M in the previous example, set M(3,2) refers to the number 1.0206 and M(2,3) –the number 1.5692. Row or column numbering begins with 1. Therefore, for example, the first element in matrix M is M(1,1). For several sequential elements or a row or column, the semicolon can be used. For example, M(2:3,2) refers to the numbers in the second and third rows of the column 2 in matrix M. The single colon sign addresses to the entire row or column, e.g. M(:,2) refers to the elements of all rows in column 2 and M(3,:) to those of all the columns in row 3.

In addition to addressing by two numbers (row-column), linear one-number addressing can be used. In this case, the element's place within the matrix is detected sequentially, beginning with the first element and down the first column, then continuing analogously with the second column and so forth, up to the last element in the last column. For example, for the above matrix M, the M(6) command refers to element M(2,2), M(8) to the M(4,2) while M(5:8) is the same as M(:,2), etc.

Using square brackets, you can generate a new matrix by combining an existing matrix with a vector or another matrix. These kinds of examples are presented below:

```
>> No=[1;2;3;4]                    % column vector with serial numbers
No =
   1
```

```
   2
   3
   4
>> >>M_No=[No M]                           %assembles column vector No and matrix M
M_No =
  1.0000  10.0000  1.0616  1.5643
  2.0000  20.0000  1.0308  1.5692
  3.0000  30.0000  1.0206  1.5700
  4.0000  40.0000  1.0154  1.5704
>>M_No(3,2)                        %refers to the element in 3rd row and 2nd column
ans =
  30
>>M_No(2:4,1)                    %refers to the elements in column 1 and rows 2,3,4
ans =
   2
   3
   4
>>M_No(2,1:3)                       %refers to the elements in row 2 and columns 1…3
ans =
  2.0000  20.0000  1.0308
>>M_No(3,1:3)                       %Refers to all elements in row 2 and columns 1…3
ans =
  3.0000  30.0000  1.0206  1.5700
>> M_No(2:4,2)=5.1                        %replaces the elements 2…4 in line 2 with 5.1
M_No =
  1.0000  10.0000  1.0616  1.5643
  2.0000   5.1000  1.0308  1.5692
  3.0000   5.1000  1.0206  1.5700
  4.0000   5.1000  1.0154  1.5704
```

To convert a row/column vector into a column/row one and for the rows/columns exchange in matrices, the transpose operator ‘ (quote) is applied, as illustrated:

```
>>M_tr=M'                          %M' changes rows with columns and assigns to M_tr
M_tr =
  10.0000  20.0000  30.0000  40.0000
   1.0616   1.0308   1.0206   1.0154
   1.5643   1.5692   1.5700   1.5704
```

Note that the 4x3 M matrix was transposed into a 3x4 matrix and this result was assigned to M_tr.

2.2.2. Matrix Operations

In various mathematical operations, vectors, matrices and arrays are not always used in the same way as in operations with individual variables. Addition and subtraction are executed the same as ordinary arithmetic but multiplication, dividing and exponentiation are different. In addition, there are also differences between operations with arrays and matrices. Commands to complete these operations are discussed below.

Addition and Subtraction

These two basic arithmetical operations are performed element by element in the matrices/arrays that must be equal in size, e.g. when $A1$ and $A2$ are two matrices sized 3x2 each:

$$A1=\begin{bmatrix} A1_{11} & A1_{12} & A1_{13} \\ A1_{21} & A1_{22} & A1_{23} \end{bmatrix} \text{ and } A2=\begin{bmatrix} A2_{11} & A2_{12} & A2_{13} \\ A2_{21} & A2_{22} & A2_{23} \end{bmatrix}$$

the sum or subtraction of these matrices $A=A1\pm A2$ is

$$A=\begin{bmatrix} A1_{11} \pm A2_{11} & A1_{12} \pm A2_{12} & A1_{13} \pm A2_{13} \\ A1_{21} \pm A2_{21} & A1_{21} \pm A2_{21} & A1_{23} \pm A2_{23} \end{bmatrix}$$

In addition and subtraction operations, the commutative law is valid, namely $A1\pm A2=A2\pm A1$.

Multiplication

Matrix multiplication is a more complex operation performed in accordance with the rules of linear algebra and is possible only when the number of row elements in the first matrix is equal to the number of column elements in the second matrix.

Thus, the above matrices $A1$ has size 2x3 and $A2$ - 2x3 and thus cannot be multiplied. Replace $A2$ for further explanations with another matrix, sized 3x2

$$A2=\begin{bmatrix} A2_{11} & A2_{12} \\ A2_{21} & A2_{22} \\ A2_{31} & A2_{32} \end{bmatrix}$$

The inner dimensions of the multiplied matrices $A1$ and $A2$ are equal (2x**3** * **3**x2, the inner sizes are 3, bold), and now it becomes possible to multiply $A1 * A2$.

$$A1 * A2 = \begin{bmatrix} A1_{11} & A1_{12} & A1_{13} \\ A1_{21} & A1_{22} & A1_{23} \end{bmatrix} * \begin{bmatrix} A2_{11} & A2_{12} \\ A2_{21} & A2_{22} \\ A2_{31} & A2_{32} \end{bmatrix}$$

$$= \begin{bmatrix} A1_{11}A2_{11} + A1_{12}A2_{21} + A1_{13}A2_{31} & A1_{11}A2_{12} + A1_{12}A2_{22} + A1_{13}A2_{32} \\ A1_{21}A2_{11} + A1_{22}A2_{21} + A1_{23}A2_{31} & A1_{21}A2_{12} + A1_{22}A2_{22} + A1_{23}A2_{32} \end{bmatrix}$$

If we multiply $A2$ by $A1$, their inner dimensions are equal to 2. It is not difficult to verify that the product $A1*A2$ is not the same as $A2*A1$. In other words, the commutative law does not apply to multiplication of the matrices. All this is true of course for vectors.

Various examples of addition, subtraction and multiplication of matrices using MATLAB commands are presented below:

```
>> A1=[0.5570 1.9298;1.0938 0.3152;1.9150 1.9412]             %; between rows
A1 =
   0.5570   1.9298
   1.0938   0.3152
   1.9150   1.9412
>>A2=[1.1767 2.1181 0.1385
    1.9664 0.0955 0.2914                                     %generated by pressing
    0.5136 0.8308 2.4704];                                   % Enter between rows
>>A1*A2                    % error message! A1-column and A2-row numbers are unequal
```

Error using *
Incorrect dimensions for matrix multiplication. Check that the number of columns in the first matrix matches the number of rows in the second matrix. To perform elementwise multiplication, use '.*'.

```
>>A=A2*A1                              % the A1-column and A2-row numbers are equal (=3)
A =
   3.2374   3.2073
   1.7578   4.3905
   5.9256   6.0486
>>C1=A2(1,:),C2=A1(:,2)                          %produce row/column vectors from A1& A2
C1 =
   1.1767   2.1181   0.1385
C2 =
   1.9298
   0.3152
   1.9412
>> C1*C2
ans =
```

```
  3.2073
>>C1*C2                              % multiplies column vector C1 by row vector C2
ans =
  9.1863
>> C2*C1                                       % multiplies row vector C2 by C1
ans =
  2.2708  4.0875  0.2673
  0.3709  0.6676  0.0437
  2.2842  4.1117  0.2689
```

A possible application of matrix multiplication in solving the M&T problems is a set of linear equations represented in the matrix form. For example, the set of two equations with two variables

$$\begin{aligned} a_{11}x_1 + a_{12}x_2 &= b_1 \\ a_{21}x_1 + a_{22}x_2 &= b_2 \end{aligned}.$$

may be written in compact matrix form as $AX=B$ or in full matrix form as

$$\begin{bmatrix} a_{11} & a_{11} \\ a_{11} & a_{11} \end{bmatrix} \begin{bmatrix} x_1 \\ x_2 \end{bmatrix} = \begin{bmatrix} b_1 \\ b_2 \end{bmatrix}$$

where A is the matrix $\begin{bmatrix} a_{11} & a_{11} \\ a_{11} & a_{11} \end{bmatrix}$ and X and B are the column vectors $\begin{bmatrix} x_1 \\ x_2 \end{bmatrix}$ and $\begin{bmatrix} b_1 \\ b_2 \end{bmatrix}$, respectively.

Division

The division of matrices is much more complicated than their multiplication due to the non-commutative properties of matrices. Here we only briefly touch on some features and describe MATLAB® commands for matrix division. A comprehensive explanation can be found in books on linear algebra.

Identity and inverse matrices are often used in dividing operators.

An **identity** matrix I is a square matrix whose diagonal elements are 1's and the others are 0's. The eye command (see Table 2.4) generates the identity matrix. In the case of multiplication of identity by square matrix, the commutative low is retained, namely A by I, or I by A yields the same result: $AI=IA=A$. For example

```
>>D1=[13 14 5;14 10 9;2 2 15]              % generates a 3x3 square D1-matrix
D1 =
  13  14   5
  14  10   9
   2   2  15
>>I=eye(3)                            % eye(3) generates the 3x3 identity I-matrix
I =
```

```
    1   0   0
    0   1   0
    0   0   1
>>D1*I                                                   % left multiplication
ans =
   13  14   5
   14  10   9
    2   2  15
>>I*D1                          % right multiplication, result is the same as with D1*I
ans =
   13  14   5
   14  10   9
    2   2  15
```

In case of multiplication of the identity matrix by a non-square matrix, the result is identical to multiplying this matrix by scalar 1, e.g. [4 3 6;12 1 9]*eye(3) and leads to the same product as 1*[4 3 6;12 1 9].

Matrix B is termed **inverse** to A in the case when AB (left multiplication) and BA (right multiplication) lead to the identity matrix: $AB = BA = I$. The inverse to the A- matrix is typically written as A^{-1}. In MATLAB®, this can be written in two ways: B=A^-1 or with the inv command as B=inv(A). For example, multiplying previously generated matrix D1 by its inverse D1-1 leads to the identity matrix

```
>> D1*D1^-1
ans =
    1.0000  -0.0000  -0.0000
    0.0000   1.0000   0.0000
   -0.0000  -0.0000   1.0000
```

The left, \, or right, /, division is often used when matrix manipulations are performed. For example, to solve the matrix equation $AX=B$ (introduced in the Multiplication subsection) with A as a square matrix and X and B as column vectors, left division should be used: $X=A\backslash B$. By contrast, to solve the same matrix equation but rewritten in form $XC=B$, with C as a transposed matrix of A and X and B as row vectors, the right division should be used: $X=B/C$.

For example, the set of linear equations

$$7.3x_1 + 3.0x_2 = 6.9$$
$$-x_1 + 8.7x_2 = -10.5$$

can be rewritten with coefficients to the right of unknowns x_1 and x_2

$$x_1 7.3 + x_2 3.0 = 6.9$$
$$-x_1 + x_2 8.7 = -10.5$$

These two possible forms of writing equations correspond to the two matrix forms:

$$AX{=}B \text{ with } A=\begin{bmatrix}7.3 & 3\\ -1 & 8.7\end{bmatrix},\ B=\begin{bmatrix}6.9\\ -10.5\end{bmatrix},\ \text{and } X=\begin{bmatrix}x_1\\ x_2\end{bmatrix}$$

or

$$XC{=}D \text{ with } C{=}A^{-1}{=}A'=\begin{bmatrix}7.3 & -1\\ 3 & 8.7\end{bmatrix},\ D{=}B^{-1}{=}B'=\begin{bmatrix}7 & 11.5\end{bmatrix},\ \text{and } X=\begin{bmatrix}x_1 & x_2\end{bmatrix}$$

The solutions for the two discussed forms require to use the two mentioned forms of matrix division – left and right division, the commands being:

```
>>A=[7.3 3;-1 8.7];                          % A for form AX=B
>>B=[6.9;-10.5];              % B is the column vector,form AX=B
>>X_left=A\B                      % defines X with left division
 X_left =
  1.3762
  -1.0487
>> C=A';                    % C is the inverse of A; form XC=B
>> D=B';                     % B is the row vector; form AX=B
>> X_right=D/C                   % defines X with right division
X_right =
  1.3762   -1.0487
```

An application example with matrix division is provided in Subsection 2.3.4.1.

2.2.3. Array Operations

In previously described commands for matrix operations, the rules of linear algebra are applied. However, in mechanical, tribological and other calculations there are many operations performed using element-by-element procedure. We use the term *array* instead of *matrix* when such operations are performed. In the multiplication, division, and exponentiation, these operations are conducted in the same way as addition or subtraction - elementwise. Unlike matrix operations, elementwise operations apply only to arrays of the same size. In MATLAB®, they are denoted with a point typed preceding the arithmetic operator, namely

.* (element-wise multiplication);
./ (element-wise right division) ;
.\ (element-wise left division);
.^ (element-wise exponentiation).

For example, if we have two vectors of three elements $V1=[\ V1_1\ V1_2\ V1_3]$ and $V2=[\ V2_1\ V2_2\ V2_3]$, element-by-element multiplication, division and exponentiation are performed as follows:

$V1.*V2=[\ V1_1 * V2_1 V1_2 * V2_2 V1_3 * V2_3]$, $V1./V2=[\ V1_1 / V2_1 V1_2 / V2_2 V1_3 / V2_3]$

and

$V1.\wedge V2=\left[V1_1^{V2_1}\ V1_2^{V2_2}\ V1_3^{V2_3}\right]$.

The same operations applied for two arrays

$$A1=\begin{bmatrix} M1_{11} & M1_{12} & M1_{12} \\ M1_{21} & M1_{22} & M1_{23} \end{bmatrix}$$

and

$$A2=\begin{bmatrix} A2_{11} & A2_{12} & A2_{12} \\ A2_{21} & A2_{22} & A2_{23} \end{bmatrix}$$

lead to:

$$A1.*A2=\begin{bmatrix} A1_{11}A2_{11} & A1_{12}A2_{12} & A1_{13}A2_{13} \\ A1_{21}A2_{21} & A1_{22}A2_{22} & A1_{23}A2_{23} \end{bmatrix},$$

$$A1./A2=\begin{bmatrix} A1_{11}/A2_{11} & A1_{12}/A2_{12} & A1_{13}/A2_{13} \\ A1_{21}/A2_{21} & A1_{22}/A2_{22} & A1_{23}/A2_{23} \end{bmatrix},$$

and

$$A1.\hat{}\,A2=\begin{bmatrix} A1_{11}^{A2_{11}} & A1_{12}^{A2_{12}} & A1_{13}^{A2_{13}} \\ A1_{21}^{A2_{21}} & A1_{22}^{A2_{22}} & A1_{23}^{A2_{23}} \end{bmatrix}.$$

Elementwise operators are frequently used to compute a function with an argument given as a series of values. Examples of manipulations with the array operations are:

```
>>A=[2.77 0.73;-1.35 -0.06;3.03 0.71]              % generates a 3x2 array A
A =
  2.7700  0.7300
  -1.3500  -0.0600
  3.0300  0.7100
>>B=[-0.2 1.41;-0.12 1.42;1.49 0.67]               % generates a 3x2 array B
B =
  -0.2000  1.4100
  -0.1200  1.4200
  1.4900  0.6700
>>A.*B                          % Element-by-element multiplication of A by B
ans =
  -0.5540  1.0293
  0.1620  -0.0852
  4.5147  0.4757
>>A./B                          %Element-by-element right division of A by B
ans =
 -13.8500  0.5177
  11.2500  -0.0423
  2.0336  1.0597
>>B.^2                          % each term in B is raised to the power of 2
ans =
  0.0400  1.9881
  0.0144  2.0164
  2.2201  0.4489
>> A*B                % matrix operation, do not work here - see error message
```

Error using *
Incorrect dimensions for matrix multiplication. Check that the number of columns in the first matrix matches the number of rows in the second matrix. To perform elementwise multiplication, use '.*'.

```
>>x=0:3               %generates four-element vector for the two next commands
x =
   0   1   2   3
>> y=4+exp(-2.1*x.^1.9)                    % elementwise exponentiation: x^1.9
y =
  5.0000  4.1225  4.0004  4.0000
>> y=(sqrt(x)+3)./(1.5*x.^2-1)        % elementwise division and exponentiation
y =
  -3.0000  8.0000  0.8828  0.3786
```

2.2.4. Supplementary Matrix Commands

Vectors, matrices, or arrays with some certain or random values can be generated with special commands. The

```
ones(m,n)
```

and

```
zeros(m,n)
```

commands generate matrices of m rows and n columns with 1 and 0 as all elements.

Many scientific and engineering problems, including those related to modeling in the M&T areas, descriptive statistics, material physics, testing, measurements technique, material informatics, properties of materials and various simulations, involve random numbers for which the following generators of pseudorandom numbers should be used:

```
rand(m,n)
```

or

```
randn(m,n)
```

The former command yields a uniform distribution of elements between 0 and 1 while the latter a normal one with mean 0 and standard deviation 1. For generating a square matrix, these commands can be abbreviated to rand(n) and randn(n). Examples are:

```
>>ones(3,4)                  %generates a 3x4 matrix in which all elements are 1s
ans =
   1   1   1   1
   1   1   1   1
   1   1   1   1
>>zeros(4,3)                 %generates a 4x2 matrix in which all elements are 0s
ans =
   0   0   0
   0   0   0
   0   0   0
   0   0   0
>>a=rand(2,3)                 %Uniform distr.,2x3 matrix, numbers between 0 and 1
a =
  0.8147  0.1270  0.6324
  0.9058  0.9134  0.0975
>>v=rand(1,3)                 %Uniform distr.,1x3 vector, numbers between 0 and 1
v =
```

```
  0.8147   0.9058   0.1270
>>b=randn(2,3)                    %Normal distr., 2x3 matrix with random numbers
b =
  0.5377   -2.2588   0.3188
  1.8339   0.8622   -1.3077

>>w=randn(3,1)                 %Normal distr., column vector with random numbers
w =
  0.5377
  1.8339
  -2.2588
>>d=randn(2)                      %Normal distr., 2x2 matrix with random numbers
d =
  0.5377   -2.2588
  1.8339   0.8622
```

Integer random numbers can be generated with the randi-command as shown in Table 4.

Table 4. Commands for matrix manipulations, generation and analysis (alphabetically)[1]

MATLAB® Command	Parameter/s	Description	Example (Inputs and Outputs)
char(s1,s2,s3,…)	s1,s2,s3 … are the strings not all of the same length; written in the one row.	Creates a matrix of rows, each row containing one string with a length equal to the longest of all rows; missing characters in shorter lines are added with spaces.	>>char('Screw','Gear'); ans = 2×5 char array 'Screw' 'Gear '
cross(a,b)	a and b are three- element vectors each.	Calculates the cross product of 3D-vectors **a**= $a_1\hat{i} + a_2\hat{j} + a_3\hat{k}$, **b**= $b_1\hat{i} + b\hat{j} + b\hat{k}$: **c**=**a** x **b**= $(a_2b_3 - a_3b_2)\,\hat{i}$ + $(a_3b_1 - a_1b_3)\,\hat{j}$ + $(a_1b_2 - a_2b_1)\,\hat{k}$	>>a=[2 -3 6]; >>b=[3 1 7]; >>c=cross(a,b) c = -27 4 11
det(A)	A – square matrix	Calculates the determinant of A.	>> A=[7 -9 6;19 13 11;3 2 20]; >> det(A) ans = 4783
diag(x)	x - vector	Generates a matrix with elements of x placed diagonally.	>> x=1.1:2:5.1;diag(x) ans = 1.1000 0 0 0 3.1000 0 0 0 5.1000

continued on following page

Table 4. Continued

MATLAB® Command	Parameter/s	Description	Example (Inputs and Outputs)
dot(a,b)	a and b are the *n*-element vectors each	Calculates the scalar product of two vectors: a·b=$a_1b_1+a_2b_2+\ldots+a_nb_n$; the result is a scalar.	>>a=[2.9 0.9 5.1] ; >>b=[3 1 6.1] ; >>c=dot(a,b) c = 40.7100
eye(n)	n is the square marix size	Generates a identity matrix - a square matrix with diagonal elements 1 and others 0	>> eye(3) ans = 1 0 0 0 1 0 0 0 1
length(x)	x - vector	Calculates the amount (length) of the x elements.	> x=[-3.1 5 -2.22 2]; >>length(x) ans = 4
max(a)	a - vector or matrix	Returns a row vector with maximal numbers of each column in the matrix a. If a is a vector, it returns the maximal number in a .	>>a=[6 7;2 3;8 1]; >>b=max(a) b = 8 7 >>a=[4 9 2 1 8 2]; >>b=max(a) b = 9
mean(a)	a - vector or matrix	Analogous to max but for arithmetical mean/s.	>>a=[6 7;2 3;8 1]; >>b=mean(a) b = 5.3333 3.6667 >>a=[4 9 2 1 8 2]; >> b=mean(a) b = 4.3333
median(a)	a - vector or matrix	Analogous to max but for median/s.	>>a=[6 7;2 3;8 1]; >>b=median(a) b = 5 3 >>a=[4 9 2 1 8 2]; >>b=median(a) b= 4
min(a)	a - vector or matrix	Analogous to max but for minimal value/s	>>a=[6 7;2 3;8 1]; >> b=min(a) b = 2 1 >>a=[4 9 2 1 8 2]; >> b=min(a) b = 1
num2str(a)	a - vector or matrix	Converts a single number or numerical vector/matrix elements into a string representation.	>>a=[-19.73546 5.1]; >>v=['a=' num2str(a)] v = 'a=-19.7355 5.1'
randi(imax,m,n)	imax - maximal integer value, m - number of rows, n - number of columns.	Returns an m by n matrix of integer random numbers from value 1 up to imax –maximal integer value.	>>randi(9,2,3) ans = 8 2 6 9 9 1

continued on following page

Table 4. Continued

MATLAB® Command	Parameter/s	Description	Example (Inputs and Outputs)
repmat(a,m,n)	a - matrix, m – number of rows, n - number of columns.	Generates the large matrix containing m×n copies of a.	> A=[3 4;2 5]; >>b =repmat(A,1,2) b = 3 4 3 4 2 5 2 5
reshape(a,m,n)	a - matrix, m - number of rows, n - number of columns.	Returns an m-by-n matrix whose elements are taken column-wise from a. Matrix a must have m*n elements.	>>a=[6 7;2 3;8 1] a = 5 8 1 2 7 3 >>reshape(a,1,6) ans = 6 2 8 7 3 1
sign(x)	x – vector or single number	signum or sign function; returns: 1 if the element of x >0; 0 if x=0; and -1 if x<0. x '/| x '|, when x is complex	>>X=[-11.2 3 0 5]; >>sign(X) ans = -1 1 0 1
size(a)	a – matrix	Returns a two-element row vector; the first element is the number of rows in matrix a while the second is the number of columns.	>>a=[6 7;2 3;8 1]; >>size(a) ans= 3 2
[y,ind]= sort(a,dim,mode)	a -vector/matrix; dim – dimension; mode are the 'ascend' or 'descend' sorting order ; y – ordered vector/ matrix; ind – indices a for each y.	Sorts elements of a along columns (dim=1) or along rows (dim=2) and in 'ascend' (default) or 'descend' mode order. It returns ordered y and indices of the a for each y .	>>a=[6 7;2 3;8 1] a = 6 7 2 3 8 1 >> [y,ind]=sort(a,2,'descend') y = 7 6 3 2 8 1 ind = 2 1 2 1 1 2
std(a)	a - vector or matrix.	Calculates standard deviation by $\left[\frac{1}{n-1}\sum_{i=1}^{n}\left(a_i-\mu\right)^2\right]^{\frac{1}{2}}$, with μ as a mean of *n* elements for each column of matrix a. If a is a vector, it returns the standard deviation value of vector.	>>a=[6 7;2 3;8 1]; >> std(a) ans = 3.0551 3.0551 >>a=[4 9 2 1 8 2]; >> std(a) ans = 3.3862

continued on following page

Table 4. Continued

MATLAB® Command	Parameter/s	Description	Example (Inputs and Outputs)
strvcat(t1,t2,t3,...)	t1, t2, t3, ... - strings.	Generates the matrix containing the t1, t2, t3... strings as rows.	>>t1 = 'Friction'; >>t2 = 'Wear'; >>t3 = 'Lubrication'; >>strvcat(t1,t2,t3) ans = 3×11 Char array 'Friction ' 'Wear ' 'Lubrication'
sum(a)	a - vector or matrix.	Analogous to max but for column (matrix) or row (vector) sums of elements.	>> a=[5 8; 1 2; 7 3]; >> sum(a) ans = 13 13

Note: when we reuse the rand, randn or randi command, new random numbers are generated each time. To restore the settings of the random number generator to retrieve the same random numbers as when **restarting** MATLAB®, you must enter the rng default command as shown:

```
rng default                                    % a starting generator settings
>> a=rand(2,3)          % first usage of the rand command, first random numbers
a =
  0.8147   0.1270   0.6324
  0.9058   0.9134   0.0975
>> a=rand(2,3)          % reusing the rand command leads to new random numbers
a =
  0.2785   0.9575   0.1576
  0.5469   0.9649   0.9706
rng default                               % restoring a starting generator settings
>> a=rand(2,3)     % reusing rand command restores the first generated numbers
a =
  0.8147   0.1270   0.6324
  0.9058   0.9134   0.0975
```

In addition to the previously discussed commands, there are many others that can be used for manipulation, generation and analysis of vectors, matrices and arrays, some of which are listed in Table 4.

Note: when we reuse the rand, randn or randi command, new random numbers are generated every time. To restore the random number generator settings in order to obtain the same random numbers as when restarting MATLAB®, you must enter the rng default command in the command window as in following example:

```
rng default                                    % a starting generator settings
>>a=rand(2,3)                         %first RAND usage, first random numbers
```

```
a =
  0.8147  0.1270  0.6324
  0.9058  0.9134  0.0975
>>a=rand(2,3)              %reusing the rand command leads to new numbers
a =
  0.2785  0.9575  0.1576
  0.5469  0.9649  0.9706
rng default                     % restoring a starting generator settings
>>a=rand(2,3)            %here the rand command restores the first numbers
a =
  0.8147  0.1270  0.6324
  0.9058  0.9134  0.0975
```

2.2.5. Strings as Matrix Elements

All vectors / matrices discussed so far, except for some of those in Table 4, have numerical elements even if they are written as variables or expressions (since the variables contain numerical values and the expressions are executed and the resulting values are assigned to the matrix). However, the elements of the matrix can also contain strings. A string is an array of characters, numbers and other symbols. A string is entered in MATLAB® between single quotes:

```
          %The 'Mechanics studies…' sentence is assigned to the str1 variable.
>>str1='Mechanics studies the motion, forces, and interaction of bodies'
str1 =
  'Mechanics studies the motion, forces, and interaction of bodies'
>>str2='Strength'                  %generates the 8th-element string vector str2
str =
  'Strength'
>>length(str2)                  %the str2 length is 8, the ' signs are not counted
ans =
   8
```

In computing, each character of the string is treated and stored as a number (thus the set of characters represents a vector or an array) and can be addressed as an element of a vector or array, e.g. str2(4) in the string Strength is the letter 'e'. Single quotes are not counted in the string length. Some examples with string manipulations are:

```
>>str2(6)                                    % 'g' is the 6th element of str2
ans =
  'g'
>>str2(3:5)                  % 'r', 'e' and 'n' are the 5th …7th elements of str2
ans =
  'ren'
```

```
>>str2([7 4 5])              %'t','e' and 'n' are the 7th…5th elements of str2
ans =
  'ten'
```

To display string without quotes, the disp command can be used, e. g.

```
>>disp(chr1)
Mechanics studies the motion, forces, and interaction of bodies
>> disp(str([7 4 5]))
red
```

Just as numerical rows, the strings between the rows are divided by a semicolon (;) and strings within the rows are divided by a space or comma. Each row with strings must be the same length as the longest of the rows. To achieve the character alignment between string rows, the spaces should be added to shorter strings; for example,

```
>> Name=['Piston';'Bolt';'Crankshaft']
Error using vertcat
Dimensions of arrays being concatenated are not consistent.
```

This error message appears due to inequality in length of the strings: The number of the characters in the word Piston and in Bolt is 6 and 4 while in Crackshaft – 10. For the right matrix generating, the four spaces should be added after Piston and six after Bolt; causing all the strings to be of the same length. Thus, they can be successfully entered and displayed.

```
>> Name=['Piston    ';'Bolt      ';'Crankshaft'];
>> disp(Name)
Piston
Bolt
Crankshaft
```

In order to determine a longer string, calculating the number of spaces added to each row of a column is a tedious procedure for the user. To avoid this, use the char or strvcat command as shown in Table 4.

2.2.6. About Table Outputting

The disp and fprintf commands can display single numbers as well as vectors, matrices and captions (Subsection 2.1.6). Therefore, these commands can be used to output data in tabular form.

First, we show how it can be done with the disp command. For example, gear teeth height and thickness data with captions as per Table 3 should be displayed using the following commands:

```
>>M = [10 1.0616 1.5643;20 1.0308 1.5692;30 1.0206 1.5700;40...
1.0154 1.5704]                    % Generates matrix M with data from Table 3
M =
```

```
   10.0000   1.0616   1.5643
   20.0000   1.0308   1.5692
   30.0000   1.0206   1.5700
   40.0000   1.0154   1.5704
>>disp('            Table'),disp('   N     H, mm    S, mm'),disp(M)
               Table
    N     H, mm    S, mm
   10.0000   1.0616   1.5643
   20.0000   1.0308   1.5692
   30.0000   1.0206   1.5700
   40.0000   1.0154   1.5704
```

First two disp commands show captions and the third output numbers of the M matrix.

Note: All disp commands must be written on the same command line in order to display the table in the presented view.

Now use the fprintf command. This command permits a formatted output when, for example, the N values can be displayed without decimal digits while H and S with two decimal digits. The table can be presented with the two following commands:

```
>>fprintf('Table\n   N   H,mm   S,mm\n'),fprintf('%5.0f %6.2f %6.2f\n', M')
         Table
   N    H,mm   S,mm
   10   1.06   1.56
   20   1.03   1.57
   30   1.02   1.57
   40   1.02   1.57
```

The first fprintf command shows two caption lines and the second displays numbers of the M matrix. These fprintf commands must be written on the same command line in order to display the table in the view presented.

Note:

- The **fprintf command prints matrix rows as table columns**. Considering this, in the example above, the 4x3 M matrix was transposed to the 3x4 matrix (written as M') and the conversion format for the outputting values was written for three columns;
- The \n symbols (back slash and letter n without space) must be written at the end of the column format.

2.2.7. Application Examples

2.2.7.1. Forces Applied to a Bracket and Its Equivalent Force

Three forces F_1, F_2, and F_3 are applied to a bracket in the *xy* plane. In Cartesian coordinates, each of the ***F*** forces can be given via its coordinate components F_x and F_y in the vector form

$$\boldsymbol{F}=F_x\mathbf{i}+F_y\mathbf{j}=F_{mg}\,(\cos\theta\mathbf{i}+\sin\theta\mathbf{j})$$

where θ is the angle (in respect to the axis *x*) of the vector F_{mg} called magnitude (vector length)[2].

The equivalent (otherwise termed total), force F_{eq}, its length and angle can be determined using the equations

$$\boldsymbol{F}_{eq}=\sum_i \boldsymbol{F}_i$$

$$F_{eq\,mg}=\sqrt{F_{eq\,x}^2+F_{eq\,y}^2}$$

$$\theta=\tan^{-1}\frac{F_{eq\,y}}{F_{eq\,x}}$$

where *i* denotes the serial number of the force applied to the bracket, *i*=1, 2 and 3 in this case.

The equivalent force vector magnitude can be calculated also with the dot (·) product (see the dot command in Table 4) as

$$F=\sqrt{F_{eq}\cdot F_{eq}}$$

Problem: Calculate the equivalent force on the bracket and define its angle and magnitude with the two form above for the applied force magnitudes F_1=350, F_2=400 and F_3=550 N and the angles θ_1=15°, θ_2=−25 ° and θ_3=120 ° respectively. Display the defined F with the fprintf command in vector form F=XXX.Xi+XXX.Xj where the X symbol designates position where the defined components of F vector should be printed.

The following steps should be applied to solve the problem:

- Enter values of the force magnitudes and angles F_1, F_2, F_3, θ_1, θ_2 and θ_3;
- Calculate each F_{mg} as a two-element vector;
- Calculate the equivalent force magnitude *F* in two possible ways – as a square root from the sum of its squared components or as square root from the dot product of two F_{eq} vectors and show both defined values;

- Use the fprintf command to display the elements of the F_{eq} vector in a format with one decimal number; to print in one row; transpose the two-element row vectors F_{eq} to the column vector.

```
>>F1M=350;F2M=400;F3M=550;%                                         % degrees
>>Thet1=15;Thet2=-25;Thet3=120;                                      % Newton
>>F1=F1M*[cosd(Thet1) sind(Thet1)];                    %vector F1: [F1x F1y]
>>F2=F2M*[cosd(Thet2) sind(Thet2)];                    %vector F2: [F2x F2y]
>>F3=F3M*[cosd(Thet3) sind(Thet3)];                    %vector F3: [F3x F3y]
>>Feq=F1+F2+F3;                                    %vector Feq as [Feqx Feqy]
>>FeqM1=sqrt(Feq(1)^2+Feq(2)^2);                            % magnitude Feq
>>FeqM2=sqrt(dot(Feq,Feq));                % magnitude Feq with another way
>>FeqM1,FeqM2,Feq
>>fprintf('Equivalent Force: Feq=%5.1fi+%5.1fj\n',Feq')
FeqM1 =
 582.5978
FeqM2 =
 582.5978
Feq =
 425.5972 397.8533
Equivalent Force: Feq=425.6i+397.9j
```

2.2.7.2. Generating Table with Names and Numbers

The linear temperature expansion coefficients α of some technological/industrial materials are presented in Box 1.

Box 1.

Material	Linear Temperature Expansion, $\alpha \cdot 10^{-6}$, m/(m K)
Polyethylene	108
Titanium	0.5
Polypropylene	54
Magnesium	26
Aluminum	23
Steel	12
Silicon Carbide	450
Titanium	2.77
Copper	16

Problem: Generate and display the two-column array in which the first column contains the material names while the second columns contains the linear temperature expansion coefficients.

To execute, use the following steps:

- Generate a string column with the material names as they written in the first table column, note that all name should have the same number of characters as the longest column;
- Generate a column with the linear temperature expansion data from the second table column;
- Combine these two columns into the same matrix. Note that all matrix elements must be of the same type, e.g., if strings are written in one column, then another column must also contain strings or vise-versa; in our case, the num2str command (Table 4) can be used to transform the numerical α data into strings;
- Use the disp command to display the table header ' Material Expansion$\cdot 10^{-6}$, m/(m K) ';
- Use the disp command to display a two-column array with material names and measure names;
- Add the spaces for the table header and for each of the material name to center table captions above the displayed columns.

The commands are:

```
>>Name=char('Polyethylene    ','Titanium','Polypropylene',…
'Magnesium','Aluminum','Steel','Silicon Carbide','Titanium',… 'Copper')
>>alpha=[108 0.5 54 26 23 12 450 2.77 16]';
>>Table=[Name num2str(alpha)];
>>disp(' Material  Expansion·10^-6 m/(m K)'),disp(Table)
```

```
 Material  Expansion·10^-6, m/(m K)
Polyethylene      108
Titanium            0.5
Polypropylene      54
Magnesium          26
Aluminum           23
Steel             12
Silicon Carbide   450
Titanium          2.77
Copper            16
```

2.2.7.3. Transformation from Polar to Cartesian Coordinates

Conversion from polar r,θ – coordinates to Cartesian x,y-coordinates is a frequently need in M&T calculations. This can be performed with the following equations:

$x = r\cos\theta$ and $y = \sin\theta$

where r and θ are the distance and directional angle from the reference point.

Problem: Generate the *x* and *y* coordinate matrices for *r* and θ in the range 0.1 … 1 and 0 …π/2 respectively, when there are the four *r*- points and three θ-points; each of the x and y matrices must have three rows and four columns.

The steps are as follows:

Generate a string column as follows:

- Using the linspace command, generate a 1x4 row vector with *r*-coordinates in the range 0.1 …1;
- Using the linspace command, generate a 1x3 row vector with θ-coordinates in the range 0 …π/2;
- Calculate the Cartesian coordinates *x* and *y* for the assigned values of the polar coordinate *r* and θ.

Note: in order to get 3x4 *x* and *y* matrices by multiplication of the two vectors, the first vector should be 3x1 and the second 1x4 by their sizes, but the product *r·cos* θ [1x4]*[1x3] cannot be performed because the number of columns of the first vector is not equal to the number of rows in the second. To perform the multiplication, it is necessary to convert the first vector from 1x4 row-vector to the 4x1-column vector. In this case, multiplication is possible, but we get the 4x3 matrix while a 3x4 matrix is required. Now let's try to change the places *r* and *cos* θ, namely *cos* θ ·*r*. In this case, we have the product [1x3]*[1x4], which also cannot be calculated. However, converting the first term from vector 1x3 into 3x1 vector, we get the product [3x1]*[1x4], which can be performed and leads to a matrix of the desired size - 3x4.

The commands are:

```
>>r=linspace(0.1,1,4);                                          %generates 1X4 vector r
>>Th=linspace(0,pi/2,3);                                       %generates 1X3 vector Th
>>Th=Th';                  %converts Th from 1x3 to 3x1 for further multiplication
>>x=cos(Th)*r                                    % cos(Th)*r possible as [4x1]*[1x3]
>>y=sin(Th)*r                                             % ibid.(but for sin(Th))
x =
  0.1000  0.4000  0.7000  1.0000
  0.0707  0.2828  0.4950  0.7071
  0.0000  0.0000  0.0000  0.0000
y =
       0       0       0       0
  0.0707  0.2828  0.4950  0.7071
  0.1000  0.4000  0.7000  1.0000
```

2.2.7.4. Instantaneous Positions of the Engine Piston Pin

The position *x* of the engine piston pin with respect to the angle of rotation of the crankshaft can be calculated as

$$x = r\cos\theta + \sqrt{l^2 - r^2\sin^2\theta}$$

where *r* and *l* are the lengths of the crank arm and connecting rod, respectively, and θ is the angle of rotation of the crank arm.

Problem: calculate the piston pin instantaneous position at ten rotation angles in range $0\ldots2\pi$ for the crank and rod lengths 0.12 and 0.25 m respectively. Present the results as a two-column table with header 'Angle, degree Position, m' where the θ and x values are represented in the first and second table columns. Use the disp command for displaying 'Angle,grad Position, m' caption, and the fprintf command for displaying the θ and x values with, respectively, two and four decimal digits in the resulting table.

The solution steps are:

- Assign r and l values, and generate a ten-element row vector with θ values from the range $0\ldots2\pi$, the latter using the linspace command;
- Calculate vector x by the above expression; use elementwise operation for $\sin^2\theta$.
- Join the two separate row vectors θ and x into a two-row matrix called Table;
- Use the disp command with the table header and the and the fprintf command with the required format of the θ and x numbers; note that the fprintf command outputs rows of the matrix as columns.

The commands are:

```
>>r=0.12;l=0.25;                                                    % in m
>>Thet=linspace(0,2*pi,10);                       % generates 1X10 vector Thet
>>x=r*cos(Thet)+sqrt(l^2-r^2*sin(Thet).                 ^2%calculates position x
>>Table=[Thet;x];                    %collects two row vectors in two-row matrix
>>disp('Angle,grad Position, m'),fprintf(' %7.2f  %7.4f \n',... Table) % disp and
fprintf in one line for interactive entering
Angle,grad Position, m
   0.00    0.3700
   0.70    0.3297
   1.40    0.2411
   2.09    0.1674
   2.79    0.1338
   3.49    0.1338
   4.19    0.1674
   4.89    0.2411
   5.59    0.3297
   6.28    0.3700
```

2.3. FLOW CONTROL

A simple sequence of commands is a computing program. Commands are implemented in the order in which they are written. However, in many cases, the order of execution of the commands must be changed. Examples include, when calculations must be repeated with new parameters or when it is necessary to select and calculate a variable with one of several expressions, each of which is accurate within a different area. In the latter case, the area should be checked to use the correct computing expression. In other situations, when the result should be calculated with the required accuracy, it may be necessary to repeat the commands several times until the error in the answer diminishes to the required size. Flow control is applied to realize such processes. MATLAB® uses special commands (usually called conditional statements) for these purposes. These commands direct the computer to choose which command to execute next using for this some relational, logical operators and conditional statements. The most common flow control commands are described below.

2.3.1. Commands for Relational and Logical Operations

Flow control operations are carried out using relational and logical commands. Both types of commands compare the pairs of values or statements but the former operates mainly with numerical values while the latter with logical (Boolean) values.

Relational Operators

The commands performing a pairwise comparison of values are called relational or comparison operators. The result of applying such an operator is written as 1 (true value) or 0 (false value), e.g., the expression x<5.1 results in 1 if x less than 5.1, while in 0 if otherwise.

The relational operators are:

< (less than),
> (greater than),
<= (less than or equal to),
>= (greater than or equal to),
== (equal to),
~= (not equal to).

Note: two-sign operators must be written without spaces between the first and second signs, e.g., two equal signs must be written without space between them: ==.

When a relational operator is applied to a matrix or array, it performs elementwise comparisons. In this case, the result is the array of 1's where the relation is true (the size of the array is the same as the size of the compared matrix) and 0's where it is not. If one of the compared operands is scalar while the other is a matrix, the scalar is matched against every element of the matrix. The ones and zeros are logical data types, not numerical, although they can be used in arithmetical operations in the same way as numerical data.

Some examples are:

```
>>3*4==24/2                                % the answer is 1 since 3*4=24/2=12
ans =
 logical
  1
>>sin(pi/2)~=0               % the result is 1 as sin(pi/2)=1 and it is not '0'
```

```
ans =
 logical
  1
>>M=[7.1 -9.4 11.3;-6 5 -3.9;7 -16 -1.8]          %Produces a 3x3 matrix M
M =
   7.1000  -9.4000  11.3000
  -6.0000   5.0000  -3.9000
   7.0000 -16.0000  -1.8000
>>B=M<=0                          % Assigns 1s and 0s to B as results of M<=0
B =
 3×3 logical array
  0  1  0
  1  0  1
  0  1  1
>>M(B)
                           % Selects elements M<=0, using logical B as numerical
ans =
  -6.0000
  -9.4000
 -16.0000
  -3.9000
  -1.8000
>>M(M<0)=0                 %Assigns 0s to the elements of M that are less than 0
M =
  7.1000     0      11.3000
  0          5.0000  0
  7.0000     0       0
```

Logical Operators

These operators are designed for operations with logical expressions, e.g. $(x>3)\&x(<\pi)$, and operate with the true or false logical values. Similar to relational operators, they can be used as addresses in another vector, matrix or array; see, for instance, the last two example commands.

There are three logical operators: & (logical AND), | (logical OR) and ~ (logical NOT). Like the relational operators, they can be used as arithmetical operators with scalars, matrices and arrays. Comparison is carried out elementwise and yields logical 1 or 0 when the result is true or false, respectively. There are also equivalent logical commands: and(A,B) is equivalent to A&B, or(A,B) – to A|B, not(A,B) – to A~B. The results of operations with logical operators correspond to the rules of Boolean algebra and are logical 1 or 0. The actual result depends on the order in which logical operations are performed. The order is determined by the so-called precedence rules, which we set according to their priority, from highest to lowest: parentheses, exponentiation, NOT (~), multiplication/division, addition/subtraction, relational operators, AND (&), OR (|). The desired order of execution each of operator can also be set/changed using parentheses.

Some examples are:

```
>>x=-4.2;                                                    % assign -4.2 to x
>>-2<x<-3             %runs from left to right:-2<-4.2 is false and 0<-3 is 0
ans =
 logical
  0
>>x>-5&x<-1                                 % x>-4 is true (1); then the & leads to 1
ans =
 logical
  1
>>~(x<5)                                  % x<4 runs first, is true (1), then ~1 is 0
ans =
 logical
  0
>>~x<5
%~x runs 1st and is true as x is nonzero, then 1<4 is true
ans =
 logical
  1
```

Among the MATLAB® logical functions is the find command, which in its simplest forms reads as

```
i=find(x)
```

or

```
i=find(A>c)
```

where in the first case, i is a vector of the place addresses (indices) where nonzero elements of x are located, and in the second case, i is the vector of addresses of those A elements that is greater than c. In the second form of this command, any of the communication operators can be used, namely, <,> =, etc. These forms of the find command use the linear addressing form (subsection 2.2.1). For example

```
>>V=[0 3 6.1 7.2 -7.4315 0 -2.6];                            %assigns vector V
>>i1=find(V)                    % returns the addresses of the non-zero V-elements
i1 =
   2   3   4   5   7
>>i2=find(V<3)                    % returns the addresses of the V with elements <3
i2 =
   1   5   6   7
>>M=[0 3 6.1;7.2 -7.4315 3.1; 0 -2.6 1];                  %assigns 3x3 matrix M
>>i=find(M<3)                  %returns the LINEAR addresses of M with elements <3
```

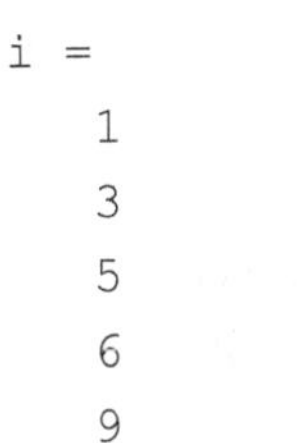

```
i =
    1
    3
    5
    6
    9
```

2.3.1.1. Application Example: Screening of the Yield Strength of Some Metals

The yield stress of some metals in annealed state appears in Box 2.

Box 2.

Material	Yield Strength, MPa
Aluminum	35
Brass	75
Copper	69
Iron	130
Nickel	138
Steel	180
Titanium	450
Molybdenum	565

Problem: Use relational and logical operators to form the three lists of materials with the following yield strengths: a) less than 80; b) between 80 and 200; and c) greater than 200. Calculate number of metals in each or group. For each of these groups, display the yield strength together with names of the metal.

The steps are as follows:

- Assign the material name and density values as they appear in the table.
- Input the names of the material as character matrix and its yield strengths as a numeric column vector, each with the names Material and Yield. The Material names have different lengths but each row of the matrix must be the same length. To do this it is most simple to use the char command that adds automatically the space signs to the names shorter than longest in the list of the names. In addition, to create a gap between each material name and its yield strength, one space character can be written at the end of each name.
- Assemble the two separate vectors Material and Yield in the one matrix with the Material_Yield name. Note that these vectors are of different types: the Material vector comprises strings while the Yield vector comprises numbers. Therefore, it necessary to transform the numeric-type vector to the string-type using the num2str command.

- Use the find command (logical operator) to identify the row indices of the Yield vector that designate the locations where the yield strengths are less than 80 MPa; assign them to the d_less80 vector.
- Calculate the amount of materials with yield strengths of less than 80 MPa by the sum command with the condition yields<80 (relational operator); this operand creates logical ones (that can be used as the numerical values) where the condition is true and zeros in other cases, therefore the sum of this numbers gives the amount of the materials with .
- Find in the Metal_Density matrix and display the rows containing name of the materials and its yield strengths; this can be executed using the y_less80 vector with row addresses and the colon (:) operator denoted all the column indices.
- The row indices, amount of materials with yield strengths between 80 and 200 and above 200 MPa, and names of the yields should be defined and displayed in the same way as yield strengths of less than 80 MPa.

The commands used to solve this problem are:

```
>>Material=char('Aluminum','Brass','Copper','Iron','Nickel',…
'Steel ','Titanium','Molybdenum');                    %assigns vector with names
>>Yield=[35 75 69 130 138 180…
     450 565]';                                      %assigns vector with numbers
>>Material_Yield=[Material num2str(Yield)];                         %creates matrix
>>y_less80=find(Yield<80);                  %finds addresses where yields are <80
>>n_y_less80=sum(Yield<80)                                    %calculates the amount
n_y_less80 =
   3
>>disp(Material_Yield(y_less80,:))                         %displays names and yields
Aluminum  35
Brass    75
Copper   69
>>                % beginning from this, the commands are as above but for the
>>                            % yields of 80…200 MPa and greater then 200 MPa
>>y_between80and200=find(Yield>=80&Yield<=200);
>>n_y_between80and200=sum(Yield>=80&Yield<=200)
n_y_between80and200 =
   3
>>disp(Material_Yield(y_between80and200),
ans =
  'Iron    130'
  'Nickel  138'
  'Steel   180'
>>y_above200=find(Yield>200);
>>n_y_above200=sum(Yield>200)
n_y_above200 =
```

```
    2
>>disp(Material_Yield(y_above200,:))
Titanium 450
Molybdenum565
```

2.3.2. The If Statements

There are various conditional statements to manage the order of the execution of commands. One important among them is the if statement, which has three basic forms: **if ... end**, **if ... else ... end** and **if ... elseif ... else ... end**. Each if construction should terminate with the word end. The words of each form of statement appear on the screen in blue. The if statement forms, their designs and description are shown in Table 5.

Table 5. Different forms of the If statement

The If Statement Form	Design	Description
if ... end	*if conditional expression* ... ... MATLAB® command / s ... *end*	Executes the inner MATLAB® command/s when the conditional expression is true
if ... else ... end	*if conditional expression* ... ... MATLAB® command / s ... *else* ... ... MATLAB® command / s ... *end*	Executes the command/s that is/are located between the words if and else in case the conditional expression is true; otherwise, the command/s placed between else and end are executed.
if ... elseif ... else... end	*if conditional expression1* ... ... MATLAB® command / s ... *elseif conditional expression2* ... ... MATLAB® command / s ... *else* ... ... MATLAB® command / s ... *end*	Executes the command/s that is/are located between the words if and elseif in case the first conditional expression is true; otherwise, when the second conditional expression is true the command/s located between elseif and else is/are executed, and if this expression is false the commands between else and end is/are executed.

In this table, the conditional expression can contain the relational and/or logical operators; for example, a<=v1&a>=v2 or b ~ = c.

When the if conditional statement is typed and entered in the Command Window next to the prompt >>, the new line (and additional lines, after pressing enter) appears without the prompt until the final end is typed and entered.

An application example with the if statement is presented at the end of this chapter (Subsection 2.3.4.2).

2.3.3. Loops in MATLAB®

Another commands used in flow control are the commands for loops that allow repeat one or a group of commands. In loop, each command/s re-execution is termed a pass. There are two loop commands in MATLAB®: for … end and while … end. These words appear on the screen in blue. Similar to the if statement, each for or while construction must terminate with the word end.

The loop statements are written in a general form in Table 6.

Table 6. Commands for loop generation.

Design	Description
$for\ k = [initial : step : final]$ $\left.\begin{matrix}\ldots\\\ldots\\\ldots\end{matrix}\right\} MATLAB^{®} command / s$ end	Executes k -times the internal command/s; k can be specified as sequence of numbers with the colon command or by numbers written in the square brackets.
$while\ conditional\ expression$ $\left.\begin{matrix}\ldots\\\ldots\\\ldots\end{matrix}\right\} MATLAB^{®} command / s$ end	Executes the internal command/s repeatedly while the conditional expression is true.

The commands written between for and end are repeated k times; k is a number that is changed in every loop pass or by the addition of the step-value or by getting the next value in the brackets. This process continues until k reaches or exceeds the final value.

The square brackets in the expression for k (Table 6) mean that k can be assigned as a sequence, for example in such way k=[2.41 -1.16 1:2:6], where first k is 2.41, then -1.16, then 1, 3, and 5.

If there are only colons in k, the brackets can be omitted, e.g., k =1:3:10. Immediately after the last pass, the command following the loop is executed. Sometimes, the for ... end loops can be replaced with the matrix operations, which are actually superior as for…end loops work slowly. This advantage is negligible for short loops with a small number of commands but appreciable for large loops with numerous commands.

The while … end loop is used when the number of passes is not known in advance. In each pass, MATLAB® executes the commands written between the while and end and repeats the passes until the conditional expression is true. An incorrectly written loop may continue indefinitely, for example

```
>>b=2;
>>while b >0                    %b is always >0, so the loop continues indefinitely
   c=sqrt(b)
end
c =
  1.4142
...
```

This result (c= 1.4142) appears repeatedly on the screen. To interrupt the loop, press the Ctrl and C keys simultaneously.

To illustrate the described loop functions, consider an example with two function for … end and while … end with both used to calculate the exponential function e^x via the Taylor series at $\sum_{n=0}^{\infty}\frac{x^n}{n!}$ for x=1.0472. Assume n=0, 1, …, 7 in case of for … end, and for the while … end assume that the conditional expression requires $\frac{x^n}{n!}$ greater than 0.0001.

The solutions are:

```
>>x=1.0472;                    % defines x for the both loop-statement examples
>>                             %             Calculating e^x with the for … end loop
>>                %Note: amount of terms in series is counted below by k from 0
>>                             %                  thus total number of terms is n+1
>>s=0;                 % sets s = 0 for further summation of the series' terms
>>n=6;                 % defines the last given number of the terms: n+1=6+1=7
>>for k=0:n                                                  % for ... end loop
  s=s+x^k/factorial(k);                     % s increases by x^k/k! each pass
end% the end of the for ... end loop
>>s% displays the resulting s that is the e^x
s =
  2.8493
>>                           %           Calculating e^x with the while … end loop
>>s=0;                 % sets s = 0 for further summation of the series' terms
>>k=0;                                             % sets the term counter k to 0
>>term=x^k/factorial(k);                  %calculates the first series' term(k=0)
>>while term>=.0001                                        % while ... end loop
   s=s+term;                                     % s increases by term each pass
   k=k+1;                                                 % new term number of k
   term=x^k/factorial(k);                            % calculates new term value
end% the end of the while...end loop
```

```
>>fprintf('exp(x)=%7.4f n=%i\n',s,k)                %displays result: e^x and n
exp(x)= 2.8496 n=8
```

The first example examines the for … end loop. At the beginning of the loop, in the first pass, the s value is zero. During this pass, the first term (its k number is 0) is calculated and added to the s. In the second pass, k= k +1=0+1=1, the second term of the series is calculated and added to the previous s value. This procedure is repeated up to k=n =5. After this, the cycle ends with the obtained value displayed by typing and entering of the variable names. In the case of the for … end loop, the number of passes must be fixed.

The second example examines the while … end loop and describes a bit more complicated solution. In this case, the condition expression must be specified for ending the cycle. It is assumed here that the value of the kth term may not be greater than 0.0001 (this number represents the accuracy of the calculated value). As in the previous case, in the first pass, the s and k values are equal to zero (k is sometimes called a counter); the first term value is calculated for k=0 before the start of the loop. In the first pass, the first term of the series is checked if it is greater than 0.0001. If it is true, the term added to the sum s. The new term is checked if it is greater than 0.0001. If this condition is true, the next pass is started for the next term calculation. If it is false, the loop ends. The fprintf command would then display the obtained s value and number of times the term was added to s, which represents the number of terms of the series used. If the value is false, the loop ends. The fprintf command displays the resulting value of e^x and number of loop members used to obtain this value. As the latter number is an integer, the conversion character i is used for its displaying.

Note:

- the while … loop does not summarize a value that does not meet the required conditional statement. As a result, the last value of the counter k is increased by 1 if the starting k-value was 1; in the example above the starting k=0, therefor the subtraction is not required;
- the for … end and while … end loops and if statements can incorporate additional loops and/or if-statements. The order and quantity of these inclusions are not limited and are predetermined only by calculation purposes.

2.4. APPLICATION EXAMPLES

2.4.1. Friction Force Versus Contact Area at the Nanometer Scale: Defining Coefficients of the Linear Fit

The fitting of the observed data with some mathematical expression is a widely used technique in M&T sciences to describe possible relationships between the dependent y and independent x variables. For example, a laboratory study of the nano-tribological behavior a material (Enachescu et. al., 1999) shows the dependence between the friction force F and contact area s at the nanometer scale. The processed data is:

Friction force F =3, 4, 5, 6 and 7 nN

Area of contact, s =0.77, 1.00 1.29 1.51 and 1.71 nm^2

This data can be described by the linear equation (termed also as linear regression) $y=a_1+a_2x$ in which y is the friction force F, x is the contact area s and the coefficients a_1 and a_2 can be obtained from the following set of equations

$$a_1 n + a_2 \sum_{i=1}^{n} x_i = \sum_{i=1}^{n} y_i$$

$$a_1 \sum_{i=1}^{n} x_i + a_2 \sum_{i=1}^{n} x_i^2 = \sum_{i=1}^{n} x_i y_i$$

where x and y are the force and density, respectively, and n is the number of the observed values.

This set of equations can be represented in the two matrix forms $AX=B$ or $XA=B$. In the first case:

$$\begin{bmatrix} n & \sum_{i=1}^{n} x_i \\ \sum_{i=1}^{n} x_i & \sum_{i=1}^{n} x_i^2 \end{bmatrix} \begin{bmatrix} a_1 \\ a_2 \end{bmatrix} = \begin{bmatrix} \sum_{i=1}^{n} y_i \\ \sum_{i=1}^{n} x_i y_i \end{bmatrix}$$

while in the second case

$$\begin{bmatrix} a_1 \\ a_2 \end{bmatrix} \begin{bmatrix} n & \sum_{i=1}^{n} x_i \\ \sum_{i=1}^{n} x_i & \sum_{i=1}^{n} x_i^2 \end{bmatrix} = \begin{bmatrix} \sum_{i=1}^{n} y_i & \sum_{i=1}^{n} x_i y_i \end{bmatrix}.$$

Problem: Define the a-coefficients with left- and right-division; print the result as a linear equation with the relevant coefficients a_1 and a_2.

The following steps are to be taken:

- Generate the y and x row vectors with the observed data;
- Generate a 2x2 matrix A of the first matrix form- the matrix rows are the rows of the first matrix form equation; the `sum` command can be used for sums;
- Generate a column vector B with the sums on the right-hand side of the first matrix equation;
- Use left division $A \backslash B$ to calculate the a-coefficients;
- Display the coefficients in the written linear equation using the `fprintf` command;
- Use right-hand division A/B for which transform the column vector B into the row vector. The right division in this case is simply a verification of the previous solution;
- Display the coefficients directly in the written linear equation using the `fprintf` command.

The commands for the solution are:

```
>>x=3:7;                                                % contact area in nm^2
>>y=[0.77 1.00 1.29 1.51 1.71];                        % friction force in nPa
>>A=[length(x) sum(x);sum(x) sum(x.^2)];                 % A for the form Ax=B
>>B=[sum(y);sum(y.*x)];                   % column vector B for the form Ax=B
>>a=A\B             %calculates and displays a-coefficients for the form Ax=B
a =
  0.0610
  0.2390
>>aa=B'/A               %calculates and displays a-coeffic. for the form xA=B
aa =
  0.0610  0.2390
>>            %  The fprintf is used below for showing a-coefficients with >>%
3 decimal digits
>>                    %      Output of the equation obtained for the form Ax=B
>>fprintf('\n  The equation is F=%5.3f+%5.3fs\n',a(1),a(2))
The equation is F=0.061+0.239s
>>                    %      Output of the equation obtained for the form xA=B
>>fprintf('\n  The equation is F=%5.3f+%5.3fs\n',aa(1),aa(2))        %displays
equation
The equation is F=0.061+0.239s
```

Note: In general case, to solve the equations with accordance to the second matrix form XA=B, the matrix A and the vector B should be transposed by the quote operator ('); in the above example this was not necessary since A=A' for the set of two linear equations.

2.4.2. Steady-State Temperatures of the Rectangular Plate

The dimensionless steady-state temperatures of a rectangular *a*x*b* plate can be calculated with the following expression (Hsu, 1984)

$$T = \frac{4}{\pi}\sum_{i=1,3,5,\ldots}^{n}\frac{\sinh\left(i\pi\alpha\eta\right)\sin\left(i\pi a\xi\right)}{i\sinh\left(n\pi\alpha\right)}$$

where $\eta = x/a$, $\xi = y/b$, $\alpha = a/b$, *a* and *b* are the plate lengths in the *x* and *y* directions, respectively.

Problem: Obtain the temperatures T using the <u>for … end</u> loop for ξ and the <u>sum</u> command for summarizing elements of the series (Σ) within the loop.

Take α=a=*b*=1, n=11, η=1. Display a two-column table with ξ and T–columns and with heading 'Coordinate Temperature'. Use the **disp** commands for heading and the **fprinf** command for the numeric values.

The following steps must be taken:

- Assign the given values 1, 1, and 11 to the eta, alfa, and n variables, respectively, and the 0, 0.1, …1 values to the vector xi;
- Organize the for … end loop, for instance, as follows for j=1:length(xi) … end; within the loop, place the vector n=1:2:11 and then the T expression using the indices for xi and T, the element-wise operators, and the sum commands;
- Assemble the xi and T vectors into a two-row matrix named Table for further outputting;
- Display results with the disp commands for the resulting table head ' Y, nondim T, nondim', and the fprinf command for the numeric xi and T values with one and three decimal digits respectively.

The commands are:

```
>>eta=1;a=1;n=11;                                                   % assignments
>>xi=0:.1:1;                                          % generates the vector xi
>>for j=1:length(xi)                    %length(xi) returns theelements number ofxi
  i=1:2:n;                                % generates vector of i for the sums in T
  T(j)=4/pi*sum(sinh(i*pi*a*eta).*sin(i*pi*a*xi(j))./...
    (i.*sinh(i*pi*a)));                      % xi(j) - for the new xi in each pass
end                                                     % ends of the for … end loop
>>Table=[xi;T];                                 % two-row matrix for the fprintf
>>disp(' Y, nondim T, nondim'),fprintf('  %4.1f    %6.3f\n',...
  Table)                      % write in one line for the interactive mode input
Y, nondim T, nondim
  0.0     0.000
  0.1     1.147
  0.2     0.964
  0.3     0.984
  0.4     1.044
  0.5     0.947
  0.6     1.044
  0.7     0.984
  0.8     0.964
  0.9     1.147
  1.0     0.000
```

2.4.3. Radial Thickness of the Piston Ring

In the design of piston rings, the radial thickness t_r (in mm) of the compression piston rings is calculated by the equation:

$$t_r = D\sqrt{\frac{3P_w}{\sigma}}$$

where D is the cylinder bore diameter, mm; P_w is the gas pressure, N/mm^2 and σ - allowable bending stress, MPa (Sobhy M., n.d.).

Problem: Calculate the thickness of the piston ring with diameter 130 mm and for P_w =0.025 … 0.042 N/mm^2 and σ=85 ...110 MPa; take the four values of P_w and five values of σ. Solve the problem in two ways: with and without the for… end loops (using the vectors only). Present the results in a table in which each row shows the thickness for all σ-values while the row corresponds to one specific pressure; use the fprintf command.

The following steps realize both possibilities:

- Assign the value of D and generate the P_w and σ row vectors;
    - Calculate t_r (with the expression above) in the two for… end loops: the external for Pw and the internal – for σ; before the external loop the t_r zero matrix can be generated to reduce computer calculations due to matrix size changing during each loop pass.
    - Display the calculated Pw values with the fprintf in which the obtained values are presented with single digits after the decimal point.
    - Realize now the second possibility, without loops. To do this, rewrite the expression under the square root in the form $3P_w(1/\ \sigma)$ so that the division in brackets comes first and is followed by the matrix multiplication; ($1/\sigma$) produces a row vector with size 1x5. For the inner dimension equality, transpose (with the sign ') the 1x4 row vector P_w to the 4x1 column vector. The next multiplication by the scalar 3 does not change the vector size. Finally, the product of the [4x1]*[1x5] matrices is the 4x5 matrix with the M_2 values that must be the same as in the first case calculated.

```
>>D=130;n_Pw=4;n_sigma=5;                                        % assignments
>>Pw=linspace(0.025,0.042,n_Pw);             %generates row vector Pw, N/mm^2
>>sigma=linspace(85,110,5);                 % generates row vector sigma, MPa
>>                 %          Calculations tr with the for ... end loop
>>tr=zeros(n_Pw,n_sigma);                  %preallocated matrix for further calc.
>>for i=1:n_Pw                                             % the external loop
  for j=1:n_sigma                                          % the internal loop
    tr(i,j)=D*sqrt(3*Pw(i)/sigma(j));                        % calculates tr
  end                                                 % ends the internal loop
end                                                   % ends the external loop
>>fprintf('%4.1f %4.1f %4.1f %4.1f %4.1f\n',tr')                  % displays tr
 3.9 3.7 3.6 3.5 3.4
 4.3 4.1 4.0 3.9 3.8
 4.7 4.5 4.3 4.2 4.1
 5.0 4.8 4.7 4.5 4.4
>>      %         Calculations tr without loop with the matrix manipulations
>>tr=D*sqrt(3*Pw'*(1./sigma));                  % 1./sigma first then 3*Pw'*
>>fprintf('%4.1f %4.1f %4.1f %4.1f %4.1f\n',tr')                  % displays tr
 3.9 3.7 3.6 3.5 3.4
 4.3 4.1 4.0 3.9 3.8
 4.7 4.5 4.3 4.2 4.1
 5.0 4.8 4.7 4.5 4.4
```

2.5. CONCLUSION

The MATLAB® desktop and its main windows are presented and used in interactive calculations. Primary operations, elementary and specific commands, flow control statements, and looping are examined by the manifold of simple examples from the M&T areas. Discussed commands used to solve the followed problems:

- Stiffness of the threaded bolt, friction factor of the pipe surface,
- Stress intensity factor near a centrally located crack,
- Number of teeth of a spur gear, adhesive force in contact between two spheres,
- Forces applied to the bracket, momentary position of the engine piston pin,
- Yield strength screening of some metals,
- Fitting coefficients determination for the friction force – contact area data,
- Radial thickness of the piston ring.

In general, the material of this chapter demonstrates the ability to perform calculations and solve various M&T problems using the MATLAB® commands discussed here.

REFERENCES

Bhushan, B. (2013). Introduction to Tribology (2nd ed.). John Wiley & Sons Inc. doi:10.1002/9781118403259

Burr, A. H., & Cheatem, J. B. (1995). *Mechanical analysis and design* (2nd ed.). Prentice Hall.

Burstein, L. (2020). *A MATLAB® primer for technical programming in material science and engineering*. Elsevier-WP.

Enachescu, M., van den Oetelaar, R. J. A., Carpick, R. W., Ogletree, D. F., Flipse, C. F. J., & Salmeron, M. (1999). Observation proportionality between friction force and contact area at the nanometer scale. *Tribology Letters*, *7*(2-3), 73–78. doi:10.1023/A:1019173404538

Hsu, H. P. (1984). *Applied Fourier Analysis*. Harcourt Brace Jovanovich.

Rooke, D. P., & Cartwright, D. J. (1976). *Compendium of stress* intensity factors. HMSO Ministry of Defense. https://en.wikipedia.org/wiki/Stress_intensity_factor#cite_note-rooke-4

Single, J. E., & Mischke, C. R. (1989). *Mechanical engineering design* (5th ed.). McGray-Hill.

Sobhy, M. (n. d.). *9 Piston Rings*. Retrieved from https://www.academia.edu/26077072/9_Piston_rings

ENDNOTES

1 Based on tables in chapter 2 in Burstein, 2020.

2 The traditional notation is ||*F*||, not used here to facilitate understanding.

APPENDIX

Table 7. List of Examples, Problems, and Applications discussed in the chapter

No	Example, Problem, or Application	Location, Subsection
1	Displaying coefficients of friction with the fprintf command.	2.1.6.
2	The stiffness of a threaded bolt.	2.1.7.1
3	The friction factor of the pipe surface.	2.1.7.2.
4	Stress intensity near a centrally located crack.	2.1.7.3.
5	The minimal number of teeth of a spur gear.	2.1.7.4.
6	Adhesive force in the contact between two spheres.	2.1.7.5.
7	Vector representation of nominal thread diameter and thread pitch data.	2.2.1.
8	The matrix representation of data on the height, number and thickness of the gear teeth.	2.2.1.
9	Forces applied to a bracket and its equivalent force.	2.2.7.1.
10	Generating table with names and numbers.	2.2.7.2.
11	Transformation from polar to Cartesian coordinates.	2.2.7.3.
12	Instantaneous positions of the engine piston pin.	2.2.7.4.
13	Screening of the yield strength of some metals.	2.3.1.1.
14	Calculation the exponential function via the Taylor series.	2.3.3.
15	Friction force versus contact area at the nanometer scale: defining coefficients of the linear fit.	2.4.1.
16	Steady-state temperatures of the rectangular plate.	2.4.2.
17	Radial thickness of the piston ring.	2.4.3.

Note, some small examples, mostly related to non-M&T issues, are not included in the list.

Chapter 3
Script-, Function-Files and Program Managing

ABSTRACT

The Editor window for writing scripts and user-defined functions are presented, as well as the Live Editor window for writing live scripts and functions. All commands, regular and live scripts, and functions are explained by examples from the mechanics and tribology (M&T) fields. After that, the application examples are given; they include the stress unit converters, computing of the stress factor of a shaft with a transverse hole, gear warm K-parameter calculations, installation, and operation stresses on the piston ring.

INTRODUCTION

All studied in the preceding chapter commands, its managing were performed interactively; in this mode of working with MATLAB® the commands not stored and should be re-entered in the Command Window each time when calculations must be repeated. This is not convenient and is a disadvantage of the interactive mode; another disadvantage is that the any corrections of the entered command is possible only on the executable line. If any of the consecutive commands has been corrected, all predecessors along with this and subsequent commands must be repeated to obtain the correct result. This situation is uncomfortable, and the reader who has studied the preceding chapter attentively has undoubtedly experienced it. The solution of this problem is to write a list of commands (termed program), save them to a file and run it if necessary. MATLAB® has two types of such files, called script and function files. Their creation, storage, and execution as well as M&T-oriented examples are discussed below.

DOI: 10.4018/978-1-7998-7078-4.ch003

3.1. SCRIPT FILES AND EDITOR WINDOW

3.1.1. Editor Window, Creating, Saving and Running a Script File

The list of commands written and executed in the order in which they are written is called a script and is a program. The Editor window is used to type a script, after which the script must be saved in a file and run. In case you need to introduce corrections or new commands, they can be entered directly into the file. Saved files have the extension '.m' and termed m-files. To open the Editor window enter the edit command in the Command Window. Another way is to click the New Script icon located on the toolstrip in the File section of the Desktop Home tab. After this, the Editor window appears. The view of the window detached from the Desktop is shown in Figure 1. The window contains toolstrip at the top and a large blank field in the rest of it. The commands should be typed in this field. Like to the Desktop toolstrip (see Subsection 2.1.1), this toolstrip includes the EDITOR, PUBLISH, and VIEW tabs. The EDITOR tab (default appearance) is commonly used for writing/editing commands, saving/ opening/debugging, and running.

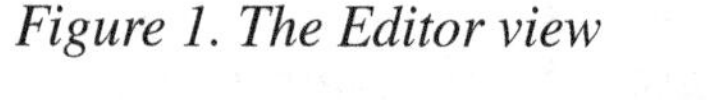

Figure 1. The Editor view

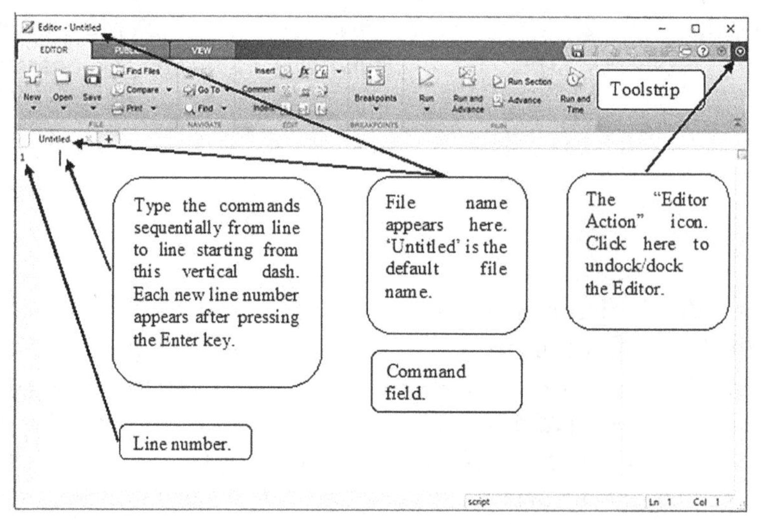

The script commands should be typed line by line in the blank field of the Editor window, starting from the place shown in Figure 1. Two or more commands can be placed on the same line; commas or semicolons should separate these commands. A new line is available after pressing the Enter key. Each line with command/s has/ve a serial number; the number appears automatically with opening the line. A typical script file written in the Editor window is shown in Fig.3.2. The file is named EditorView_ScriptExample and calculates spur gear chordal thickness. The first four lines of this file are the explanatory comments they precede by the comments sign % and in appear green. Comments are not a part of the execution and do not get a dash against the number. Two further lines content commands designed to calculate and display the tooth thickness *s*, in mm, according to a given modulus *m* and the number of teeth *Z*. For better legibility, the commands in the Editor appear in black.

Figure 2. View of the Editor window containing the script file and the M-Lint message

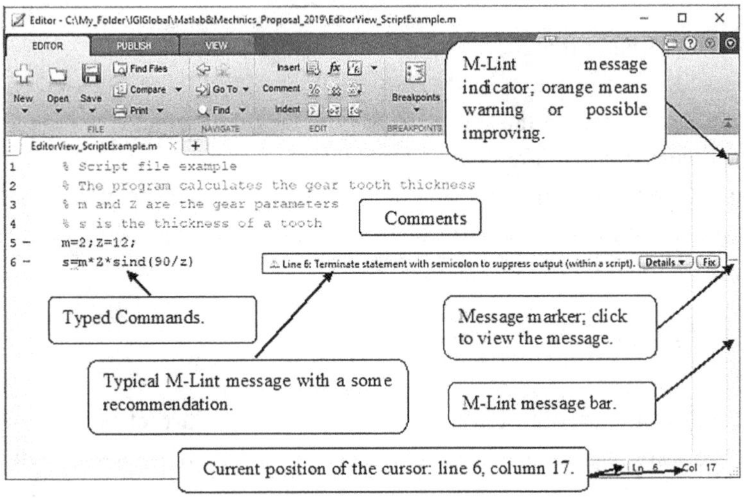

The right vertical line of the Editor Window frame is a bar on which the message markers are displayed - horizontal colored strokes. These are messages from the code analyzer (also called the M-Lint analyzer). This analyzer detects possible errors, comments on them and recommends corrections to improve program performance. The small square ☐ at the top of the message bar indicates the presence or absence the errors and/or makes some warnings. When the indicator is green, it means no errors, warnings, or script improvements are needed; a red indicator – syntax errors detected; an orange indicator – warnings or possibility of improvements (but no errors). The place where an error is detected or where there is a possibility of improvement is underlined / highlighted and a horizontal colored dash appears in the message bar. Moving cursor to this dash or to the marked place, we can obtain the error/ warning message; such a message can be seen in Figure 2. Not every warning recommendation should be considered, for example the recommendation shown in this figure (addition of a semicolon) need not be executed because we want to display the resulting value of the cell volume (semicolon suppresses display of the value). The code analyzer operates in the Editor window by default; it can be disabled by un-signing the 'Enable integrated warning and error messages' check box in the 'Code Analyzer' option of the 'Preferences' window that can be opened by clicking the Preferences line in the Environment section of the MATLAB® Desktop Home tab. The opened 'MATLAB Code Analyzer Preferences' window contains many other default settings that can be adjusted.

Saving the Script File

The commands written in the Editor window must be saved in a file. To perform this, select the 'Save As ...' option of the Save button (from the FILE section of the EDITOR tab); enter the desired pass to the file location and name of file respectively into the path- and 'File name:' fields of the 'SelectFile for Save As' window. The default location of the saving file is the 'MATLAB' folder of the 'My documents' directory. Any other directory can be chosen from the directory/file tree that appears on the left-hand side of the 'Save File for Save As' window.

When naming a file, follow these guidelines:

- The name cannot be longer than 63 characters;
- It is desirable that the name begins with a letter;
- The name should not repeat the names of the MATLAB® commands/functions, user-defined functions, predefined variables, or other saved files;
- the mathematical operation signs (e.g. +, -, /, \,*,^) should not be used within the file name;
- it is strongly not recommended to introduce dots, commas, spaces, or other punctuation marks into the file name.

About the Current Folder

The directories and files of the present folder are displayed in the Current Folder window located to the left of the Desktop. Additionally, the path to the used folder is shown in the current folder field located just under the Desktop toolstrip – see Figure 3. To set a desired, non-default folder in the Command Window, e.g. the folder with m-file previously created and saved with Editor, the following operations should be performed:

Figure 3. The 'Current Folder' window, current folder toolbar, 'Browse for folder' button, and opened 'Select a new folder' dialog box.

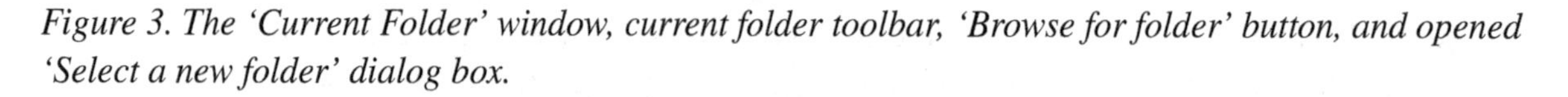

- Click the 'Browse for folder' icon located on the left-hand side of the current directory field (in the toolstrip bottom), after that the 'Select a new folder' dialog box appears;
- Select the line with the desired directory name and then click the 'Select Folder' button; selected directory appears in the current directory field. The current folder that was selected with the Desktop should be the same as in the Editor window.
- For example, the Figure 2 shows the EditorView_ScriptExample.m file stored next to the Examples subfolder with the following path to him C:\Users\leoni\OneDrive\Documents\MATLAB.

Running Script File

The developed script file should be run in order to execute it and display the calculation results. For executing a script file:

- The available files in current folder should be checked firstly, and if the file absences in the installed folder, an appropriate directory with this file should be set;
- The file name (without m-extension) should be typed in the Command Window and then entered; alternatively the 'Run' icon on the Editor window toolstrip can be clicked.

Following this, to run the EditorView_ScriptExample script, we need to check/install the C:\Users\leoni\OneDrive\Documents\MATLAB path by the described way, and then to type and enter the file name in the Command Window:

```
>> EditorView_ScriptExample                          % enter to run the script file
s =
    3.1326
```

3.1.2. Input the Variable Values From the Command Window

Whenever it needs to use the script file for recalculations with new parameters, the new values should be typed into the script, saved then, and run. To avoid the alterations of the script file the input command should be used, this command has forms:

```
Numeric_Variable = input('Text to be displayed ')
Character_Variable =input('Text to be displayed ','s')
```

where 'Text to be displayed' is a message that displays in the Command Window and enables to assign: a number to the Numeric_Variable, or a string to the Character_Variable; 's' signs that inputting text are string and it can be written without single quotes. To input a string containing more than one line, the \n specifier should be entered at the end of the each line.

After the script file is run, when the input command is initiated, the text written in single quotes is displayed on the screen and the user should type and enter a number or string; the values are assigned to the Numeric_Variable or to the Character_Variable depending on the used command form.

As example for using this command create script file that convert US (Imperial) density units, lb_m/ft^3 - pound mass per cubic foot, into international system SI standard units, kg/m^3, by the expression $d_{SI} = 16.0185 \cdot d_{US}$:

```
%                                          density: lbb/ft^3 to kg/m^3 convertor
d_US=input('Enter density in lbm/ft^3, d = ');
d_SI=16.0185*d;
fprintf('\n Density in kg/m^3 is %10.4f \n',d_SI)
```

The commands should be typed into the Editor window and saved in the m-file; name the file as density_convertor. After entering this file name in the Command Window, the input command prompt 'Enter density in lbm/ft^3, d =' appears on the screen. Type now a density value (in lb_m/ft^3) and press enter; the inputted density value is converted to kg/m^3 and the 'Density in kg/m^3 is' string together with the obtained value displayed on the screen.

Running command, appeared prompts, inputted pressure in pounds per square inch, displayed string, and defined density in kg/m^3 are:

```
>> density_convertor
Enter density in lbm/ft^3, d = 172
 Density in kg/m^3 is  2755.1820
```

The input command enables to input the vectors and matrices; this can be performed by typing the numbers into the brackets. Use for example the psi2Pa script file to convert just two densities:

```
>> density_convertor
Enter density in lbm/ft^3, d = [172 63.9]
 Density in kg/m^3 is  2755.1820
 Density in kg/m^3 is  1023.5821
```

3.2. USER-DEFINED FUNCTIONS AND FILES

3.2.1. Function Creation

In classical mathematics a function f is relationship between an inputted, independent, e.g. x, and an outputted, dependent, e.g. y, parameters (called also arguments or variables). The simplest form of a function is $y=f(x)$. The right and left parts of the function can be sets of variables, e.g. $\boldsymbol{u}=f(x_1,x_2,x_3)$ with $\boldsymbol{u}$ as a three-element vector. In terms that are more commonly used in programming, parameters on the right side of the function can be called input parameters, and on the left side, output parameters. When f (x) is specified as a mathematical expression and the input parameters are assigned, the output parameters can be calculated. Many commands discussed in the preceding chapter were written in the function form, e.g. log(x), sqrt(x), sin(x), cos(x), exp(x), etc.; this form enables to use of commands in simple direct calculations or in complex expressions during interactive or programmatic actions by entering the name of the function with the corresponding argument. In addition to the available MATLAB® functions, it is possible to create any new function and then reuse it with arbitrary input values and in different programs; such functions are termed 'user-defined'. Not only an individual expression, but also a complete program can be defined as a function and saved as a function file. This is especially useful in cases where there is no desirable MATLAB® function for the required calculations.

In general, the function should comprise the following parts: function definition, help lines with explanations, and a body with program commands. These function components should be written sequentially in the Editor window, as in the example shown in Figure 4.

Figure 4. Typical view of the function file with a user-defined function.

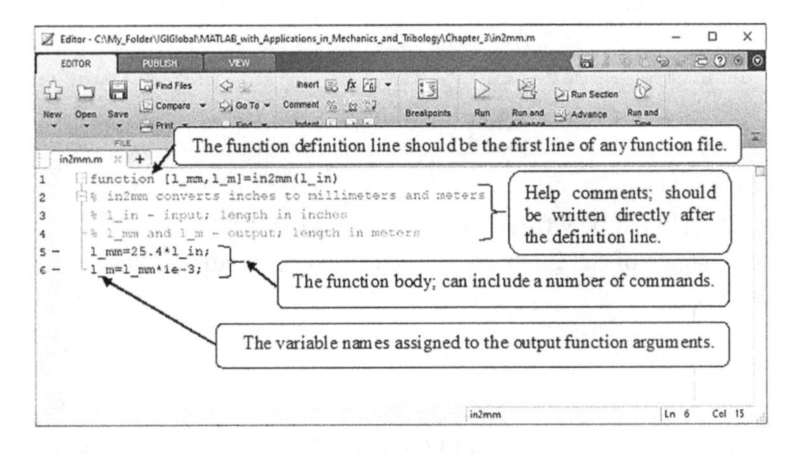

The function and its file are named in2mm; and it converts the lengths given in inches to the lengths in millimeters and meters (SI units). According to the first definition line, the function has: one input parameter – the vector l_in with lengths in inches that should be converted and two output parameters – the converted lengths in millimeters and in meters.

The requirements and recommendations regarding the each part of the user-defined function are as follows.

Function Definition Line

The line containing the function definition should look like this:

```
function [output _parameters]=name(input_ parameters)
```

The first word of the user-defined function must be the word function and after typing it is highlighted in blue. The name is the function name that is placed to the right of the ' = ' sign and it must be given in accordance with the same rules as for variable names (Subsection 2.1.4). The input_parameters denotes a comma-divided variables which values should be transferred into the function; the output _ parameters is a list of those we want to process and derive from the function. The input parameters must be written between the parentheses and the output parameters - between square brackets. In the case of a single output parameter, the brackets can be dispensed with.

The input and output parameters of the function can be completely or partially omitted. Various possible views of the function definition line are given in the following examples:

```
function [A,B]=fun1(a,b,c,d) - full record, function named 'fun1' with four
                               input (a, b, c, and d) and two output (A and B)
                               parameters;
function A= fun2(a,b) - full record, function named 'fun2' with two input (a
                        and b) and one output (A) parameters;
function [A,B,C]= fun3 - function named 'fun3' with three output parameters
                         (A, B, and C), no input parameters;
function fun4(a,b,c) - function named 'fun4' with three input arguments (a, b,
                       and c), no output parameters;
function fun5 - function named 'fun5' with no input and output parameters.
```

Note:

- The 'function' word should be written in lower-case letters;
- The amounts and names of the input and output function arguments can differ from those given in the examples above.

Help Comments

The help lines containing the comments should be located just after the function definition line. The first line of the help should contain the function name and a short description of the function (this line is displayed by the lookfor command when it searches the information throughout all the MATLAB paths), e.g. typing lookfor in2mm (this function is shown in Figure 4) in the Command window yields:

```
>>lookfor in2mm
in2mm - converts inches to millimeters and meters
```

All comments that you wrote in the function help are displayed when the help command with a name of the function is entered, for example:

```
>> help in2mm
  in2mm converts inches to millimeters and meters
  l_in - input; length in inches
  l_mm and l_m - output; length in meters
```

Note, the help lines are not obligatory for the user-defined function, and can be omitted.

Hereafter, to be short, the user-defined function help part we will in no more than two lines: line of function explanation and line with an example run command.

Function Body, Local and Global Variables

The body of a function should comprise one or more commands for actual calculations. Between these commands, frequently at their end, should be placed the assignments to the output parameters; e.g., in the example in the Figure 4 the output parameters are l_mm and l_m thus the two commands of this function calculate and assign defined values to these parameters. When the function is executed, the actual values must be assigned to the input arguments, otherwise the commands of the function body cannot be performed.

Another used-defined function/s can be included in the function body; such functions are called subfunctions.

The variables in the user-defined function are local and relevant only within this function. That means, the variables are not saved and are no remain in the workspace after the completion of the function. To share some or all of the function variables with other function/s, we should make them accessible outside the function, which can be perform with the global command:

```
global variable_1 variable_2 …
```

The global command with one or more space-divided variable names (designated here as variable_1 variable_2 …) must be written within the function before the variable/s is/are firstly used and should be repeated also in other function/s where it/s is/are intended to be used.

3.2.2. Function File

Before running, the function written in the Editor window should be saved in a file. This is realized exactly as for the script file: select 'Save As' line of the Save option from the File section of the Editor tab and then enter the desired location and name of the file. It is strongly recommended to name the file by the function name, e.g. the in2mm function should be saved in a file named in2mm.m.

Examples of function definition lines and corresponding names of the function files:

```
function [F,tau]=friction_force(Mu,u,c) - the function file should be named
                                           and saved as  friction_force.m;
function screw(m,n) - the function file should be named and saved as screw.m;
function  [sigmax,sigmay]=stiffness  - the function file should be named and
                                        saved as  stiffness.m;
function  mydiffusion - the function file should be named and saved as
                        mydiffusion.m.
```

The name of a function file comprising additional sub-functions should match the name of the main function (from which the program starts).

3.2.3. Running a User-Defined Function

The user-defined function stored in a file can be launched from another file/program or from the Command Window as follows: the function file definition line should be typed without the word 'function' and to the input parameters should be assigned their values. The latter can be done by pre-assigning the values to the input parameters or strictly into the function launch command, replacing the variables with their values. For example, the in2mm function file (Figure 4 in chapter 3) can be run with the following command:

```
>>[l_mm,l_m]=in2mm([5.7 6.2])                        % l_in replaced by its values
l_mm =
  144.7800  157.4800
l_m =
    0.1448    0.1575
```

Alternatively, with pre-assignment of the input variables:

```
>> l_in=[5.7 6.2];                          % pre-assigns l_in with its values
>> [l_mm,l_m]=in2mm(l_in)                       % launches the in2mm function
l_mm =
  144.7800  157.4800
l_m =
    0.1448    0.1575
```

A user-defined function or its output parameters can be used in interactive or program calculations. For example, to calculate a cube volume $v=l^3$ in m^3 with the length cube edge length l given in inches, l=12 in, we should convert the inches to meters with the in2mm function and then calculates l^3; the following commands should be typed in the Command Window:

```
>>l_in=12;                   % assigns to l_in the length value  in inches
>>[l_mm,l_m]=ih2mm(l_in);                           % converts from inches to meters
>>V_cube = l_m^3          % uses the l_m output parameter of the in2mm V_cube =
    0.0283
```

Comparison of Script and Function Files

It is usually difficult for a beginner to understand the differences between script and function files. Indeed, most of the real problems can be solved using an ordinal script-type file. For a better understanding of two file forms presented in the chapter, their similarities and differences are outlined below.

- Both types of files are created using the Editor and saved with the extension m, as m-files.
- User-defined function, containing in the function file, has the function definition line as their first line; this feature is absent in a script file.
- The name of the function file must match the function name; this requirement does not make sense for the script file, since there is no definition line in the latter.

- Function files can receive/return data through the input/output parameters of the function; script files do not have this possibility and values of their variables should be assigned directly within the file or inputted, for example, with the `input` commands.
- Only the script files use the variables that were previously defined and are in the workspace.
- Only function files can be used as functions in other user-defined functions, or simply in the Command Window.

Ultimately, the user must decide for himself which file is preferable to develop in order to solve the problem.

3.3. INTERACTIVE SCRIPT AND FUNCTION PROGRAMS: LIVE EDITOR

As previously discussed, a program saved in a script or function files should be written in the Editor window, and then their launch and display of the results are performed in the Command Window. Thereby, two different windows should be used to work with these files. Moreover, when these programs generate any plots they used for this an additional graphical window (see Chapter 4). All this is not convenient, especially for non-programmers, it would be more suitable to use the same window for recording programs and for displaying results. Therefore, in 2016 year MATLAB® announced the new tool termed Live Editor that allows you to produce interactive script programs including commands, explanatory text and images together with numerical and graphical results of program executions. Starting with the R2018a version, the Live Editor features were expanded, and now the user-defined functions can also be created with this editor. Here we bring a brief description of the Live Editor and show creation of the live script file containing a M&T- oriented example.

3.3.1. Launching the Live Editor

To launch the Live Editor - click the New Live Script button located in the File section of the MATLAB® Desktop menu. The Live Editor window appears - Figure 5.

Figure 5. The Live Editor window

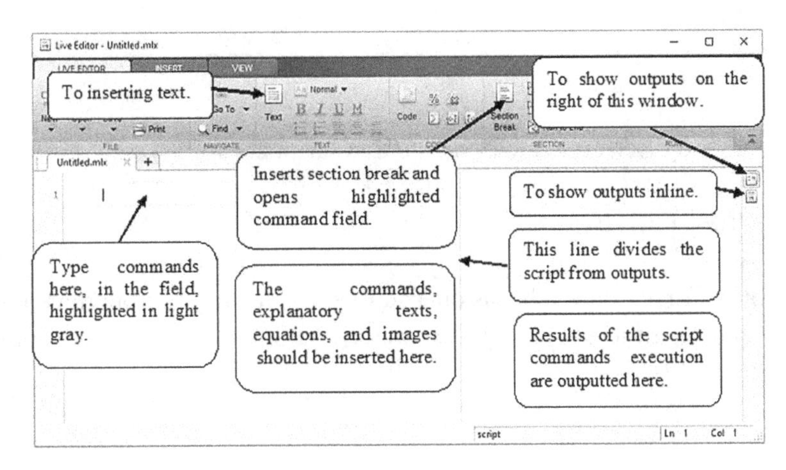

The toolstrip comprises three tabs – Live Editor, Insert, and View. The Live **Editor** tab contains the sections with buttons that use to edit the scripts. The Insert tab contains functions to insert texts, images, equations, code examples, section breaks, controls, and others to inserting items into the script. The View tab includes various buttons to the live script output. Below we describe those of the tabs buttons that are commonly used for creating actual live script.

The Live Editor window contain the vertical dividing line (default) that separates the window into two parts: the left - for entering commands, explanatory texts, equations, and images, and the right – to show output numerical and graphical results. To show outputs inline, not on the right, the "Show outputs inline" button should be clicked. In this case, the dividing line disappears and the resulting numbers and plots are outputting directly below the appropriate command/s.

3.3.2. Creating the Life Script With the Live Editor

To demonstrate the Live Editor use, write, for example, the program presented in Figure 2 for the gear tooth thickness.

Problem: Write the live script with the text and code parts that calculate and output inline the gear tooth thickness for the modulus m=2 and teeth number Z=12.

The program LiveEditorEx1 that solve this problem is presented in Figure 6.

Figure 6. Live Editor view with a live script program.

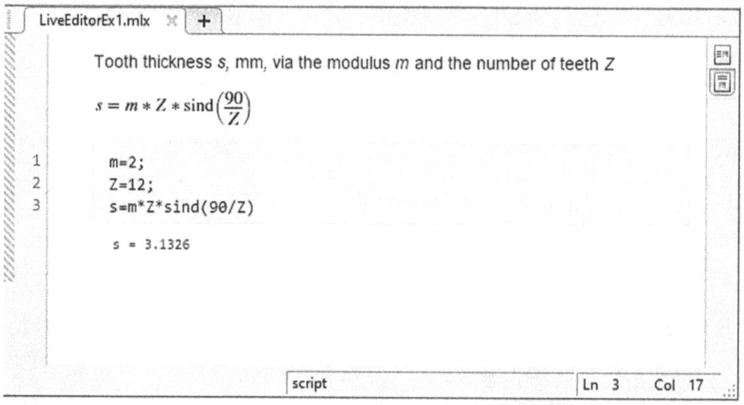

LiveEditorEx1.mlx

Tooth thickness s, mm, via the modulus m and the number of teeth Z

$$s = m * Z * \mathrm{sind}\left(\frac{90}{Z}\right)$$

```
1  m=2;
2  Z=12;
3  s=m*Z*sind(90/Z)
```

s = 3.1326

script | Ln 3 | Col 17

The following steps were performed to build the live script above.

Step 1. Inserting the Title/Text

The Text button of the Live Editor tab was clicked to insert the text "Tooth thickness s, mm, via the modulus m and the number of teeth Z". This text was typed into the white space appearing above the light gray command field.

Step 2. Inserting the Equation in the Text Field

The equation button of the 'Insert' tab, was to insert the solving equation as text. The equation was typed into the "Enter your equation" field appeared under the inserted image. After selecting the Equation option, in the toolstrip appears additional tab named Equation; the tab allows inputting the necessary symbols and mathematical structures to typing the equations.

Step 3. Entering Commands for Thickness Calculations

The following commands were typed within the highlighted command filed.

```
m=2;
Z=12;
s=m*Z*sind(90/Z)
```

Step 4. Selecting the Output Inline Option

The output to the right of the dividing line appears by default. To output directly under the command, the Output Inline button of the Lay Out section in the Insert tab was clicked to display thickness as in Figure 6.

Step 5. Running the Created Live Script

To run the live script, the Run button located in the LiveEditor tub of the toolstrip should be clicked. After some time that slightly greater than in case the simple script running, the results was outputted after the last command that was inputted for this without the ending semicolon.

Step 6. Saving Created Live Script

The script was saved in a file by clicking the Save button on the toolstrip and giving file name in the opened "Select file for Save as" window; the saved script file gets the *mlx* extension. So the life script file shown in the Figure 6. is named LiveEditorEx1.mlx.

For publishing, the script can be rearranged in the pdf-file by selecting the "Export to pdf …" line in the popup menu of the Save button.

3.3.3. SUPPLEMENTARY INFORMATION FOR USING THE LIVE EDITOR

Separate Sections in the Live Program

Sometimes, in especially when the program is long, it is better to divide the commands into separate sections. To do this, before writing the text and commands, the Section Break button (located in the Text section of the Insert tab) should be clicked. Where you want to finish this section and start a new one, you need to click the Section Break button again. To run section of the script, click the blue bar (vertical line of the Live Editor' frame) to the left of the section or alternatively click the run section button located in the "Section" section of the "Live Editor" tab.

Creating Live Function

Described above steps produce a live script. To create a live function, select the Live Function line in the popup menu of the New button located in the Home tab of the MATLAB® Desktop. The Live Editor opens with the untitled function (default) – Figure 7.

Figure 7. Live editor opened for creating the live function. The Untitled.mlx is the default file with example of function calculating z by the two variables x and y.

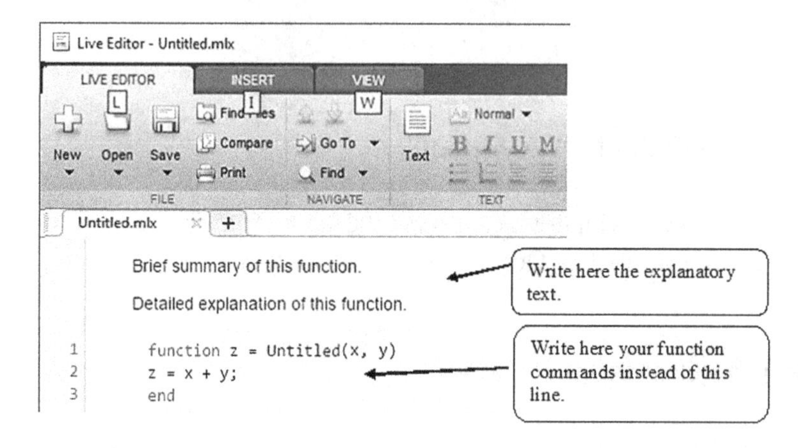

Note, the function in the Live Editor should has the end as its last line; the latter is accepted for so-called nested functions that are not studied in this book.

Two ways can be used to run the live function:

- Enter the function name with the input parameter values in the Command Window, in this case the answers are displayed in this window;
- Type the function name with the input parameter values in the Live Editor window before the live function definition line, in this case the answers are displayed in the Live Editor window before the function definition line (see subsection 3.4.5).

Opening the Existing Script as a Live Script and Existing Function as a Live Function

The existing m-file containing script can be opened as a live script mlx-file. For this, when the file opened in the Editor, click on the Save button of the Editor tab, select Save As line in the popup menu and in the "Save as type" field of the opened window select the "to MATLAB Live Code Files (*.mlx)" option and click Save. Opening/saving a script as a live script creates a copy of the original script. The comments of the script are converted into the texts of the generated live script.

The existing m-file containing function can be open as a live function mlx-file in the same way as for a script file - selecting the "to MATLAB Live Code Files (*.mlx)" option (see above).

Text Formatting Options

To format existing text the following Text section option of the Live Editor should be used, some of them are:

- Text Style: Normal, Heading 1, or 2, or 3, Title;
- Alignment: Left, Center, Right;
- Lists: Numbered list, Bulleted list;
- Standard Formatting: Bold, Italic, Underline, Monospace;
- Change Case (select the text and right click): uppercase, lowercase.

About the Interactive Controls

For interactive control of variables, the Live Editor has the sliders, dropdowns, check boxes, and buttons that can be added to live script or function. To insert a slider - click the Control button of the Insert tab and select Numeric Slider line; then specify the minimum, maximum, and step values of the variable.

For example, in discussed above live script, the liner for tooth number Z looks like Z= 29 10 100 ; . The value to the left is the current value of the Z. In case the drop-down control was selected, thus the command for Z looks like -  . The highlighted value is the current Z value.

Note: Live Editor is a new and more suitable tool for interactive programming than the regular MATLAB® Editor, but life scripts and programs work much slower than regular scripts and functions, and has restrictions on the use of certain classes of variables. All this makes it difficult to use the programs created with the Live Editor for problems with a large amount of data and/or with a large number of commands. Thus the user should choose the editor focusing on his specific problem.

3.4. APPLICATION EXAMPLES

3.4.1. Convertor for Stress Units

In the international system of units SI the stress is measured in newton per square meter, N/m^2, called pascal, Pa. In Imperial units stress is pounds-force per square inch, psi. The 1 psi in Pa calculates with the following relation 1, psi=6894.757, Pa, and 1Pa in psi is 1,Pa=0.0001450377.

Problem: Compose a script program that converts from psi to Pa and from Pa to psi. The program should request what the conversion you want and asks the input in psi or Pa in accordance to your choice. Save the program in the Ch_3_ApExample_1 file

The program that solved this problem is:

```
%                                              psi to Pa and Pa to psi conv ertor
%                                    type 1 for psi to Pa or type 2 for Pa to psi
%                           psi or Pa can be single number or vector, e.g. [29 30 32]
n=input('Enter 1- for psi to Pa \n   or 2- for Pa to psi\n  ');
if n==1
    psi=input('Enter stress in psi ');
    Pa=psi*6894.757;                                                  % psi to Pa
    Results=[psi;Pa];                                % result as two row matrix
    fprintf('\n %7.1f psi  is %7.1f in Pa\n',Results)
elseif n==2
    Pa=input('Enter stress in Pa ');
    psi=Pa*1.450377e-4;                                               % Pa to psi
    Results=[Pa;psi];                                % result as two row Matrix
    fprintf('\n %10.1f Pa  is %5.1f in psi\n',Results)
end
```

When the script file name Ch_3_ApExample_1 entered into the Command Window, the program displays the two-line prompt asks to enter 1 or 2 for psi to Pa or Pa to psi respectfully. After entering the number 1 or 2 by your choice, the "Enter stress in …" prompt appears. Type the stress/es value/s as single number or as a vector (in square brackets with spaces between the numbers); after the Enter key pressed the stresses are calculated and displayed in the entered and displayed units with one digital number after the decimal point.

The following shows the Command window when the file is running and the psi to Pa conversion is required:

```
>> Ch_3_ApExample_1
Enter 1- for psi to Pa
   or 2- for Pa to psi
 1
Enter stress in psi[29 30 32]
   29.0 psi  is 199948.0 in Pa
   30.0 psi  is 206842.7 in Pa
   32.0 psi  is 220632.2 in Pa
```

In case of the Pa to psi conversion:

```
>> Ch_3_ApExample_1
Enter 1- for psi to Pa
   or 2- for Pa to psi
 2
```

```
Enter stress in Pa [0.1e6 0.2e6 1e6]
  100000.0 Pa  is  14.5 in psi
  200000.0 Pa  is  29.0 in psi+
 1000000.0 Pa  is 145.0 in psi
```

3.4.2. The Stress Concentration Factor: Shaft With a Transverse Hole in Bending

The stress concentration factor K_t (for the bending a shaft of the diameter *D* with a transverse hole of diameter *d*) is calculated by the expression (Norton, 2010)

$$K_t = 1.58990 - 0.63550 \log \frac{d}{D}$$

Problem: Create program-function that calculates the stress concentration factor for the following rations of the *d*/*D* = 0.05, 0.1, 0.15, 0.2, 10.25, and 0.3. The program should consist from a main user-defined function with name Ch_3_ApExample_2 and subfunction named ShaftStress. The main function has not parameters, includes the d/D data, calls to the subfunction and displays results in two *d*/*D* and K_t columns. The subfunction includes the *d*/*D* and K_t as its input and output parameters respectfully, and he calculates the shaft stresses and transmits t to the main function.

The program solving the problem is written below.

```
function Ch_3_ApExample_2                          % main function, definition line
%                calculates the stress factor of the shift withhole in bending
%                                                       To run: >> Ch_3_ApExample_2
dD=.05:.05:.3;                                                % define the d/D values
Kt=ShiftStress(dD);                              % runs the subfunction ShiftStress
Results=[dD;Kt];                            % organize the output as two-row matrix
fprintf('d/D=%4.2f  Kt=%4.1f\n',Results)                       % outputs d/D and Kt
function Kt=ShiftStress(dD)                            % subfunction, definition line
Kt=1.58990-0.63550*log10(dD);                                       % calculates Kt
```

This program contains two user-defined function – the main function ApExample_3_2 that has not input and output parameters in its definition line, and the ShiftStress subfunction which has one input and output parameters each.

Inside the main function, the *d*/*D* values are assigned to the dD variable, K_t values are obtained using the ShiftStress subfunction, and the results are displayed with the fpringf function, which respectfully displays *d*/*D* and K_t values with two and one decimal digits. The ShiftStress subfunction inputs dD value and outputs Kt values that are calculated using the expression above.

The program run command, and the displayed numbers - inputted and calculated - are:

```
>> ApExample_3_2
d/D=0.05  Kt= 2.4
d/D=0.10  Kt= 2.2
d/D=0.15  Kt= 2.1
d/D=0.20  Kt= 2.0
d/D=0.25  Kt= 2.0
d/D=0.30  Kt= 1.9
```

3.4.3. Gear Worm Parameter

The lead angle λ (in degrees) of a worm gear and the ratio β (dimensionless) of the number of the worm teeth to the gear teeth number are related by the following expression (based on Spotts & Shoup, 1998):

$K = \beta/\sin\lambda + 1/\cos\lambda$

where K is a gear-worm parameter.

Problem: Compose the live script named ApExample_3_3 that calculates K for λ in the range 1° … 40° and β in the range 0.02 … 0.30. Use the "Numeric slider" for giving λ and β values and the "Output Inline" option to display the resulting K values. Give also explanatory text with the K relationship.

With the Live Editor window open, the steps to the solution are:

- For the further inline output, click the Output Inline button located within the View tab at the Layout section;
- Click the Text button in the Text section of the Live Editor tab and type an explanatory text in the white field (see Figure 8);
- Select the Equation line in the popup menu of the Equation button located in the Equation section of the Insert tab and type the K-equation into the "Enter your equation" field; use for this the Symbols (λ and β buttons) and Structures (Fraction button) sections of the appeared Equation tab;
- Click the Code button in the Code section of the Live Editor tab and type the 'beta=' command in the highlighted field; after that select the Numeric slider line in the popup menu of the Controls button located into the Insert tab.
- The line with three boxes Min, Step, and Max appears; type the β-values 0.02, 0.01, and 0.3 in the appropriate boxes;
- In the next code line, type the 'lambda='and select again the Numeric slider line in the popup menu of the Controls button located into the Insert tab; ; type the λ-values 1, 1, and 40 in the Min, Step, and Max boxes respectively;
- In the next code line type the command to the K-calculations
- Save the program with the Save button the same as it was explained for the regular Editor (subsection 3.1.1.).

- Set the beta and lambda sliders indicators by moving each of them to the desired λ and β values; the calculated K-value appears immediately below the last command; any changes in the indicator/s location/s lead to the recalculation and the appearance of a new K-value.

The live script created by the steps described is shown in Figure 3.8

Figure 8. The Live Editor window with the life script program for calculations of the gear worm parameter

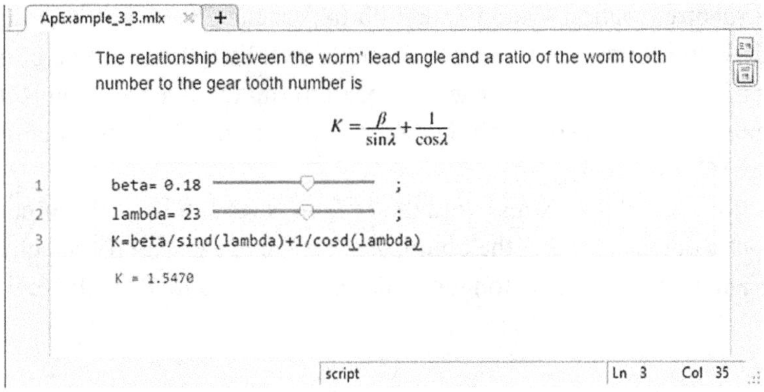

ApExample_3_3.mlx

The relationship between the worm' lead angle and a ratio of the worm tooth number to the gear tooth number is

$$K = \frac{\beta}{\sin\lambda} + \frac{1}{\cos\lambda}$$

```
1  beta= 0.18 ;
2  lambda= 23 ;
3  K=beta/sind(lambda)+1/cosd(lambda)

   K = 1.5470
```

script | Ln 3 | Col 35

3.4.4. Piston Ring. Installation and Operation Stresses

The installation stress S_a and operating stress S_w (both in MPa) on a piston ring can be calculated by the following expressions (based on MAHLE GmbH., 2016))

$$S_a = \frac{8}{3\pi}\frac{E\left(a_1 - t_y\right)\left(m_1 - m\right)}{\left(d_1 - a_1\right)^2}$$

$$S_w = \frac{8}{3\pi}\frac{Et_y\left(m - s_1\right)}{\left(d_1 - a_1\right)^2}$$

Where t_y is the radial distance from the neutral axis the ring running face, E – Young's modulus of the ring material, m – free gap, s_1 – gap in the installed state, d_1 –cylinder diameter, a_1 – ring radial dimensions, m_1=8a_1 – installation opening.

Problem: write the live script named Ch_3_ApExample_4 that obtains the stresses on piston ring at E=1.1·10^6 MPa, t_y=2.25·10^{-3} m, m=1.54·10^{-3} m, s_1=2.6·10^{-3} m, d_1=130·10^{-3} m, and a_1=4.45·10^{-3} m. The script should include function that calculates S_a and S_w values by the received the given parameters and transferring results to the script for displaying them with the disp command.

The steps to the solution are:

- Select the inline output for the final result displaying; click for this the Output Inline button located within the View tab at the Layout section;
- Click the Text button in the Text section of the Live Editor tab and type an explanatory text in the white field (see Figure 9);
- To type the stress expression into the text field, select the Equation line in the popup menu of the Equation button located in the Equation section of the Insert tab and type the *K*-equation into the "Enter your equation" field; in the appeared Equation tab use the Symbols section and the popup menu of the Structures section ('Show more' button) of the appeared Equation tab;
- Click the Code button in the Code section of the Live Editor tab and type commands for assignments of the t_y, E, m, s_1, d_1, and a_1; write command to run the live function RingStress that calculates m_1, S_a, and S_w; add in the one line the two output commands for displaying text and resulting stresses (Figure 9);
- Write the function named RingStress with the t_y, E, m, s_1, d_1, and a_1 as its input parameters and S_a and S_w as output parameters; enter the commands for the m_1, s_a, and S_w calculations.
- Save the program with the Save button the same as it was explained for the regular Editor (subsection 3.1.1.).

The solution written in the Live Editor is shown in the Figure 9:

Figure 9. The Life Editor window with the live script program for calculations of the stresses on a piston ring.

Ch_3_ApExample_4.mlx

The installation stress *Sa* and the operating stress *Sw* on a piston ring are

$$S_a = \frac{8}{3\pi}\frac{E(a_1 - t_y)(m_1 - m)}{(d_1 - a_1)^2}$$

$$S_w = \frac{8}{3\pi}\frac{Et_y(m - s_1)}{(d_1 - a_1)^2}$$

where $m_1 = 8a_1$

```
1   E=0.1e6;
2   ty=2.25e-3;
3   m=1.54e-3;
4   d1=0.13;
5   a1=4.45e-3;
6   s1=0.26e-3;
7   [Sa,Sw]=RingStress(E,ty,m,d1,a1,s1);
8   fprintf('Stresses on Piston Ring\n   Sa,MPa    Sw,MPa\n'),disp([Sa Sw])

    Stresses on Piston Ring
       Sa,MPa    Sw,MPa
      403.5086   15.5088

9   function [Sa,Sw]=RingStress(E,ty,m,d1,a1,s1)
10          m1=8*a1;
11          Sa=8/(3*pi)*E*(a1-ty)*(m1-m)/(d1-a1)^2;
12          Sw=8/(3*pi)*E*ty*(m-s1)/(d1-a1)^2;
13      end
```

Note, to output the explanatory text strongly above the resulting numbers the fprint and disp commands must be written in the same line as in the case of ordinary interactive calculations.

3.4.5. A Two-Stage Gear Train Reduction Speeds, Teeth Numbers, and Torques

A two-stage, compound reverted gear train as per Figure 10 should be designed (Budinas & Nisbett, 2011). The gear numbers are given on this figure, the expressions that should be used for gear speeds ω (rev/min), teeth number, N (units), torques T (N·m), and gear ratios (units) are:

- $e = \frac{\omega_5}{\omega_2} = \frac{N_2}{N_3}\frac{N_4}{N_5}$ - train value, $\omega_2 = \omega_{in}$ and $\omega_5 = \omega_{out}$ are the input and output rotation speeds respectively that should be specified
- $\frac{N_2}{N_3} = \frac{N_4}{N_5} = \sqrt{e}$ – teeth ratio for equal reduction on the both gear train stages,
- $m = \frac{1}{\sqrt{e}}$ – gear to pinion teeth ratio,

  $N_2 = \frac{2k}{(1+2m)\sin^2\varphi}\left(m + \sqrt{m^2 + (1+2m)\sin^2\varphi}\right)$ – number of teeth on the pinion, for 20 degree of the pressure angle, φ,
- $N_3 = mN_2$ – number of teeth of the third pinion,
- $N_4 = N_2$ - number of teeth of the fourth pinion,
- $\omega_5 = \frac{N_2}{N_3}\frac{N_4}{N_5}\omega_{in}$ - output speed, if this value is not within the specified limits then try again the calculation with a new initial ω_5 selected within the required limits,
- $T_2 = \frac{60 \cdot 745.7H}{2\pi\omega_2}$ – torque on the input shaft,
- $T_3 = T_2\frac{\omega_2}{\omega_3}$ – torque on the median shaft,
- $T_5 = T_2\frac{\omega_2}{\omega_5}$ – torque on the output shaft.

where *H* is the power to be delivered, hp; m is the teeth ratio, units; *k*=1 for studied case of the full-depth teeth; $_{in}$ - is the lower index that designates the input speed; the numbers 2, 3, 4, and 5 designate the number of gear units.

Problem: write a live program named Ch_3_ApExample_5 that calculates the above gear train parameters. The program should include the command to run the life function and the live function itself named gear, which has not output parameters, and has the following input parameters: ω_{in} (input speed), ω_{out} (output speed), $\triangle$ (possible error in the ω_{out}), *H*, and *k*. The values for input are: ω_{in} = 1750 rpm, ω_{out}= 85 rpm, $\triangle$=3 rpm, H=20 hp, and k=1. The live function should include figure representing the solved gear train and display the calculated value with the appropriate header using the disp command.

The steps to the solution are:

- Select the inline output for the final result displaying; click for this the Output Inline button located within the View tab at the Layout section;
- Click the Text button in the Text section of the Live Editor tab and type an explanatory text - Typical two-stage gear reduction - in the white field (see Figure 10);
- Write the command that will call the live function gear that will be written immediately after this line, the function call is the name of function with parameter values to be transmitted to the function;
- Click the Image button in the Insert tab of the Live Editor and in the appeared 'Load image' box select the file containing the gear train image, this file you should prepare in advance;
- Write down the line of definition of the transfer function with the input parameters required in the statement of the problem (Live Editor should add the final end to the entered word function) and then type the help line with a brief explanation of the purpose of the function;
- Write the calculating commands correspondingly to the above equations, after each part of calculations – teeth, speeds, and torques – add the appropriate disp commands to display the results;
- In the final part write the if ... else ... end statement to compare the calculated and required output speed values, write in this statement strings comment the comparing results.

The Live Editor program with the obtained results is shown in Figure 10; since the relatively large size of the program, the Live Editor window is not copied here with its menu and borders, but only with its contents.

Typical two-stage gear reduction

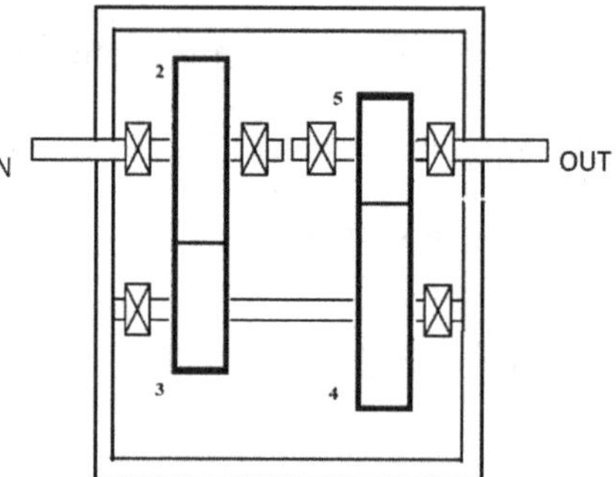


```
gear(1750,85,3,20,1)
Teeth, unit
    N2     N3     N4     N5
    16     72     16     72
Speeds, rpm
       omega_2       omega_3       omega_4       omega_5
         1750         388.9         388.9         86.42
Torques, N.m
      torque_2      torque_3      torque_5
       81.382        366.21         1648
The output gear speed is acceptable
```

```
function gear(om_in,om_out,om_del,H,k)
% gear - calculates the gear teeth numbers, speeds, and torques
om2=om_in;om5=om_out;
e=om5/om2;
m=1/sqrt(e);
phi=20*pi/180;
% Teeth calculations
disp('Teeth, unit')
N2=2*k/((1+2*m)*sin(phi)^2)*(m+sqrt(m^2+(1+2*m)*sin(phi)^2));
N2=ceil(N2);                                % teeth, rounded to inf
N3=fix(m*N2);                            % teeth, rounded down to 0
N4=N2; N5=N3;
format shortg
disp('    N2     N3     N4     N5')
disp([N2 N3 N4 N5])
% Speed calculations
disp('Speeds, rpm')
om5_cal=(N2/N3)*(N4/N5)*om2;
om3=round(N2/N3*om_in,1);
om4=om3;om5=om5_cal;
disp('       omega_2       omega_3'      omega_4       omega_5')
disp([om2 om3 om4 om5])
% Torque calculations
disp('Torques, N.m')
T2=H/om2*745.7/(2*pi)*60;
T3=T2*om2/om3;T5=T2*om2/om5;
disp('      torque_2      torque_3       torque_5')
disp([T2 T3 T5])
% Output speed check
om5_up=om_out+om_del;om5_low=om_out-om_del;
if om5_cal<=om5_up&om5_cal>=om5_low
    disp('The output gear speed is acceptable')
else
    disp('The output gear speed is unacceptable, try with
another initial output speed')
end
format
end
```

3.5. CONCLUSION

Two code editors, regular and live, were explored and used to creation, saving, and running the program scripts and functions. Appropriate commands and manipulations were studied. The effectiveness of working with programs created and saved by the editor is illustrated at first with simple examples from the M&T field and then with more complicated problems.

The developed programs solve the following M&T problems:

- stress unit conversion: psi to Pa and Pa to psi,
- stress concentration factor for the shaft with a transverse hole in banding,
- gear worm parameter,
- piston ring installation and operation stresses,
- two-stage gear train reduction speeds, teeth numbers, and torques.

It is noted that the interactive, Live Editor, and the regular, Editor, should be used for program creating program to repetitive calculations. Each time one of the editors can be used for the actual problem solution. For example, problems requiring a long time solution, a regular Editor is more suitable, in many other cases – the Live Editor.

For better understanding the script and function files their comparison are presented. It is concluded that the user needs to decide for himself which type of file is preferable to develop to solve the problem.

REFERENCES

Budinas, R. G., & Nisbett, G. K. (2011). *Shigley's Mechanical Engineering Design* (9th ed.). McGraw-Hill.

MAHLE GmbH. (2016) *Cylinder Components Properties, Applications, Materials* (2nd ed.). Springer. https://link.springer.com/content/pdf/bfm%3A978-3-658-10034-6%2F1.pdf

Norton, R. L. (2010). *Machine Design. An Integral Approach* (4th ed.). Prentice Hall.

Spotts, M. F., & Shoup, T. E. (1998). *Design of Machine Elements*. Prentice Hall.

APPENDIX

Table 1. List of examples, problems, and applications discussed in the chapter

No	Example, Problem, or Application	Subsection
1	US density units to SI convertor.	3.1.2.
2	Convertor inches to milimeters and meters.	3.2.1., 3.2.3.
3	Gear tooth thickness calculations using the Live Editor.	3.3.2.
4	Convertor for stress units.	3.4.1.
5	The stress concentration factor: shaft with a transverse hole in bending.	3.4.2.
6	Gear worm parameter.	3.4.3.
7	Piston ring. Installation and operation stresses.	3.4.4.
8	A two-stage gear train reduction speeds, teeth numbers, and torques.	3.4.5.

Note, some small examples, mostly related to non-M&T issues, are not included in the list.

Chapter 4
Basics of Graphics With Applications

ABSTRACT

The basic, special, and additional commands for generating two- and three-dimensional graphs are presented. It describes formatting commands for inserting labels, headings, texts, and symbols into a plot, as well as color, marker, and line qualifiers. Graphs with more than one curve and graphs with two Y axes are discussed. The possibilities of creating multiple plots on one page are shown. All the commands studied are presented with examples from the field of mechanics and tribology (M&T). At the end of the chapter, applications are given; they illustrate how to generate 2D and 3D graphs for engine piston velocity, power screw efficiency, engine oil viscosity, and a number of other M&T problems.

INTRODUCTION

It is widely accepted to visualize the results of observations, tests, or theoretical and engineering calculations in the form of graphs or diagrams. Such practice are widespread in engineering and science in general and in mechanics and tribology in particular. MATLAB® provides many commands for such purposes. Available commands allow you to generate two- (sometimes called XY or 2D), three-dimensional (XYZ or 3D), and some science-oriented graphics.

Two-dimensional graphics commands allow you to generate various linear, nonlinear, and semi- or logarithmic graphs, columns, histograms, pie charts, and scatterplots to name a few options. Two or more curves can be plotted on the same graph, while several plots can be presented in a separate Figure window. Produced graphs can be formatted by commands or interactively to generate the desired line style or marker shape, thickness or color, with the addition of a grid, text, caption, and legend.

DOI: 10.4018/978-1-7998-7078-4.ch004

3D graphics commands used to represent data with three variables. There are various tools for visualizing three-dimensional data. These tools allow you to build plots with spatial lines or meshs and surfaces, as well as various geometric shapes and images. Like 2D graphics, 3D graphics can be formatted using commands or interactively from the drawing window.

The most important commands that are used for two- and three-dimensional images will be presented below. The following description is based on the assumption that the reader has carefully studied the previous chapters of the book. Therefore, commands are written in most cases without explanatory comments (%), and the necessary comments are given directly in the text of the chapter.

4.1. TWO-DIMENSIONAL PLOTS: CREATION AND FORMATTING

4.1.1. One Curve on the 2D Plot

The most important basic command used to build XY is the plot command, the simplest forms of which are:

```
plot(y)
```

or

```
plot(x,y)
```

where x and y are two vectors of equal length, the first being used for horizontal, x, while the second is for the vertical, y, axes.

The first form of this command draws line by the points at which the y-vector values are the y-coordinates of the points to be plotted, and the y-vector value places (indices) are the x-coordinates. The second form of the plot command requires the both vectors – x and y with the x, and y-coordinates of each of the points.

After inputting the plot command, the curve y(x) is created in the MATLAB® Figure window. By default, a linear scale graph is generated with a blue solid line between the unmarked points.

For example, the wear w of a micro-milling tool was measured at different sliding distance, s. The obtained data is: w=24, 35, 38, 64 and 90.1 μm, s = 24.5, 41, 83, 115, and 126 m. To present this data on a plot with the x-axis as the sliding distance and the y-axis as wear, we must type the following commands in the Command Window:

```
>> w=[15 24 35 38 64];
>> s=[0 24.5 41 83 115];
>> plot(s, w)
```

After entering these commands, the Figure window is opened with the $w(s)$ plot as shown in Figure 1.

Figure 1. Figure window with the wear w of a milling tool, default settings.

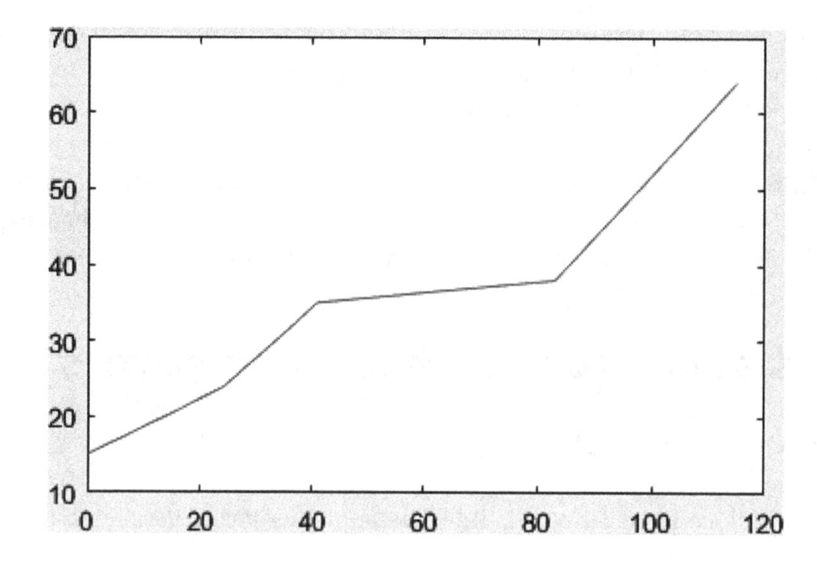

The line style, marker type, thickness, color and other parameters of the graph were performed by default. To change the default setting these parameters may be optionally added to the plot command just after the x and y identifiers:

```
plot(x,y,'Line_Specifiers','Property_Name','Property_Value',...)
```

where the Line_Specifiers parameter/s determine/s the line type, marker symbol and color of the plotted lines as per Table 1; the pair 'Property_Name' – 'Property_Value' represents the one of the possible curve property: the Property_Name is a string with name of the desired property; and the Property_Value is a number or a quoted string that specifies the property itself - see Table 2. There may be more than one property pairs.

In tables 1 and 2, the ylabel command was used, which will be explained later in subsection 4.1.4.1. The line specifiers, property names and its values (when the value is a string) are typed in the plot commands in inverted commas. The symbols within the specifiers and property name -value pairs can be written in any order, with the option of omitting any of them. Omitted properties will be assigned by default.

Below are some additional examples of the plot commands with various Line Specifiers and Name-Value properties:

- plot(y,'-m') generates the magenta solid line between the unmarked points, with x-equidistant point coordinate (addresses of the y-values located in the vector y) and y - as y-coordinates.
- plot(x,y, 'o') generates the points (not connected by a line) with x,y –coordinates marked by the circle.
- plot(x,y,'y') generates the yellow solid (default) line that connects by the points marked by default ('.').
- plot(x,y,'-h') generates the blue (default) solid line that connects the points marked with the six-pointed asterisks.
- plot(x,y,'k:d') generates the black dotted line with points marked with the diamond.

Table 1. The 'Line_Specifiers' *for the plot command.*

Specifier symbol (marker type)	Graphical representation	Specifier symbol (line or color)	Graphical representation

* (asterisk)	`>> y=1;`
	`>> plot(y,'*')` 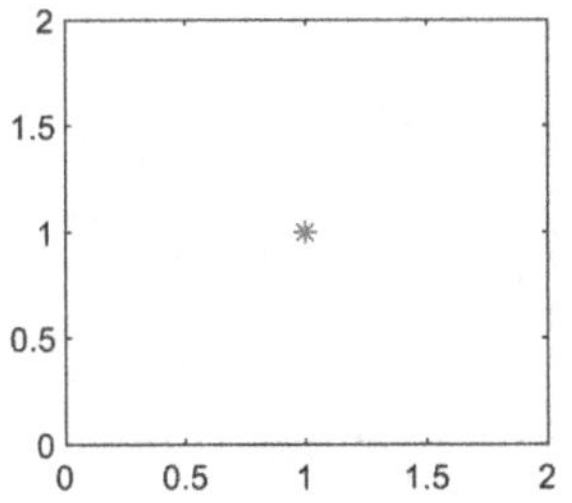
		-. (dash-dot)	`>> y=[24 38 38 64];`
			`>> plot(y,'-.')`
			`>> ylabel('w')` 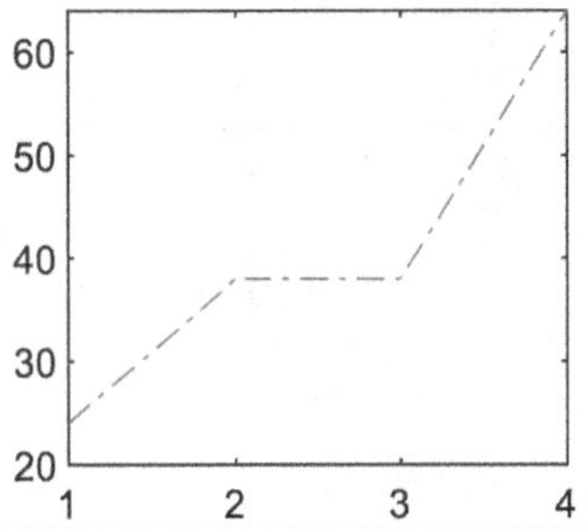

o (circle)	`>> y=1;`
	`>> plot(y,'o')` 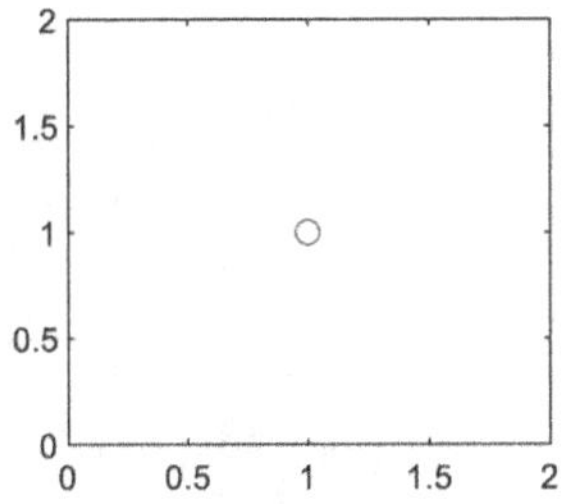
		-- (dased; two minuses without space between them)	`>> y=[24 38 38 64];`
			`>> plot(y,'--')`
			`>> ylabel('w')` 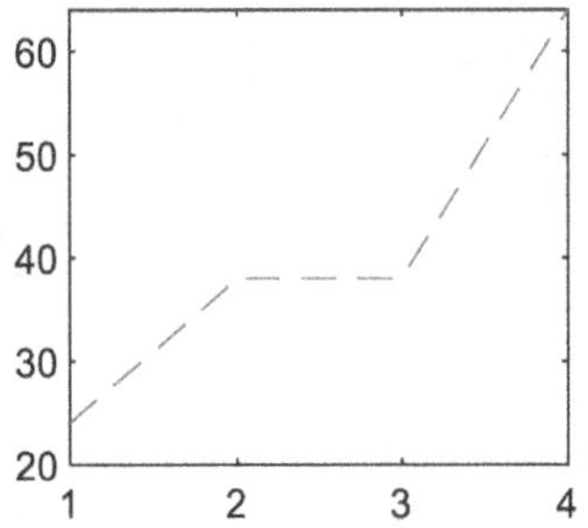

x (cross)	`>> y=1;`
	`>> plot(y,'x')` 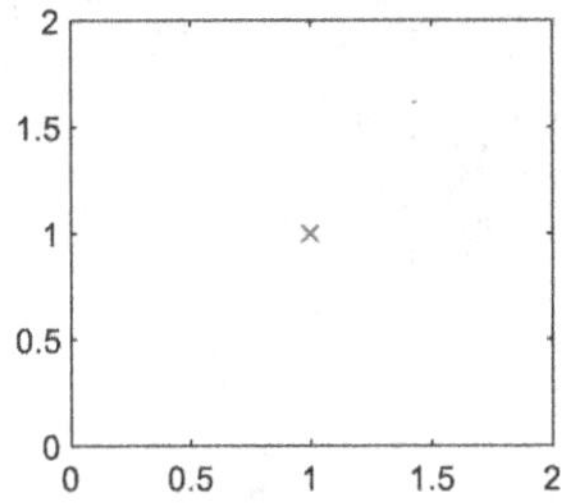
		: (dotted)	`>> y=[24 38 38 64];`
			`>> plot(y,':')`
			`>> ylabel('w')` 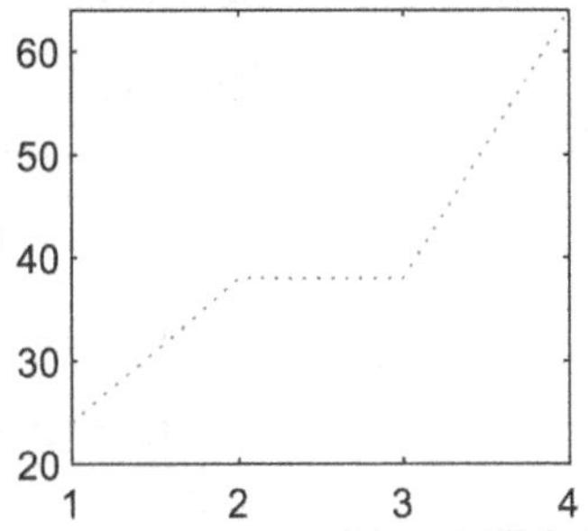

continued on following page

Table 1. Continued

d
(diamond)
	`>> y=1;`
	`>> plot(y,'d')`
		-
		(solid;
		default)
			`>> y=[24 38 38 64];`
			`>> plot(y,'-')`
			`>> ylabel('w')`

p
(five-pointed asterisk)
	`>> y=1;`
	`>> plot(y,'p')`
		k
		(black)
			`>> y=[24 38 38 64];`
			`>> plot(y,'-k')`
			`>> ylabel('w')`

.
(point)
	`>> y=1;`
	`>> plot(y,'.')`
		b
		(blue; default for single line)
			`>> y=[24 38 38 64];`
			`>> plot(y,'-b')`
			`>> ylabel('w')`

+
(plus)
	`>> y=1;`
	`>> plot(y,'+')`
		c
		(cyan)
			`>> y=[24 38 38 64];`
			`>> plot(y, '-c')`
			`>> ylabel('w')`

continued on following page

Table 1. Continued

S
(square)
	`>> y=1;`
	`>> plot(y,'s')` 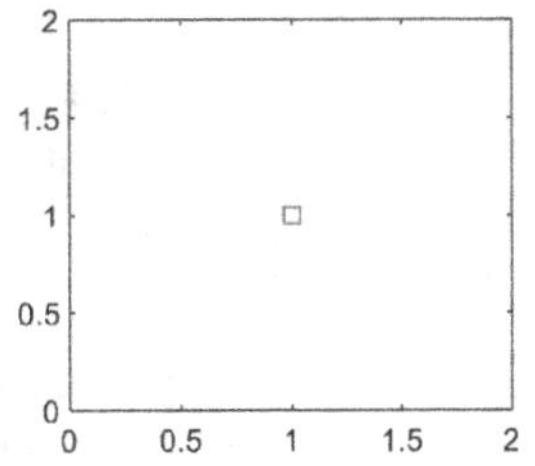
		g
		(green)
			`>> y=[24 38 38 64];`
			`>> plot(y,'-g')`
			`>> ylabel('w')` 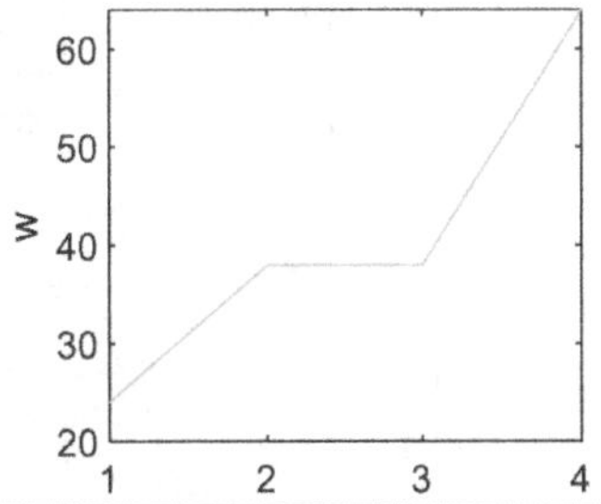

h
(six pointed asterisk)
	`>> y=1;`
	`>> plot(y,'h')` 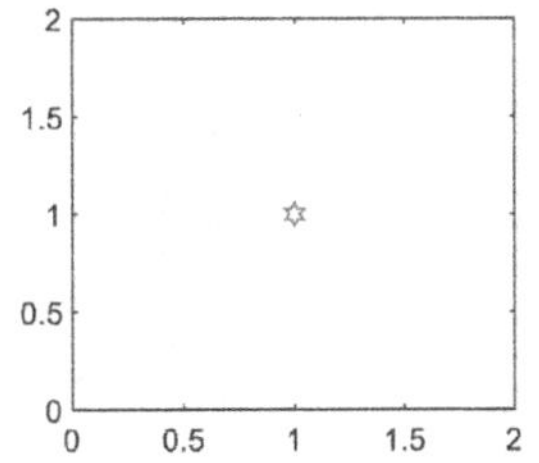
		r
		(red)
			`>> y=[24 38 38 64];`
			`>> plot(y,'-r')`
			`>> ylabel('w')` 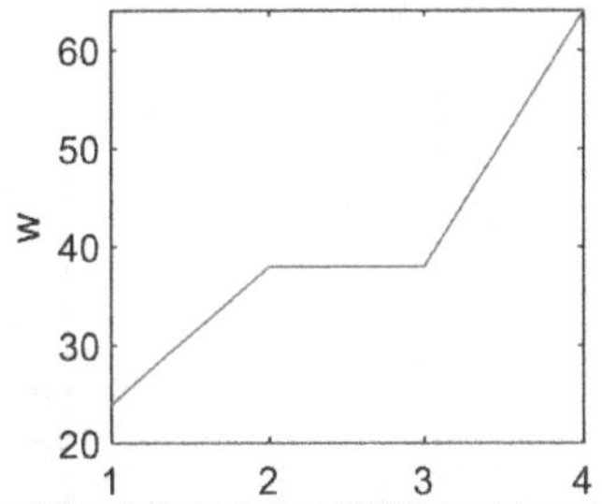

V
(inverted triangle; v-key)
	`>> y=1;`
	`>> plot(y,'v')` 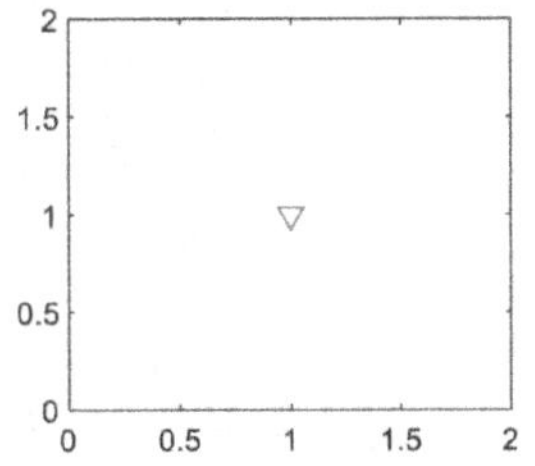
		m
		(magenta)
			`>> y=[24 38 38 64];`
			`>> plot(y,'-m')`
			`>> ylabel('w')` 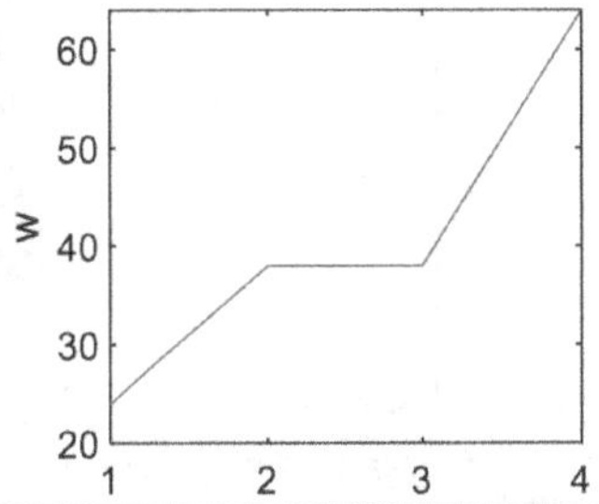

^
(triangle; upright-key)
	`>> y=1;`
	`>> plot(y,'^')` 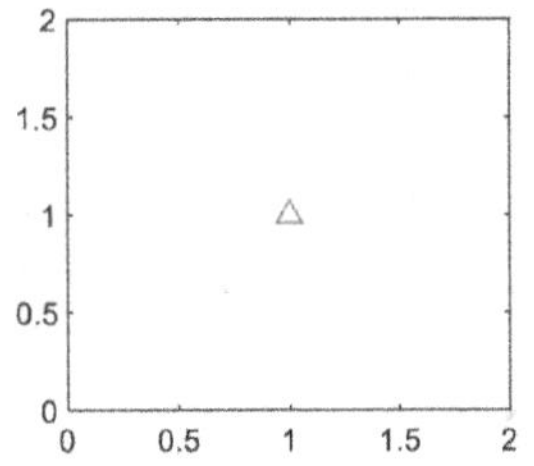
		y
		(yellow)
			`>> y=[24 38 38 64];`
			`>> plot(y,'-y')`
			`>> ylabel('w')` 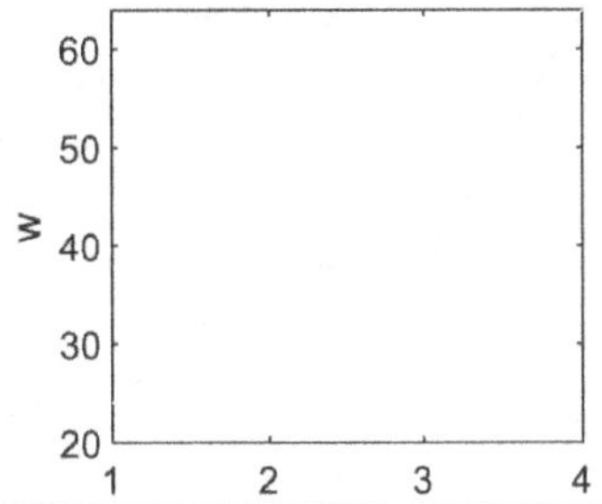

Table 2. Commonly used the Property_Name - Property_Values *pairs for the plot command*

Property name denotation	What means	Property value	Graphical representation
LineWidth or linewidth	The width of the drawn line	A decimal number that represents the line width in the points. (1 point is 1/72 inch or approximately 0.35 mm). The default line width is 0.5 points.	`>> y=[24 38 38 64];` `>> plot(y,'-b','linewidth',10)` `>> ylabel('w')` 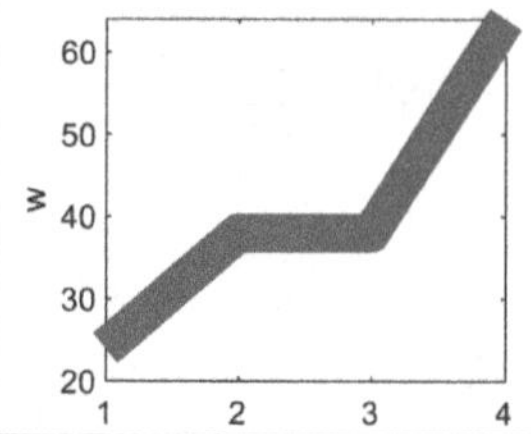
MarkerSize or markersize	The size of the marker	A decimal number in points Default value is 6. In case of the '.' marker – 1/3 of specified size	`>> y=1;` `>> plot(y,'o','markersize',20)` 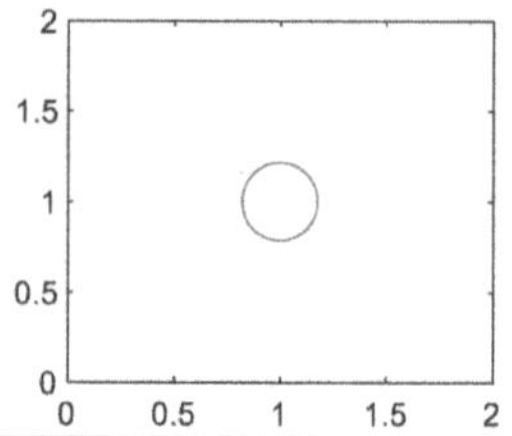
MarkerEdgeColor or markeredgecolor	Marker color for empty markers, or edge line color for filled markers	A character in accordance with the color_specifiers (Table 4.1)	`>> y=1;` `>> plot(y,'o','markeredgecolor','r','markersize',20)` 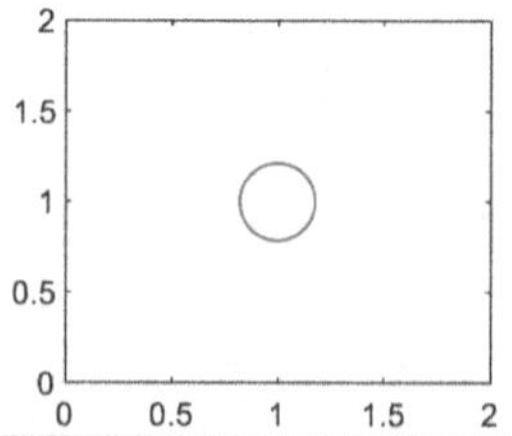
MarkerFaceColor or markerfacecolor	The fill color for markers that have closed area (e.g. circle or square)	A character in accordance with color specifiers in Table 4.1	`>> y=1;` `>> plot(1,'o','markerfacecolor','g', ... 'markeredgecolor','r','markersize',20)` 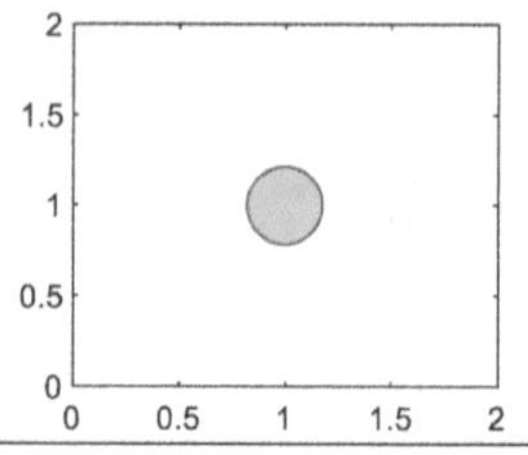

- plot(x,y,'-gx', 'markersize',10,'markeredgecolor','g') generates a green solid line (by width 0.5 points) with 10-point size marker represented by the green crosses (x).
- plot(w,s,'-ms','LineWidth',4,'MarkerSize',15,'MarkerEdgeColor','k', 'MarkerFaceColor','y') plots as magenta 4-points solid line with w, s -values marked with the 15 points black-edged yellow squares.

Using the previously entered wear-sliding distance data and entering the last of the above commands we can obtain the plot shown in Figure 2.

Figure 2. The wear-sliding distance data generated by the plot command with specifiers and property settings.

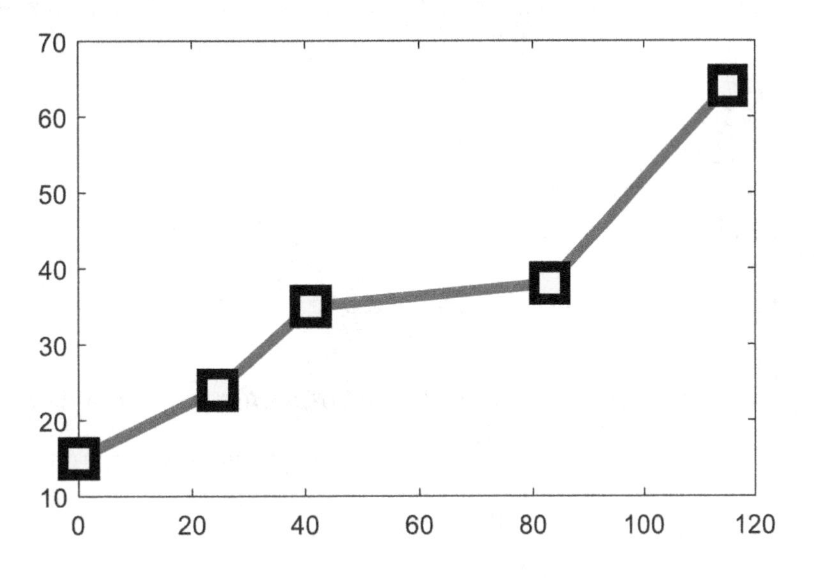

The plots should be generated by x,y-values, that can be specified in tabular form or determined from the $y(x)$ expression. In latter case, the vector of y-values should be calculated at the given vector of x-values (see Application examples, Subsection 4.4).

Note:

- To close a single Figure window, type and enter the close command in the Command Window; to close more than one Figure window, use the close all command.
- With each input of the plot command, the previous plot is deleted.

4.1.2. Multiple Curves on the 2D Plot

To generate two or more curves on the same plot, the two most popular options can be used: a plot command with more than one pairs of equal-sized x,y-vectors or use the hold on command to save the generated curve and then add a new one. Consider both of these options.

The Plot Command for Multiple Curves

The commands for generating more than one curves in the same plot have the following form:

```
plot(x1,y1,x2,y2,…,xn,yn)
```

where x1 and y1, x2 and y2, …, xn and yn are pairs of equal-length vectors containing the *x*,*y*-coordinates of the plotting points. Thts command creates plot with two, three, or n curves. For example, let's generate in a single plot two curves of the friction coefficient – normal force data for two different materials (e.g. rubber and aluminum). The normal force F_N are 1, 3, 10, 32, and 40 N. Friction coefficient $\mu = 1.2$, 0.75, 0.5, 0.37, 0.32 for one of the two substances, and $\mu = 1.55$, 1.1, 0.81, 0.69, 0.65 (dimensionless).

To generate two the μ ($F_{N)}$ curves, enter the following commands (without comments) in the Command Window:

```
>>FN=[1 3 10 32 40];                          %creates vector with normal forces
>>mu1=[1.2 0.75 0.5 0.37 0.32];                        %creates vector with mu1
>>mu2=[1.55 1.1 0.81 0.69 0.65];                       %creates vector with mu2
>>plot(FN,mu1,FN,mu2,'--k')                %creates solid and black dashed lines
```

The resulting plot with two curves is shown in Figure 3.

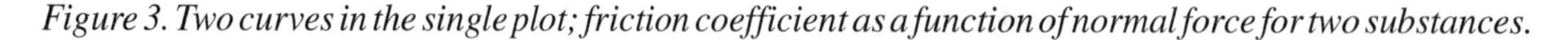

Figure 3. Two curves in the single plot; friction coefficient as a function of normal force for two substances.

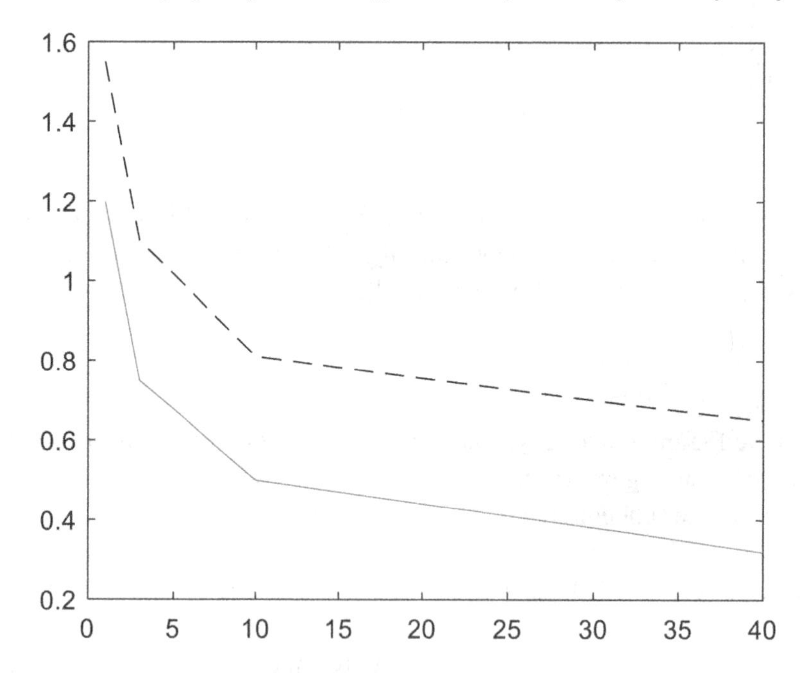

To generate more than two curves in the same plot, the new *x*- and *y*-vectors should be introduced in the plot command for each additional curve.

Note: When the *y*-data is specified as a n-column matrix (not a vector), the plot(*x*,*y*) command draws *n* curves in the same plot, each curve corresponding to *y*-values in the columns; e.g. the two curves shown in Figure 3 can be generated by entering the following commands:

```
>>mu=[mu1;mu2]'                    % joins two vectors into 2-column matrix
>>plot(FN,mu)                      % plots two columns as two mu(FN) curves
```

Using the Hold Command for Generating Multiple Curves

Another option for drawing two or more curves in the same plot is to add a new curve to an existing plot. To perform this, after the first plot creation, type the hold on command and then enter a new plot command with the new curve coordinates. To stop the hold on process, enter the hold off command, as result of which the next curve will be appeared without previously plotted curves. For example, the new series of friction cofficent values obtained for a third material of any mechanical part: μ = mu3=1.3, 1.0, 0.65, 0.52, 0.47, given at the same normal forces as the previous two μ -series, can be added to the existing graph (Figure 4.3) by additional entering the following commands:

```
>> mu3=[1.3 1.0 0.65 0.52 0.47];                    % creates new mu-vector
>> hold on                  % holds previous graph and waits for new line
>> plot(FN,mu3,':r')                       % adds dotted red line to the plot
>> hold off                                   % cancels the hold on mode
```

Generated plot is shown in Figure 4.

Figure 4. Three curves in the same plot; the 3rd curve (new friction coefficient series) produced with the hold on/off commands

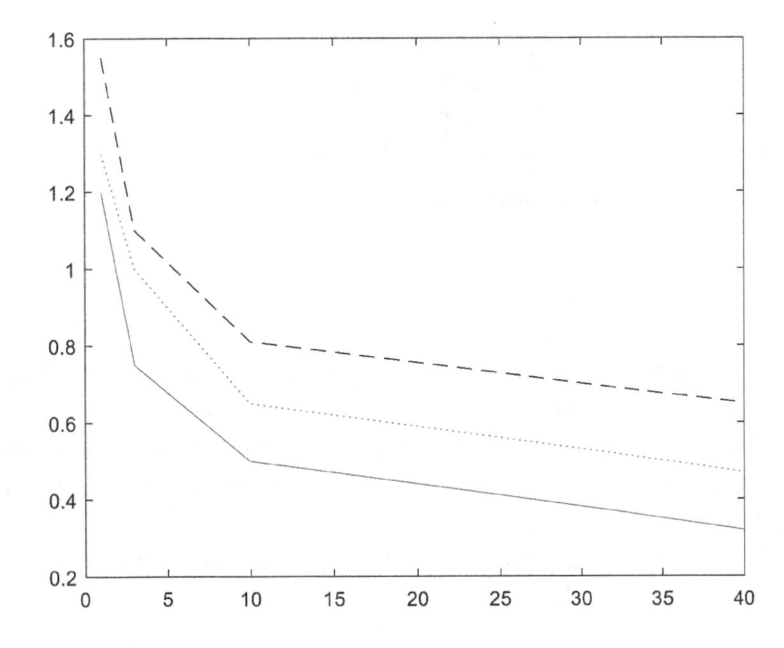

4.1.3. Several Plots in the Same Figure Window

In engineering and science, it is often necessary to place several graphs on one page or, in other words, in the same Figure window. For this purpose, the subplot command should be used; the command breaks the Figure window (and, accordingly, the page, when this window is printed) into *m*-by-*n* rectangular panes. The command can be written in two forms:

```
subplot(m,n,P)
```

and

```
subplot mnP
```

where m and n are the rows and columns of the panes and P is the current pane number whre the plot will be created (see panes and pane numbering, Figure 5, a). The P within the first command form can be written as vector containing two or more numbers of the adjacent panes, that leads to the asymmetrical arrangement of the Figure window (as, for example, in Figure 5, b).

Figure 5. The view of panes in Figure window (page) arranged in four (a) and three (b) rectangular panes.

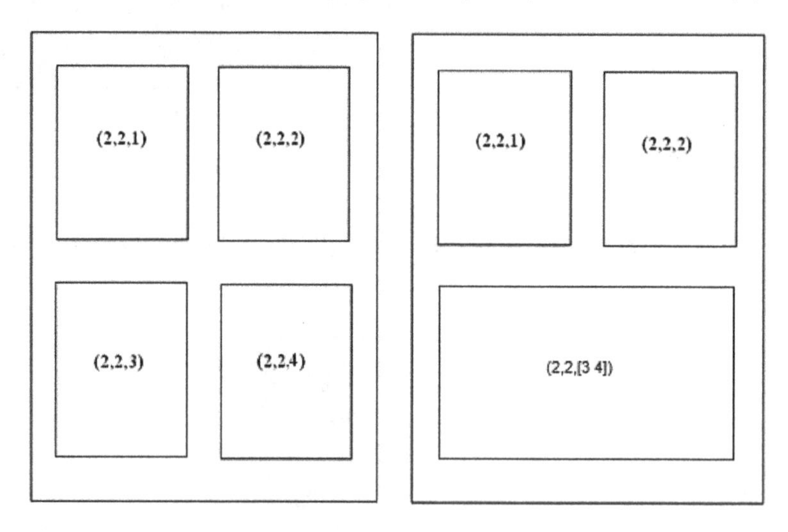

Some examples of the subplot command:

- subplot(3,2,4) or subplot 324 breaks the page into six rectangular panes arranged in 3 rows and 2 columns and makes pane 4 current.
- subplot(2,2,[2 4]) breaks the page into three rectangular panes with first and third panes of regular sizes and enlarged third pane that combines the second and fourth panes (top and bottom right panes); the enlarged pane is current.
- subplot(2,3,3) or subplot 233 breaks the page into six rectangular panes arranged in 2 rows and 3 columns and makes pane 3 current.

- subplot(2,1,1) or subplot 211 breaks the page into two rectangular panes arranged in a single column and makes the first pane current.
- subplot(1,2,2) or subplot 122 breaks the page into two rectangular panes arranged in a single row and makes the second pane current.

The plots should be placed in the current pane with the plot command. As an exemple, generate three plots on one page, placed three previously plots with one, two, and three curves of friction coefficients (as per data used for generating the Figure 3 and 4) in three panes accordingly to the Figure 5, b:

- In the top left pane, generate the $k(s)$ data plot; make two points width solid line, seven-point o-marker size, red edge and yellow face marker colors.
- In the top right pane, generate two $k(s)$ curves; leave the first line specifiers the same, mark the points of the second line with the seven-point inverted triangles, make the added line as a red dash-dotted line with a width of two points.
- In both right panes, generate three $k(s)$ curves; mark the third curve points with the seven-point triangles, make the line green and dashed with a width of two points.

The MATLAB® command for interactive generating three plots in the same Figure window:

```
>>FN=[1 3 10 32 40];                                         % vector with FN data
>>mu1=[1.2 0.75 0.5 0.37 0.32];                              % vector with mu1 data
>>mu2=[1.55 1.1 0.81 0.69 0.65];                              %vector with mu2 data
>>mu3=[1.3 1.0 0.65 0.52 0.47];              % vector with mu3 data subplot(2,2,1)
                                                             % makes pane 1 current
>>plot(FN,mu1,'-ok','LineWidth',2,'MarkerSize', 7,'markeredgecolor', 'r',
'MarkerFaceColor','y')
>>subplot(2,2,2)                                             % makes pane 2 current
>>plot(FN,mu1,'-ok',FN,mu2,'-.vr','LineWidth',2, 'MarkerSize',7,'markeredgecol
or', 'r', 'MarkerFaceColor','y')
>>subplot(2,2,[3 4])                                  % makes panes 3 and 4 current
>>plot(FN,mu1,'-ok',FN,mu2,'-.vr',FN,mu3,'--^g', 'LineWidth',2,'MarkerSize',7,
'markeredgecolor', 'r', 'MarkerFaceColor','y')
```

The resulting plot is shown in Figure 6.

Figure 6. Figure window with three frction coefficient plots.

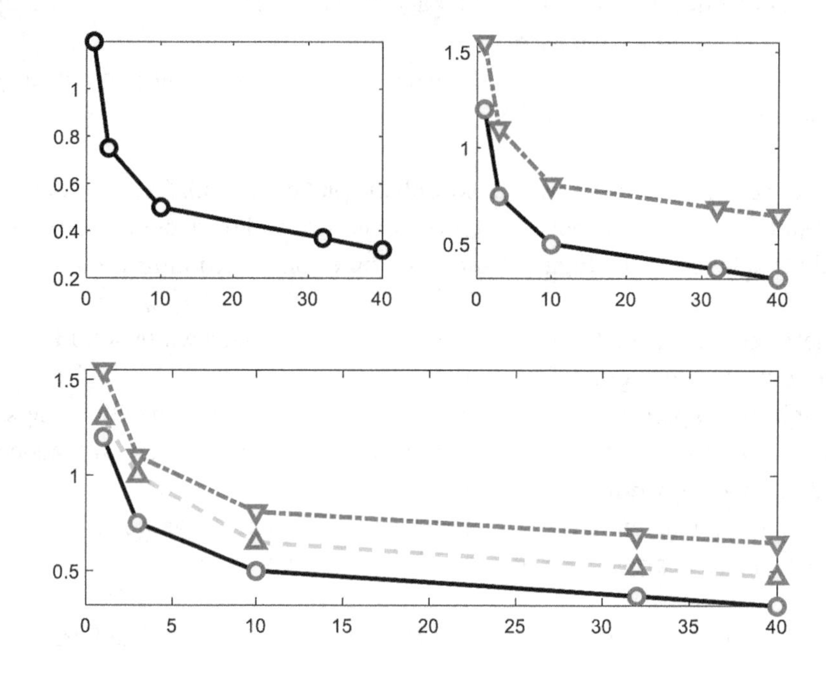

The same commands, written as live script in the Live Editor and saved with name ExSubplot, after running give inline plot as per Figure 7.

Figure 7. Live Editor window with program generated three plots in the same page.

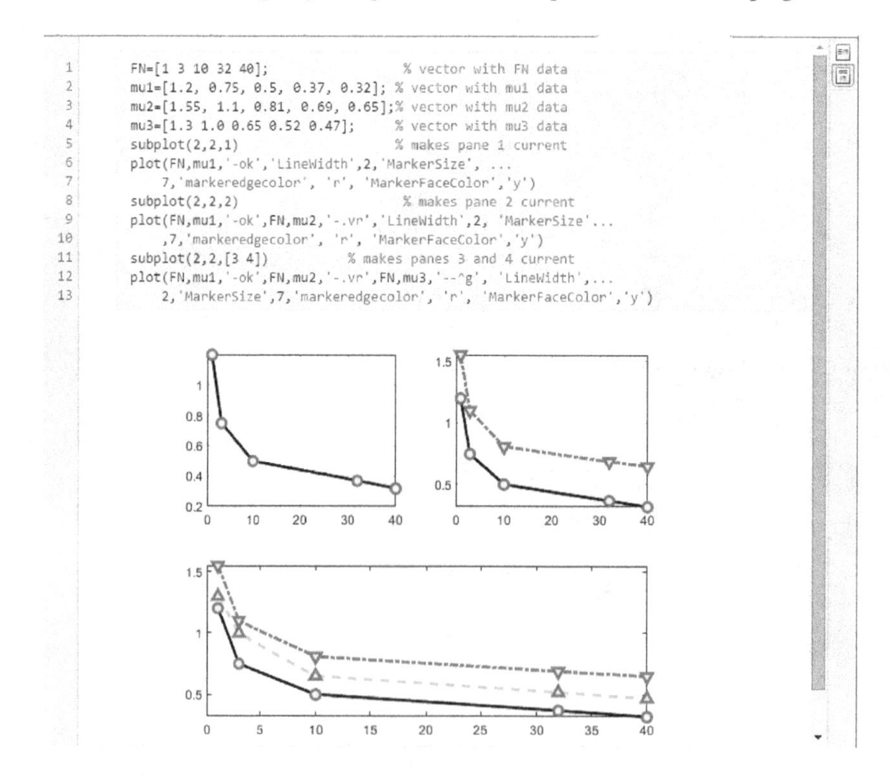

4.1.4. 2D Plot Formatting Using the Commands and the Plot Tools Editor

In general, the figure may include a grid, axis labels, suitable axes ranges, text, title, and legend, as far as vaious colors and types of curve lines. The introduction of these elements relates to graph formatting operations and possible by either using special commands or interactively using the Plot editor (see Subsection 4.1.4.1) strictly in the Figure window.

4.1.4.1. Formatting With Commands

The formatting commands are effective when the plot is created with a computer program; these commands should be entered after the plot command or when the graph has already been built. The most common commands are described below.

The Grid On/Off Command

The grid on command is applied to display the grid lines on the current plot. The grid off command hides the grid lines of the latticed plot. For example, typing grid on in the Command Window immediately after generating the Figure 4.3 by the plot command, will draw the grid lines in the figure.

The Axis Commands

There are a group of commands that controls the axis of a plot. These commands are used for the appearance, hiding, and scaling of the axes. Some of its possible forms are:

```
axis([x_min x_max y_min y_max])
axis equal
axis square
axis tight
axis off
```

The axis([x_min x_max y_min y_max]) command adjusts the x and y axes in accordance to the coordinate limits written as four-element vector in the brackets. The axis equal command sets the same scale for the x and y axes (the ratio x/y, width-to-height, is called the aspect ratio). The axis square sets the shape of the graph as square. The axis tight sets the axes limits to the range of the data to be applied to the plot; and the last command, axis off, removes the axis and background from the plot.

As an example, write script ExAxis with the sequence commands that draw the friction coefficient plot (as in Figure 4.3) without and with the axis square command. For clarity, these two plots are generated in the same Figure window with the two subplot commands in such way that the plot with two entropy lines is displayed in the left pane without any of axis commands, and the same plot is displayed with the axis square command in the right pane.

```
FN=[1 3 10 32 40];                                 % creates vector FN for x axis
mu1=[1.2 0.75 0.5 0.37 0.32];                         % creates vector mu1 for y
mu2=[1.55 1.1 0.81 0.69 0.65];                        % creates vector mu2 for y
subplot 121                      % divides window in 2 pains;1st pane is current
```

```
plot(FN,mu1,FN,mu2,'--k')                                  % generates first plot
subplot 122                  % divides window in 2 pains; 2nd pane is current
plot(FN,mu1,FN,mu2,'--k')                                  % generates second plot
axis square                                              % turns plot into a square
```

After running with the ExAxis command, the following plots are generated (Figures 8, a and b).

Figure 8. The two friction coefficient plots generated without (a) and with (b) the axis square *command*

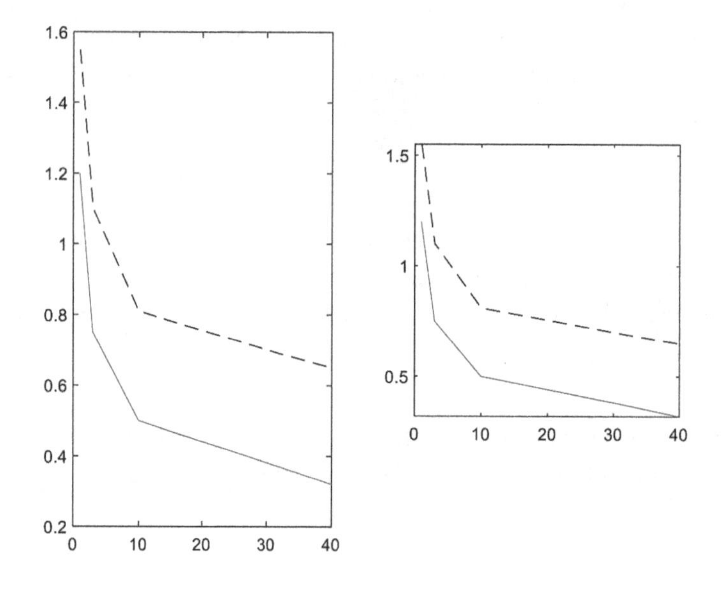

As can be seen, when the axis square command is inputted (Figure 7, b) the axes are equal in size and the plot becomes square.

The Xlabel, Ylabel and Title Commands

These commands provide the text for the *x* and *y* axes and for the top of the plot. The desired text should be written in these commands as string, between single quotes. The commands have the forms:

```
xlabel('text')
ylabel('text')
title('text')
```

The Latin and Greek letters can be included in the text. The font size of the text, its name, color, style, tilt, and some other property options can be written within each command separated by a comma from each other and from the quoted text string (see below the 'Formatting the text strings' subsection).

The gtext and text Commands

The text labels are placed in the plot with these commands; they have the following forms:

```
gtext('string/label')
text(x,y,'string/label')
```

The gtext command allows the user to put text/number/charecter written between the single quates interactively at the place that he chooses with the mouse. After this command is entered, the Figure window appears with two crossed lines; using the mouse, the user moves the crosshair to the proper point and then inputs the text by clicking the mouse button.

The second command, text, produces text/number/charecter starting from the point with coordinates x and y which the user specified in the command.

As an example, write using the'Output on right' options the ExText live script that includes commands used to construct the plot in Figure 4.2 together with the title, xlabel, ylabel, text and grid on formatting commands. The created live scrip program and the plot generated with it are shown in Figure 9.

Figure 9. Wear-sliding distance plot formatted with xlabel, ylabel, title, text, *and* grid on *commands*

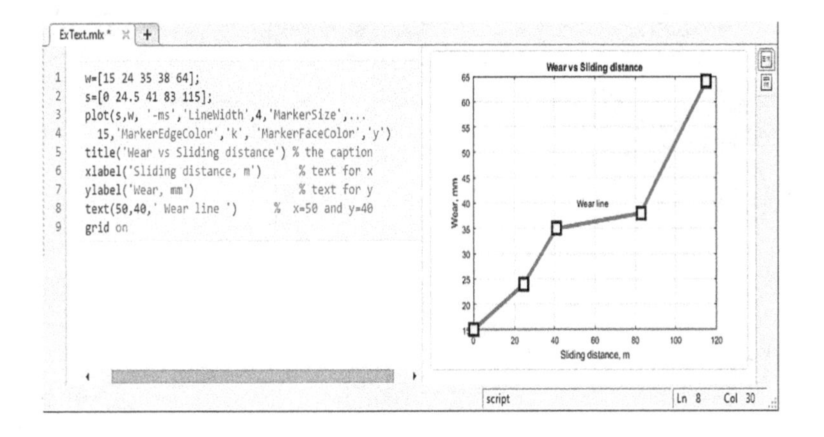

The Legend Command

This command is applied to identify each of the plotted curves; it has the following form:

```
legend('string1','string2',…,'Location', location_area)
```

The 'string1','string2',... are the explanatory text written within the frame. The 'Location' - location_area pair property is optional; location_area it specifies the area, where the explanations should be placed, for example:

- legend('line 1','line 2','Location', 'NorthWest') displays the legend frames with two line strings (line 1, line 2) placed within the plot frames on the upper-left corner;
- legend('line 1','line 2','Location', 'NorthEastOutside') displays the legend frames with two line strings (line 1, line 2) placed outside the plot frames on the upper-right corner;
- legend('line 1','line 2','Location', 'NorthOutside') displays the legend frames with two line strings (line 1, line 2) placed at the top outside the plot frames;

- legend('line 1','line 2','Location','Best') displays the legend frames with two line strings (line 1, line 2) placed inside the plot at the best possible location (having least conflict with curve/s within the plot).

Note, The default legend location (when the legend command is used without the location property,) is in the upper-right corner of the plot.

For example, input the following command to add a legend to Figure 4:

```
>>legend('Material 1','Material 2','Material 3','Location','Best');
```

After this, the plot with the legend looks like in Figure 10.

Figure 10. Plot with legend for three friction coefficient – normal force curves.

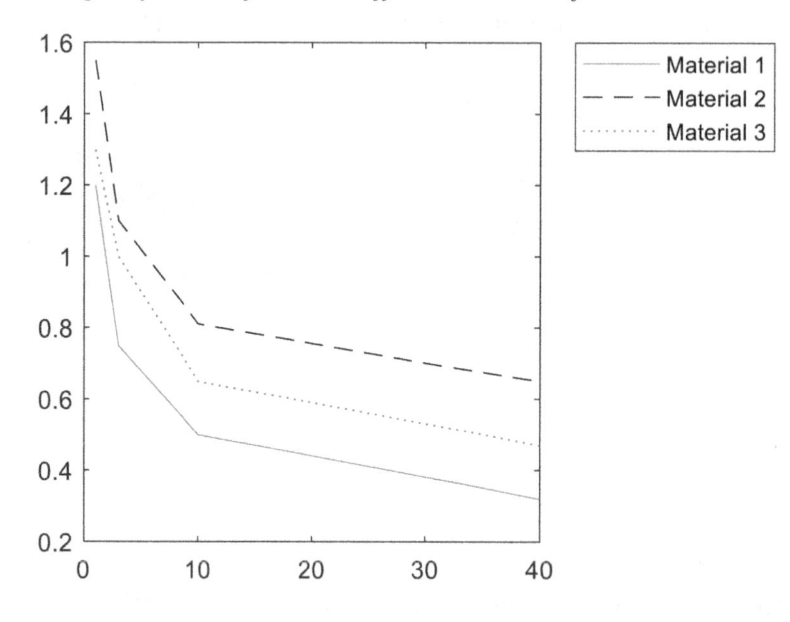

Formatting Text Strings

The text/labels/strings in the discussed commands can be formatted applying some special word/s or character/s (called modifiers) are written inside the string immediately after the backslash, in the form \ModifierName{ModifierValue} (when entered, appeared in blue). Some useful modifiers for setting the font name, style, size, color, Greek letters, or sub- and superscripts are given in Table 3.

Text can be formatted in another way by including a pair - property name and value - in the command just after the text string. As in case of the plot command (see Subsection 4.1.), to format a text, several property pairs can be included, and each of these pairs must be written after comma and with separating commas in the form 'Name', Value.

Table 3. Some modifiers for the text string/s formatting.

Modifier	Purpose	Example
\fontsize{number}	sets size of the letters of the text following the modifier	\fontsize{14} – sets the letter/s size 14
\fontname{name}	sets name of the font to the text following the modifier	\fontname{Courier New} - sets the Courier New font
\name of the Greek letter	specifies a Greek letter	\sigma sets the lowcase σ \Sigma sets the capital Σ
\b	sets the bold font to the text following the modifier	\b mechanics Displays the word "mechanics" written after the modifier as **mechanics**
\it	sets the Italic (tilted) font to the text following the modifier	\it mechanics – sets the word "substance" (written after the modifier) to be tilted: *mechanics*
\rm	sets the normal (Roman) font to the text following the modifies	\rm mechanics – displays in the Roman font the word "mechanics" written after the modifier: mechanics
_ (underscore)	subscripts	A_ d – the "_" sign sets the d-letter to be subscripted: A_d To extend this sign to more than one letter, the curled brackets are used, e.g. 12_{dec}is displayed as 12_{dec}
^ (caret)	superscripts	^oC – the ^ symbol sets 'o' to be superscripted: ^{o}C To apply this sign to more than one letters, the curled brackets are used, e.g. e^{-(x-5)} is displayed as $e^{-(x-5)}$

For example, the command

```
text(45,40,'Wear line','fontsize',14)
```

The same effect can be achieved by written the modifier name and value inside the string (as per Table 3):

```
text(45,40,'\fontsize{14}Wear line')
```

4.1.4.2. About Interactive Plot Formatting

In addition to editing the graph using commands, it is possible to edit the graph interactively using the Figure window. In the menu bar line of this window there is an assortment of buttons and other items that can be used to format a plot interactively (Figure 4.11). For beginning the plot formatting the Plot Edit button should be clicked firstly. After this you can use other buttons to format the previously created plot. Properties of axes and lines, as well as the entire figure, can be changed using the pop-up menu, summoned by clicking the Edit option in the Figure menu. The title, axis labels, texts and legend can be activated using the pop-up menu, which appears after clicking on the Insert option of the menu.

Figure 11. The menu and bar buttons for plot editing from the Figure window.

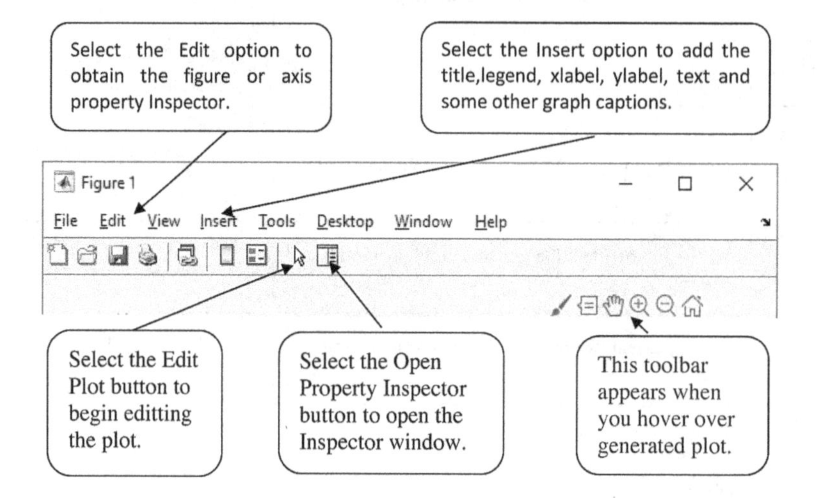

The text, legend and other plot items can be added/changed after activating 'Plot Edit' mode, by selecting the appropriate line in the Insert menu option. The window called Property Inspector is opened using the Open Property Inspector button or by double-clicking on the shifted curve, plotted point or axes; it provides means for changing or editing various characteristics of the clicked object. For more detailed information about available plot tools input the doc plottools command in the Command Window.

4.2. THREE-DIMENSIONAL PLOTS

Three main groups of commands for presenting lines, meshes, and surfaces in a three-dimensional space are available in MATLAB®. These commands along with various formatting commands are described below.

4.2.1. Presenting Line in 3D Plots

As in two-dimensional, in three-dimensional space, a line is constructed from points and a line connecting adjacent points. However, each 3D point has three coordinates here. The plot3 command is used to generate a line in three-dimensional space. The simplest shape of the command is

```
plot3 (x,y,z)

plot3(x,y,z,'Line_Specifiers','Property_Name',Property_Value,…)
```

The previously disscused grid on/off, xlabel and ylabel commands, and in addition, the zlabel commands, can be used to add grid and axis labels to the 3D plot.

For an example, design a three-dimensional plot showing a line, write for this the following program with name Ex3Dline:

```
t=linspace(0,9*pi,200);
x=sin(t);
y=t;
z=cos(t);
plot3(x,y,z,'k','LineWidth',3)
grid on
xlabel('x'),ylabel('y'),zlabel('z')
```

These commands compute the coordinates by the expressions $x=\sin t$, $y=t$ and $z=\cos(t)$, with 200 t values changing from 0 up to 9π. The plot3 command is used here, with the line color specifier 'k' (black) and the line width property pair - 'LineWidth', 3 - that increases the line width to two points (1/24 inch or approximately 1.1 mm); two last command lines generate the grid and the captions to the axes.

After running, the resulting three dimensional plot appears in the Figure window with the generated line in the form of a side helix (screw-shaped) - Figure 12.

Figure 12. Side helix generated with the plot3 command

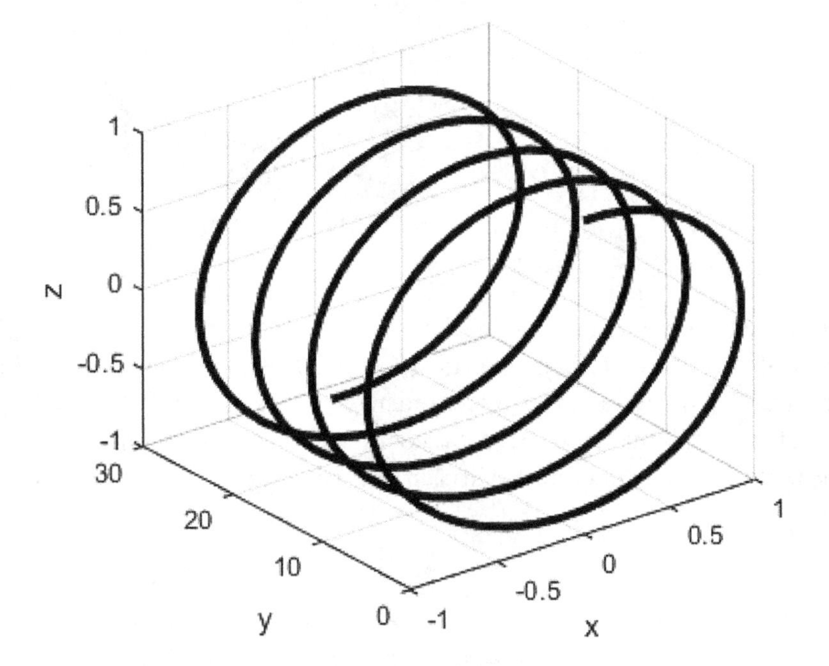

4.2.2. Representation of a Mesh in 3D Coordinates

Surfaces in MATLAB® can be drawn with the two basic commands: mesh and surf. To understand the 3D surface reproducing, it is useful to clarify who the mesh is built. In general, the mesh is comprised of the points (mesh nodes) and the lines between them. Every point in 3D space has three coordinates (x, y and z), which are used to reconstruct a surface. The x,y coordinate pairs are ordered in a rectangular grid of the x,y plane. The x,y-nodes of this grid can be represented as two size-equal matrices, one of which contains x and the other - y coordinates. For each grid x,y-node the z-coordinate should be given or calculated by any $z(x,y)$ expression and assembled then into a z-matrix.

An example of points represented in three-dimensional space is shown in Figure 13.

Figure 13. 3D points (signed by 'o') and corresponding rectangular grid in the x,y-plane (shown in blue)[1]

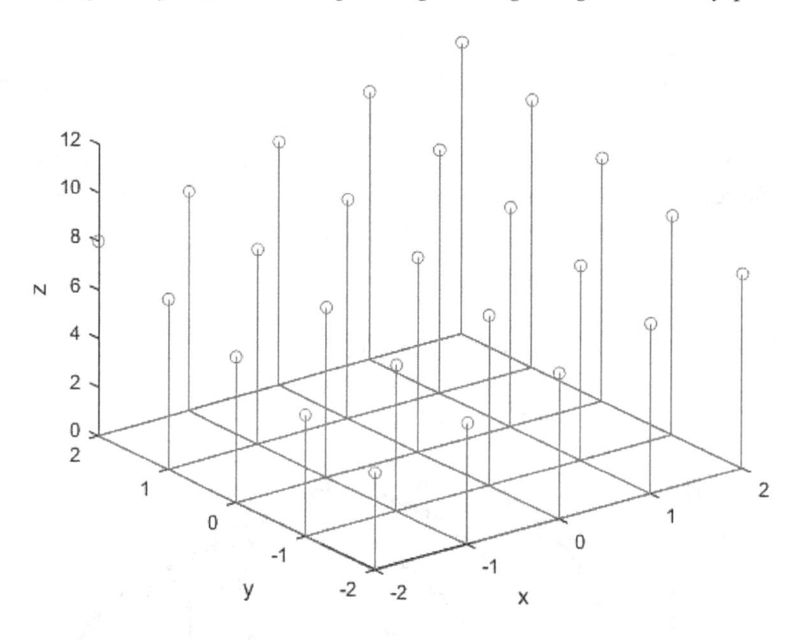

The area of the x and y coordinates for which the z-coordinates must be obtained is called the domain. In the figure above, the domain represents the orthogonal grid in the x,y-plane with the limits -2 and 2 for each of the plane axes. By writing all the x-values ordered by rows (along each iso-y line), we obtain the X-matrix; as it can be seen this matrix has the same coordinates in each column. An analogous procedure yields the Y-matrix with the same coordinates in each row. The matrices for the example in Figure 12 are:

$$X = \begin{pmatrix} -2 & -1 & 0 & 1 & 2 \\ -2 & -1 & 0 & 1 & 2 \\ -2 & -1 & 0 & 1 & 2 \\ -2 & -1 & 0 & 1 & 2 \\ -2 & -1 & 0 & 1 & 2 \end{pmatrix} \qquad Y = \begin{pmatrix} -2 & -2 & -2 & -2 & -2 \\ -1 & -1 & -1 & -1 & -1 \\ 0 & 0 & 0 & 0 & 0 \\ 1 & 1 & 1 & 1 & 1 \\ 2 & 2 & 2 & 2 & 2 \end{pmatrix}$$

The *z*-coordinates can be obtained for every *x,y* point of the domain using the element-by-element calculations. After the *X*, *Y*, and *Z* matrices are defined, the whole surface can be plotted.

The *X* and *Y* matrices can be created with a special **meshgrid** command from the specified vectors of *x* and *y*. Two simplest form of the command are:

```
[X,Y]= meshgrid(x,y)
```

or

```
[X,Y]= meshgrid(x)
```

X and Y here are the matrices of the grid coordinates that determine the division of the domain, x and y are the vectors that represent *x*- and *y* axis division respectively. The second command form can be used when the x- and y vectors are equal.

In case of Figure 12, the represented above *X* and *Y* matrices are produced with the following commands:

```
>>x=-2:2;
>>[X,Y]=meshgrid(x)
X =
   -2   -1    0    1    2
   -2   -1    0    1    2
   -2   -1    0    1    2
   -2   -1    0    1    2
   -2   -1    0    1    2
Y =
   -2   -2   -2   -2   -2
   -1   -1   -1   -1   -1
    0    0    0    0    0
    1    1    1    1    1
    2    2    2    2    2
```

When the X, Y and Z matrices are generated, a tree-dimensional mesh can be drawn. The command generating a mesh of colored lines is:

```
mesh(X,Y,Z)
```

where the X and Y matrices are defined with the **meshgrid** command for the given vectors x and y, while the third matrix, Z, is given for every X,Y – node pair or is calculated for these matrices using the given *z*(*x*,*y*)-expression.

Summarizing: to build a 3D plot you must perform the following steps. First create a grid in the *x,y* plane with the **meshgrid** command. Then calculate or define the *z* values for each *x,y* node of the grid. Finally, generate the plot using the **mesh** command.

As an example, plot the 3D mesh graph for the distance e from centroidal to neutral axis of the circular beam section (Budynas & Nissbett, 2011)

$$e = r_c - \frac{R^2}{2\left(r_c - \sqrt{r_c^2 - R^2}\right)}$$

where R is the radius of the circular cross-section; r_c – radius of the centroidal axis. To create a graphic, define R= 2.5 … 7.5 cm and r_c =7.5 … 12.5 cm, and assign ten values of R and r_c each.

The program named **ExEccentricity** that generates the 3D mesh plot reads:

```
r=linspace(2.5,7.5,10);                                  % assigns r vector
rc=linspace(7.5,12.5,10);                                % assigns rc vector
[R,Rc]=meshgrid(r,rc);                 % generates the R and Rc matrices
Rn= R.^2./(2*(Rc-sqrt(Rc.^2-R.^2)));
e=Rc-Rn;                                         % calculates the e matrix
mesh(R,Rc,e)                                     % generates the mesh plot
xlabel({'Beam cross-section';'radius, cm'})
ylabel({'Centroidal axis';'    radius, cm'})
zlabel('Eccentricity,cm')
```

The resulting plot is shown in Figure 14.

Figure 14. 3D mesh plot depicting the distance from the centroidal to the central axis of the circular beam section

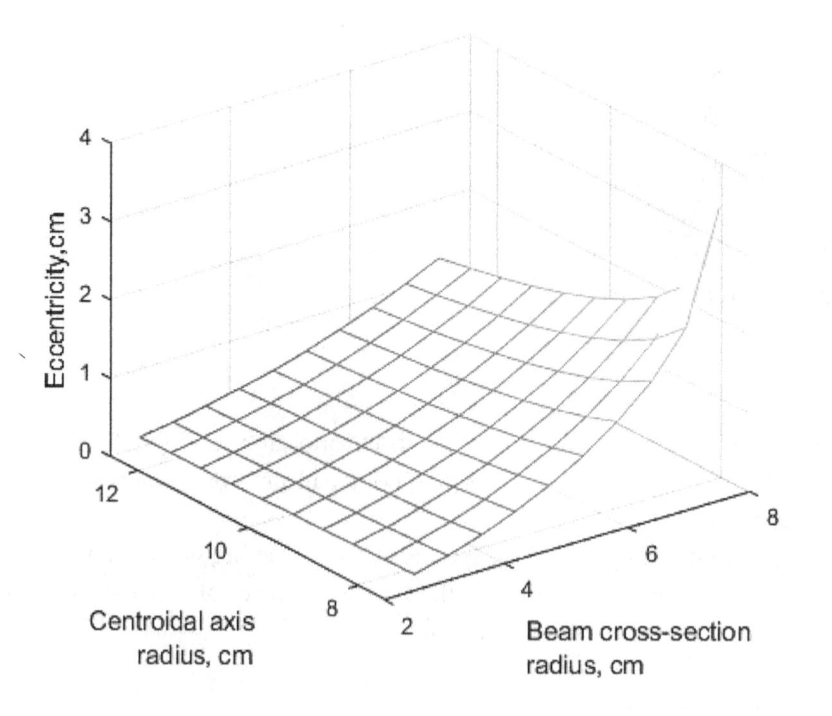

Note, to present x and y captions in two lines the curled brackets containing two rows of the strings (divided by;) are used.

4.2.3. Surfaces in 3D Plots

The mesh command described above generates a three-dimensional mesh of colored lines. The surfaces between these lines are white, not colored. To create a plot having a colored surface, the surf command is used. The form of this command is:

```
surf(X,Y,Z)
```

To illustrate the command implementation we apply the ExEccentricity program of the previous example with the surf(Rc,R,e) command entered in place of the mesh(Rc,R,e). In this case this program generates the following plot (Figure 15).

Figure 15. 3D surface plot depicting the distance from centroidal to neutral axis of the circular beam section

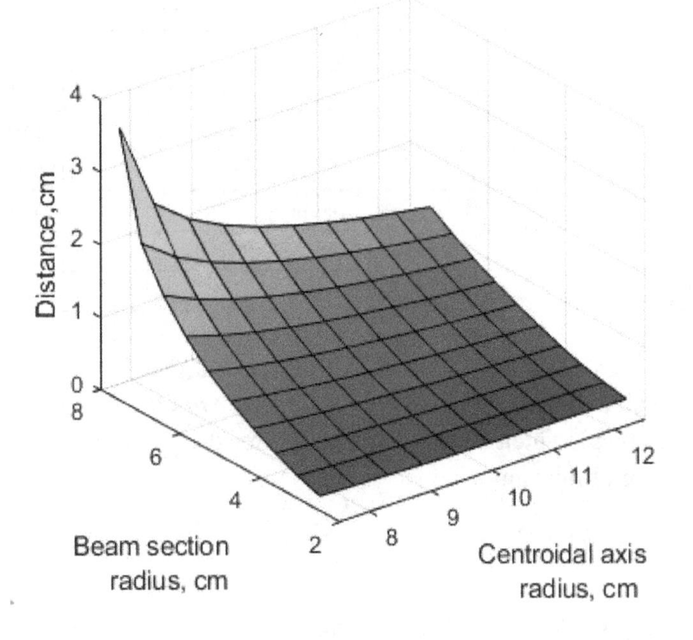

Note:

- The surf and mesh commands have also the form surf(Z) or mesh(Z). In this case, the Z matrix values are plotted versus this matix row numbers (x-coordinates) and column numbers (y-coordinates);
- When the surf and mesh commands are executed, the grid appears automatically in the plot and can be removed using the grid off command.

4.2.4. Formatting and Rotating 3D Plots

Many of the commands described for 2D plot formatting, such as grid, title, xlabel, ylabel, and axis can be used and have partially been used above to formate generated 3D plot. However, there are a variety of additional commands used for the formatting of a three-dimensional plot. Some of them are described below.

Figure Colors With the colormap Command

All previously created graphics have colors, which makes the graphics clearer and more representative. Color is an important attribute of two-dimensional and especially three-dimensional graphs. When the 3D graph commands are entered, color is generated by default according to the magnitude of the represented surface (range of z-values). Nevertheless, the user can changes/sets the color map with the colormap command:

```
colormap(c)
```

Table 4. Some typical colors and their representation with the three-number vectors

Color	Vector c	Color	Vector c
black	[0 0 0]	magenta	[1 0 1]
red	[1 0 0]	gray	[0.5 0.5 0.5]
green	[0 1 0]	copper	[1 0.62 0.4]
blue	[0 0 1]	aquamarine	[0.49 1 0.83]
cyan	[0 1 1]	gold	[0.804 0.498 0.196]
yellow	[1 1 0]	white	[1 1 1]

For example, the mesh line colors in Figure 14, assigned by default, can be changed to red with the colormap([1 0 0]) command, which should be entered after the plot creation.

Another possible form of the command is

```
colormap c_name
```

This form permits to write the name of some common color combinations (maps): the c_name may be jet, cool, summer, winter, spring, and some others. For example, the colormap summer changes the current colors to shades of green and yellow. By default the parula is used as the c_name. More complete information about the available color schemes can be defined by entering the doc colormap command in the Command Window.

The Viewpoint and Projections of the 3D Graph

All previously generated 3D plots are shown by default from a certain viewpoint. The spatial orientation of a plot relative to the viewer's eye is controlled by the **view** command that has the form

```
view(azimuth,elevation)
```

Figure 16. Viewpoint, azimuth and elevation angles in 3D graphs; positive directions are indicated by arrows

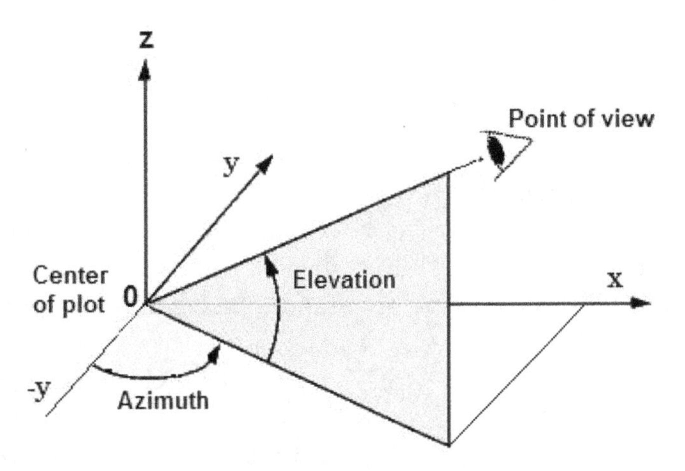

Both azimuth and elevation values must be inputted in degrees; their default values are: azimuth =-37.5°, and elevation =30°.

Various orientation angles, view projections, and examples of the view command are presented in Table 5.

Table 5. Orientation angles and corresponding view projections.

View	Azimuth, degree	Elevation, degree	Command
Top view – the *x*,*y*-projection (two-dimensional view,default)	0	90	>>view(0,90) or>>view(2)
Side view – the x,z- projection	0	0	>>view(0,0)
Side view - the y,z- projection	90	0	>>view(90,0)
Default three-dimensional view	-37.5	30	>>view(3)
Mirror view relative to default view	37.5	30	>>view(37.5,30)
Rotate the view around the x-axis by 180°	180	90	>>view(180,90)

For an example, generate four plots of various views for the previous example - the the distance from from the centroidal to the neutral axis of the section of the circular beam: the default (Figure 14), mirrored to default, the top and the front projections. The commands written as the ExView script program are:

```
r=linspace(2.5,7.5,10);                                    % assigns r vector
rc=linspace(7.5,12.5,10);                                 % assigns rc vector
[R,Rc]=meshgrid(r,rc);                     % generates the R and Rc matrices
Rn= R.^2./(2*(Rc-sqrt(Rc.^2-R.^2)));
e=Rc-Rn;                                  % calculates the distance e matrix
subplot(2,2,1), surf(Rc,R,e)                    % generates the default view
xlabel({'Centroidal axis';'   radius, cm'})
ylabel({'Beam section';'radius, cm'}),zlabel('Eccentricity,cm')
title('Default view')
axis tight                                  % sets axis limits to the data range
subplot(2,2,2), surf(Rc,R,e)                      % generates the second plot
view(37.5,30)                                      % mirroring to the default
xlabel({'Centroidal axis';'   radius, cm'})
ylabel({'Beam section';'radius, cm'}),zlabel('Eccentricity,cm')
title('az=37.5^o, el=30^o')
axis tight                                  % sets axis limits to the data range
subplot(2,2,3), surf(Rc,R,e)                       % generates the third plot
view(2)                            % azimuth =0 and elevation =90 - top view
xlabel({'Centroidal axis';'   radius, cm'})
ylabel({'Beam section';'radius, cm'}) zlabel('Eccentricity,cm')
title(' az =0^o, el =90^o')
axis tight                                  % sets axis limits to the data range
subplot(2,2,4), surf(Rc,R,e)                      % generates the fourth plot
view(0,0)                           % azimuth =0 and elevation =0 - side view
xlabel({'Centroidal axis';'   radius, cm'}) zlabel('Eccentricity,cm')
title(' az =0^o, el =0^o')
axis tight                                  % sets axis limits to the data range
```

The results obtained after running the program are shown in Figure 17.

Figure 17. Different projections of the distance from the centroidal to the neutral axis of the section of the circular beam.

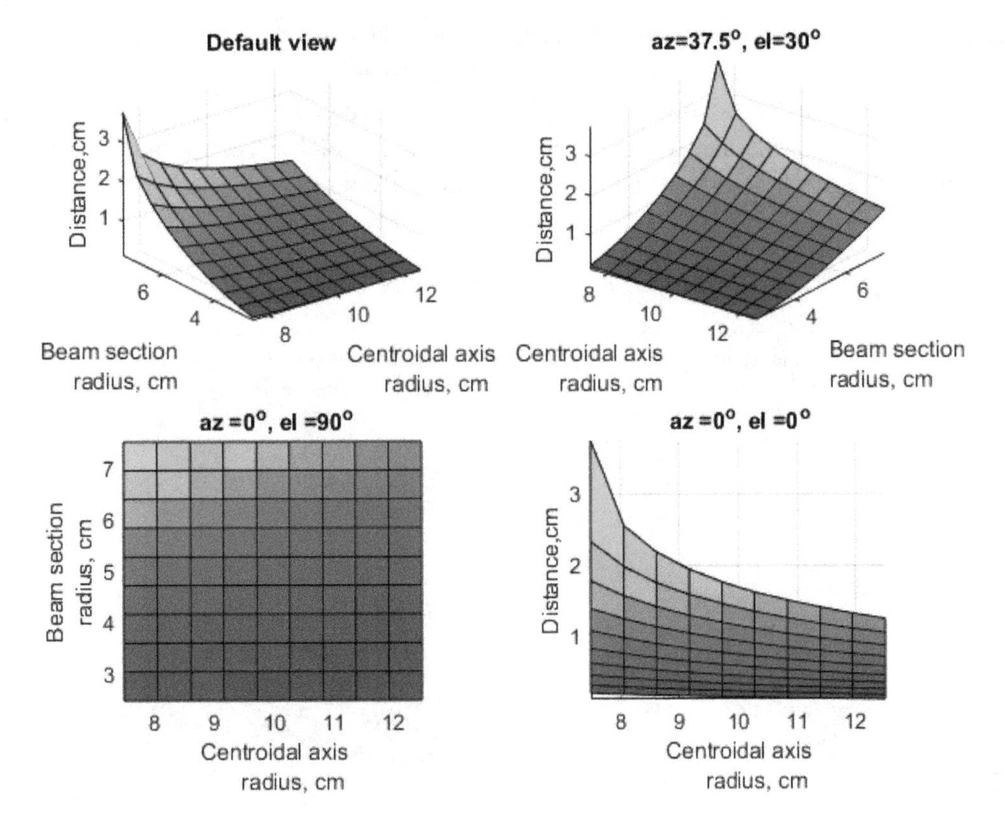

The ExView script commands perform the following actions:

- Assign values to the the r and rc vectors;
- Create matrices R and Rc in the range of the r and rc respectively using the meshgrid command;
- Calculate the *e* distance for each pair of the R and Rc values;
- Break the Figure window (page) into four panes and make the first pane current using the subplot command; generate the first plot using the surf command at default viewpoint;
- Set the axis limits for the data range with the axis tight command;
- Make the second pane current using the subplot command; generate the second plot with the surf command;
- Set viewpoint angles so that the second plot represents a mirror view of the default view with the view(37.5,30) command and set the axis limits to the data range with the axis tight command;
- Make the third pane current using the subplot command; generate the third plot using the surf command;
- Make the top view with the view(2) command and set the axis limits to the data range using the axis tight command;
- Make the fourth pane current using the subplot command; generate the fourth plot with the surf command;
- Set side view with the view(0,0) command and set axis limits to the data range using the axis tight command.

About the Rotation of the Plot

Any created plot can be rotated manually by clicking the 'rotate 3D' button, . This button, among others, appears after placing the mouse cursor on the area containing the generated plot. After clicking on the 'Rotate 3D' button, the cursor should be placed on plot and you can start the rotation. The values of azimuth, Az, and the elevation, El, angles appear in the bottom-left corner of the window. A Figure window containing a graph, Rotate 3D button, and cursor for plot rotation is shown in Figure 18.

Figure 18. The Figure window with a surface plot in rotation mode.

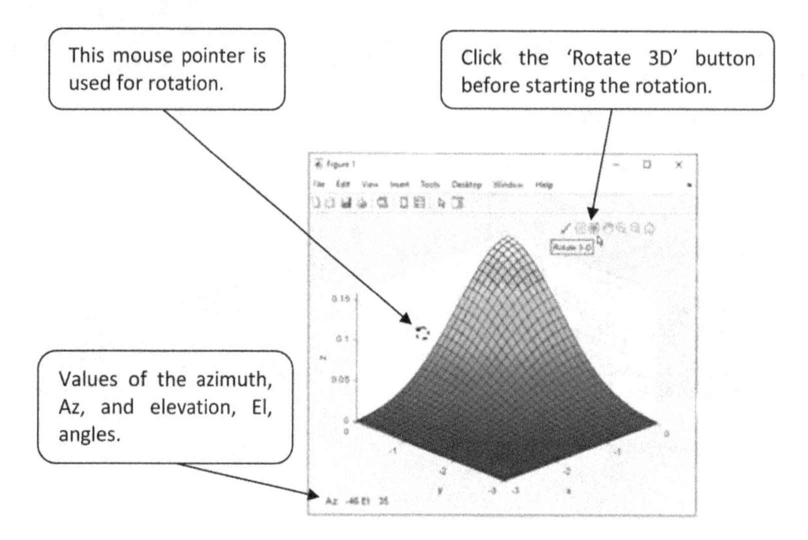

The rotation mode can also be set programmatically using the **rotate3d** on command. After activating this command, execute the following:

- Place the mouse cursor on it on the generated plot;
- By holding the left mouse button and moving the mouse, rotate the plot in accordance with the desired azimuth Az and elevation El angle values.

To interrupt the rotation mode, the rotate3d off command should be entered.

4.3. SPECIALIZED AND SUPPLEMENTARY 2D AND 3D PLOT COMMANDS

In addition to the described 2D and 3D graphs, the mechanicans and thribologists use for example logarithmical plots, plots with error boundaries for each *x,y* point, scatter plots, and many others. These types of graphs can be designed with specialized commands. Some of these graphs, together with a list of additional graphic commands, are presented briefly below.

4.3.1. Plot With Error Bars

Certain limitations of the method and/or inaccuracy of the equipment, calculated, observed or verified data are always determined with some inaccuracy. In this regard, it is desirable to plot obtained values with their error limits at each measured point. The **errorbar** command is accomplished such plotting and represents observed points with error limits. The simplest forms of this command are:

```
errorbar(x,y,l_e,u_e)
```

or

```
errorbar(x,y,e)
```

For an example, generate plot for the wear *w* of a micro-milling tool as function of the sliding distance, *s* (see data in subsection4.1.1) when lower error is 0.5w and upper error two times larger than the lower. Create for this the following script with name **ExErrorbar**:

```
w=[15 24 35 38 64];                                   % vector with wear data
s=[0 24.5 41 83 115];                      % vector with sliding distance data
l_e= 0.05*w;                                      % vector of the upper errors
u_e=l_e*2;                                        % vector of the lower errors
errorbar(s,w,l_e,u_e)                            % plots asymmetric error bars
xlabel('Sliding distance, m'),ylabel('Wea, mm')
title('w(S)data with asymmetric errorbars'),grid on
```

After running, the following plot is generated (Figure 19).

Figure 19. Wear vs sliding distance with asymmetric errors.

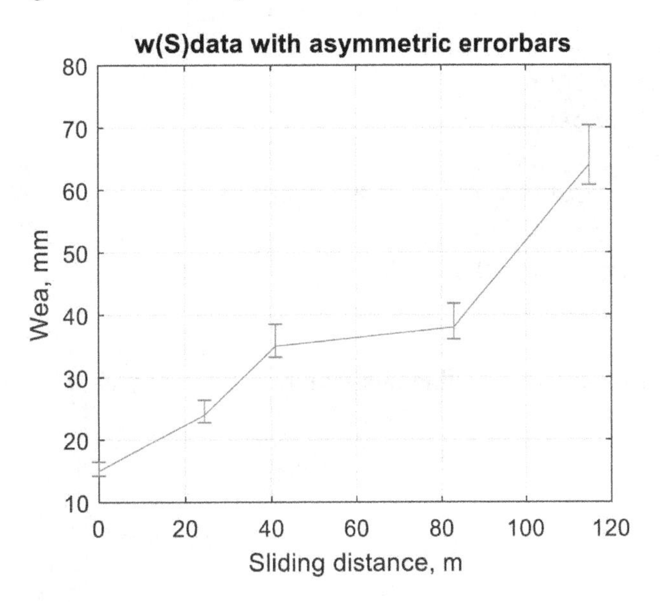

The commands formatting the line color, style or marker - the line style, color and marker specifiers can be included in the errorbar command, e.g., inputting the errorbar(w,s,l_e,u_e,'g--o') command changes the line color to green, the line style to a dashed line, and assigns the data points with circles into the plot of the Figure 19.

4.3.2. Plot With Two Y-Axes

Sonetimes two curves that need to be placed on the same graph have different *y*-scale and/or different units. In such cases, it is desirable to generate two different vertical axes in the same plot; for this, two possible forms of the yyaxis commands can be used

```
yyaxis left
```

and

```
yyaxis right
```

As an example, generate in the same plot both the kinematic viscosity and the specific volume of an engine oil as function of the temperature. The viscosity is 1684.4, 831.44, 260.69, 73.406, and 30.129 mm^2/s, at temperatures 0, 10, 30, 60, 90 °C respectively, while the specific volume is 0.001159, 0.001185, 0.001211, and 0.001239, kg/m^3 at temperatures 0, 30, 60, and 90 °C respectively. Place the left *y*-axis for kinematic viscosity and the right *y*-axis for specific volume on the graph. The program named Ex2yaxes generate the desired plot is:

```
eta=[1684.4 831.44 260.69 73.406 30.129];% vector y-left
T_eta=[0 10 30:30:90];     % vector x for the left axis
v=[0.001159 0.001185 0.001211 0.001239];% vector y_right
T_v=0:30:90;          % vector x for the right axis
yyaxis left            % activates left y-axis
plot(T_eta,eta,'o-') % left y-axes plot, circled points
xlabel('Temperature, ^oC')
ylabel('Kinematic viscosity, mm^2/s')
yyaxis right           % activates right y-axis
plot(T_v,v,'s-')   % right y-axes plot, squared points
xlabel('Temperature, ^oC')
ylabel('Specific volume, m^3/kg')
grid on
```

After running this program, the resulting plot with two different *y*-axes is (Figure 20)

Figure 20. The plot with two different y-axes representing the kinematic viscosity (left y-axis) and specific volume (right y-axis) of an engine oil.
Note: the xlabel *and* grid on *commands are entered only once since the x-axis and the grid are common to the two y-axes plot. This is also true for all other two-dimensional plot formatting commands not related to the y-axes.*

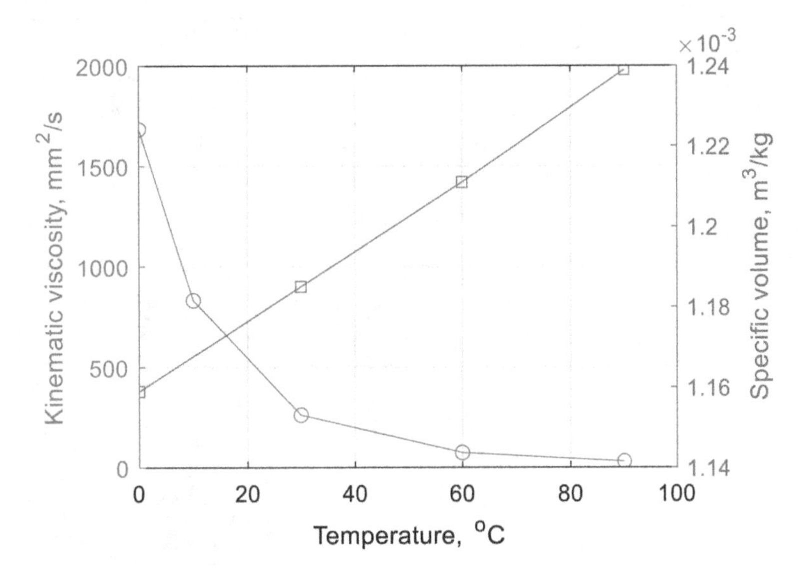

4.3.3. Supplementary Commands for Generating 2D and 3D Graphs

In addition to the graphics commands presented above, MATLAB® has many other commands for 2D and 3D plotting. By entering the help graph2d, help graph3d or help specgraph commands in the Command Window you can get a complete list of the graphical commands. Table 6 presents supplementary commands for two- and three-dimensional plotting that can be useful for graphic representation in M&T. The table shows one of the possible forms of commands, usually the simplest, with brief explanations, examples and the resulting plots.

Table 6. Supplementary 2D and 3D plotting commands.[2]

Command	Purpose	Example with generated plots

bar(x,y)	displays the vertical bars of the *y* values at the locations specified by *x*	`>>x=0.1:0.1:1.0;`
		`>>y=[1,2,1,2,5,2,5,2,3,1];`
		`>>bar(x,y)`
		`>>xlabel('Wear rate, m^3/m')`
		`>>ylabel('Frequences, units')`

barh(x,y)	displays the horizontal bars of the *y* values at the location specified by *x*	`>>x=0.1:0.1:1.0;`
		`>>y=[1,2,1,2,5,2,5,2,3,1];`
		`>>barh(x,y)`
		`>>xlabel('Wear rate, m^3/m')`
		`>>ylabel('Frequences, units')`

continued on following page

Table 6. Continued

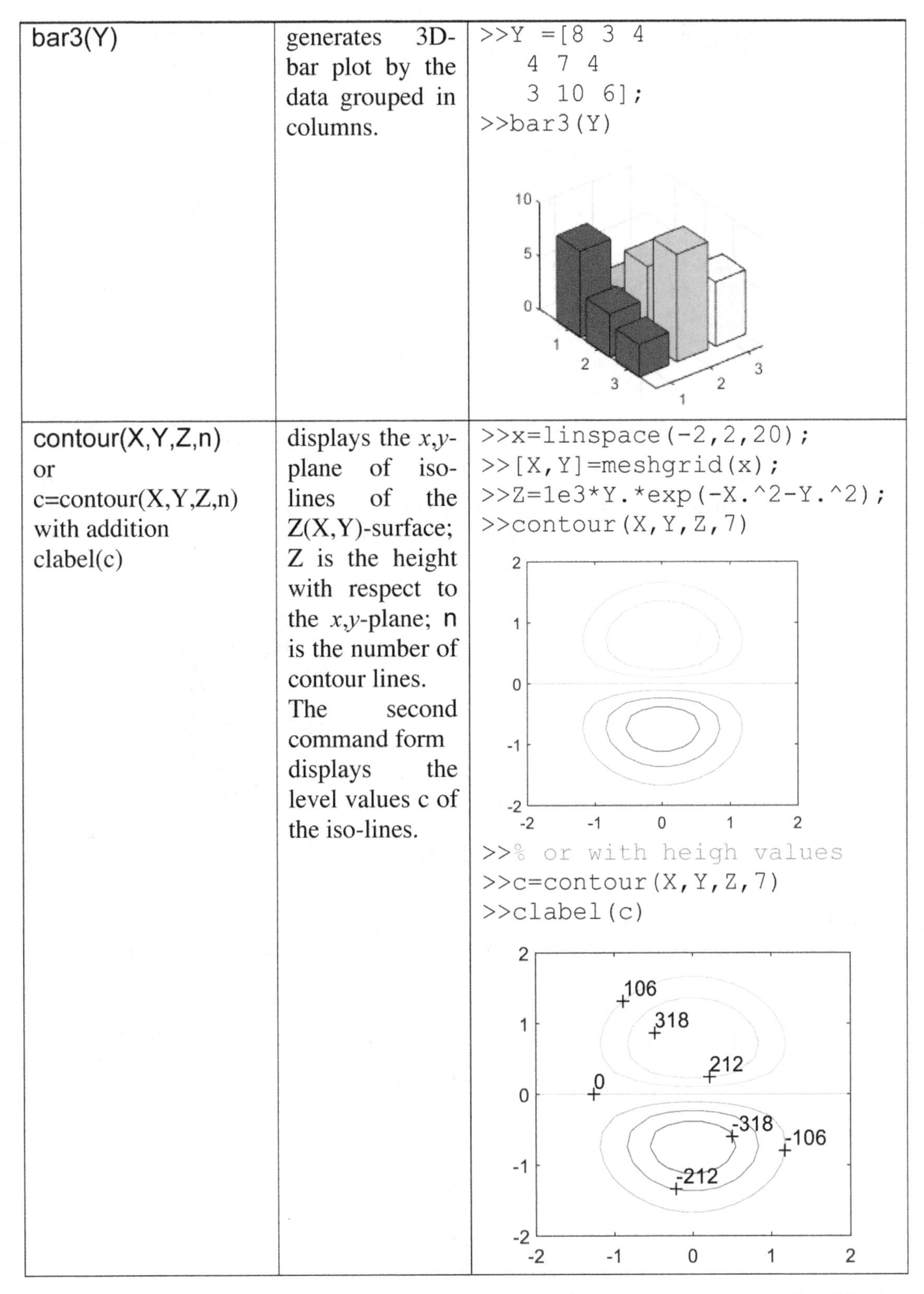

bar3(Y)	generates 3D-bar plot by the data grouped in columns.	`>>Y =[8 3 4`
		`   4 7 4`
		`   3 10 6];`
		`>>bar3(Y)`

contour(X,Y,Z,n)
or
c=contour(X,Y,Z,n)
with addition
clabel(c)
	displays the *x,y*-plane of iso-lines of the Z(X,Y)-surface; Z is the height with respect to the *x,y*-plane; n is the number of contour lines. The second command form displays the level values c of the iso-lines.	`>>x=linspace(-2,2,20);`
		`>>[X,Y]=meshgrid(x);`
		`>>Z=1e3*Y.*exp(-X.^2-Y.^2);`
		`>>contour(X,Y,Z,7)`
		`>>% or with heigh values`
		`>>c=contour(X,Y,Z,7)`
		`>>clabel(c)`

continued on following page

Table 6. Continued

contour3(X,Y,Z,n)	displays *x,y*-planes of iso-lines of the Z(X,Y) surface in 3D; n is the number of contour lines.	`>> x=linspace(-2,2,20);`
		`>> [X,Y]=meshgrid(x);`
		`>> Z=1e3*Y.*exp(-X.^2-Y.^2);`
		`>> contour3(X,Y,Z,7)`
		400 200 0 -200 -400 2 0 -2 -2 0 2

cylinder or cylinder(r)	draws an ordinary or a profiled cylinder; the profile is given by the r expression.	`>>t=0:pi/10:2*pi;`
		`>>r=atan(t);`
		`>>cylinder(r)`
		1 0.5 0 2 0 -2 -2 0 2

figure or figure(n)	generates a new figure window named as Figure 1 or Figure n where n is the number of figure	`>> figure(4)` Figure 4 Fil Ed Vie Inse Too Deskt Wind He

continued on following page

Table 6. Continued

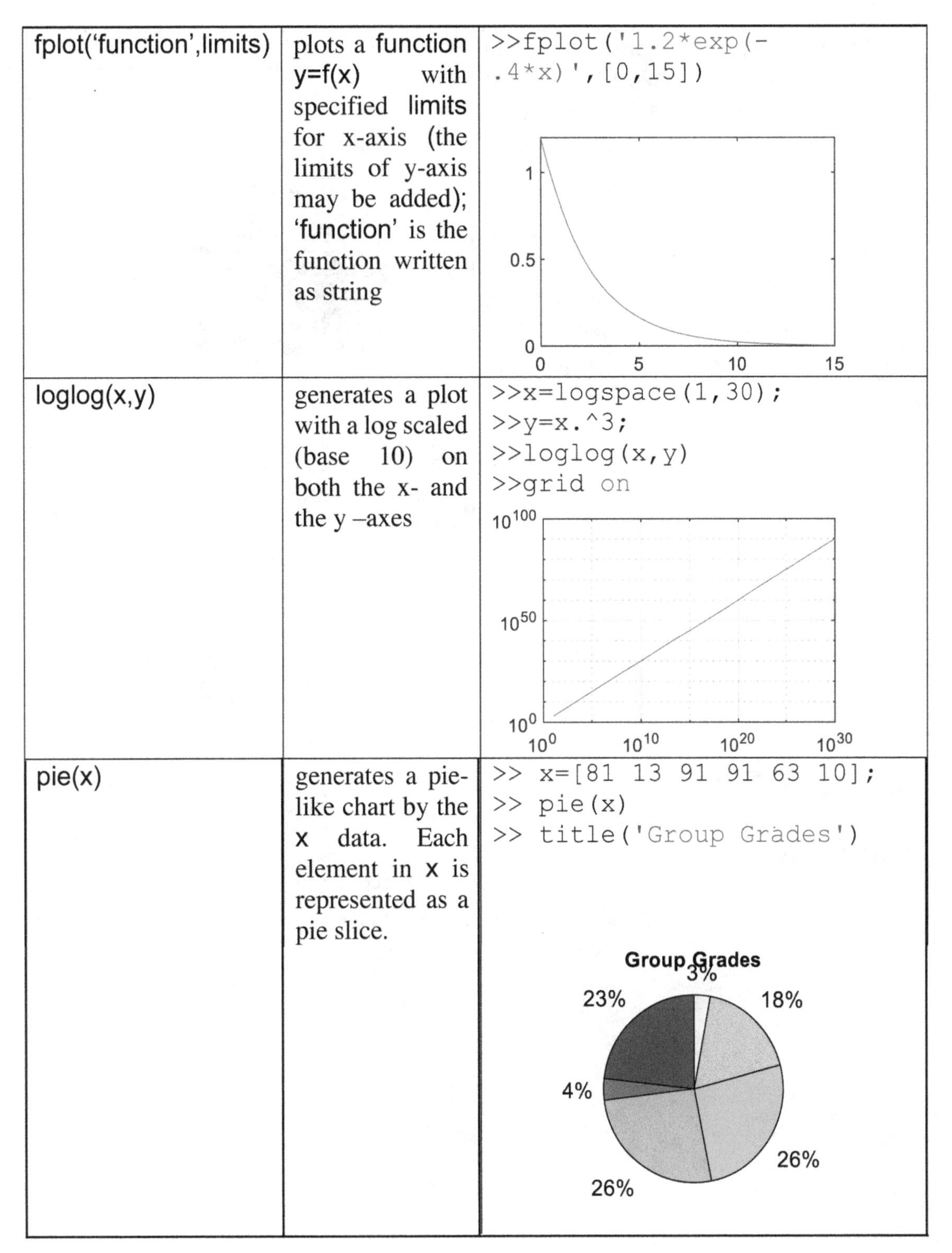

fplot('function',limits)	plots a function y=f(x) with specified limits for x-axis (the limits of y-axis may be added); 'function' is the function written as string	`>>fplot('1.2*exp(-.4*x)',[0,15])`

loglog(x,y)	generates a plot with a log scaled (base 10) on both the x- and the y –axes	`>>x=logspace(1,30);`
		`>>y=x.^3;`
		`>>loglog(x,y)`
		`>>grid on`

pie(x)	generates a pie-like chart by the x data. Each element in x is represented as a pie slice.	`>> x=[81 13 91 91 63 10];`
		`>> pie(x)`
		`>> title('Group Grades')`

continued on following page

Table 6. Continued

<table>
<tr><td>pie3(x,explode)</td><td>generates a 3D pie-like chart; explode is a vector that specifies an offset of a slice from the center of the chart; 1 denotes an offseted slice and 0 a plain slice.</td><td><pre>>> x=[81 13 91 91 63 10];
>> explode=[0 0 1 1 0 0];
>> pie(x)
>> title('Group Grades')</pre>Group Grades
3%
18%
23%
26%
4%
26%</td></tr>
<tr><td>polar(theta,rho)</td><td>generates a plot in polar coordinates; theta and rho are the angle and radius respectively.</td><td><pre>>>theta=linspace(0,2*pi);
>>rho=atan(theta);
>>polar(theta,rho)</pre>90
1.1253
120
60
0.56267
150
30
180
0
210
330
240
300
270</td></tr>
<tr><td>semilogy(x,y)
or
semilogx(x,y)</td><td>Generates with a log-scaled x- or y-axis; can incude the same specifyers as the plot command</td><td><pre>>>mu=[1300 185 38 9.6];
>>T=0:30:90;
>>semilog(T,mu,'-o')
>>xlabel('Temperature,^oC')
>>ylabel('Viscosity,mPa.s')
>>grid on</pre>10^3
10^2
10^1
Viscosity,mPa.s
0
20
40
60
80
100
Temperature, °C</td></tr>
</table>

continued on following page

Table 6. Continued

sphere or sphere(n)	generates a sphere with 20x20 (default) or nxn mesh cells respectively.	>> sphere(15), axis equal 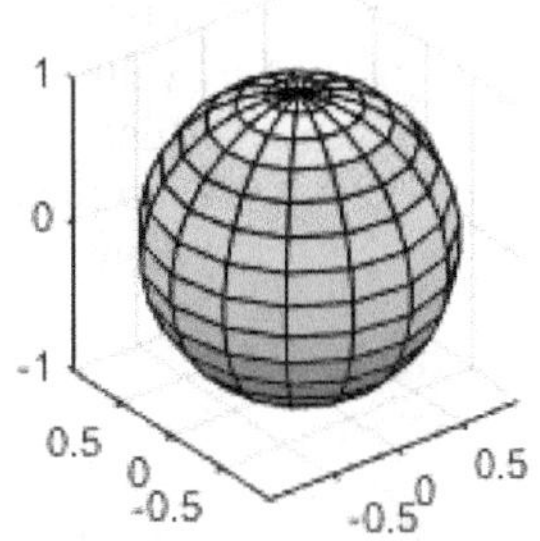
stairs(x,y)	generates a stairs-like plot of the discrete y data given at the specified x points.	>> x=2000:5:2025; >>y=[7.8 8.1 8.4 8.7 9.1 9.6]; >> stairs(x,y) >> xlabel('Year') >> ylabel('People, mln') >>axis([2000 2025 7 10]) 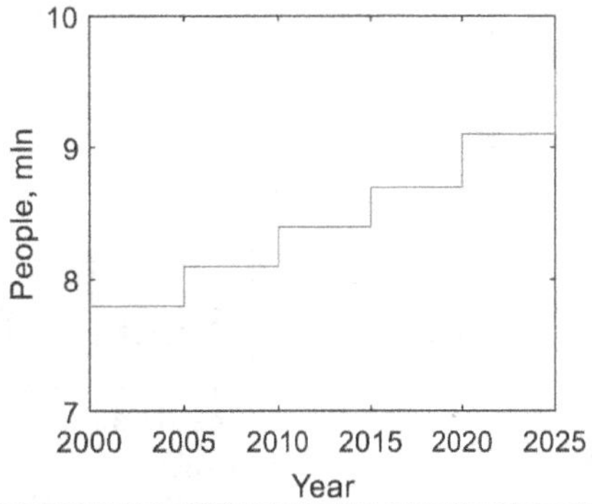
stem(x,y)	displays data as stems extending from a baseline along the x-axis. A circle (default) terminates each stem.	>> x=10:10:100; >> y=[2,5,1,2,4,2,5,2,3,1]; >> stem(x,y) >> xlabel('Grades') >> ylabel('Number of Grades') 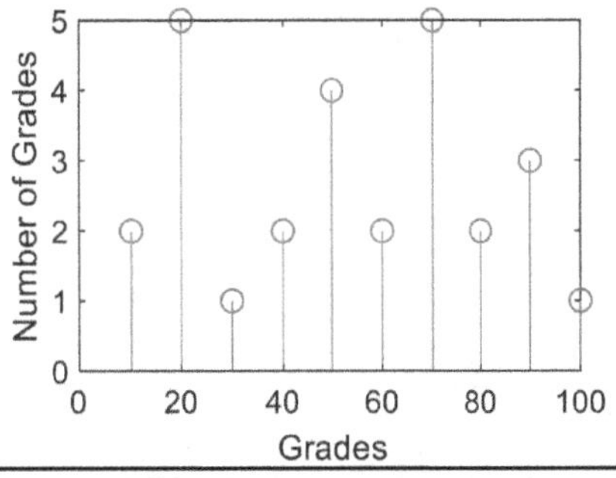

continued on following page

Table 6. Continued

stem3(x,y,z)	generates a 3D plot with stems from the x,y - plane to the z-values; a circle (default) terminates each stem.	`>> x=0:0.25:1;`
		`>> [X,Y]=meshgrid(x);`
		`>> z= (X.*Y./( X.^2+Y.^2)).^1.2;`
		`>> stem3(X,Y,z)` 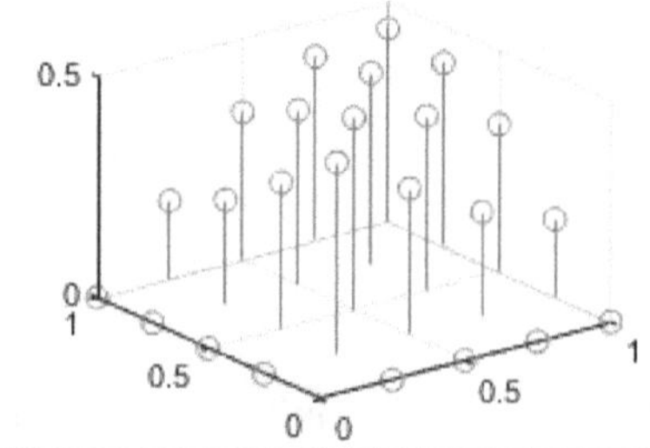
		0.5 0 1 0.5 0 0 0.5 1

surfc(X,Y,Z)	generates surface and contour plots together	`>>x=-2:.2:2;`
		`>>[X,Y] = meshgrid(x);`
		`>>Z=1e3*X.*Y.*exp(-X.^2-Y.^2);`
		`>> surfc(Z);` 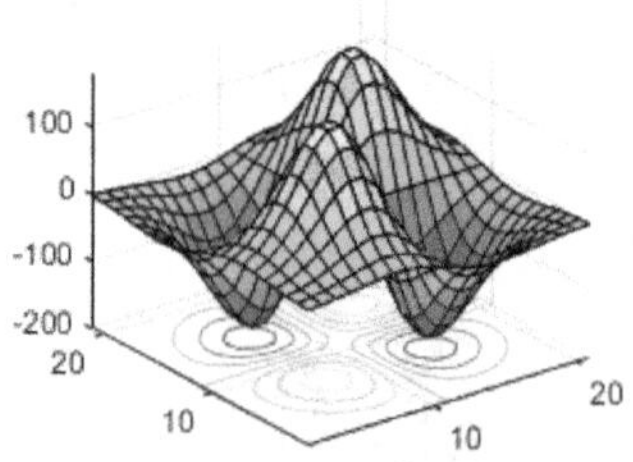
		100 0 -100 -200 20 10 10 20

waterfall(X,Y,Z)	Generates the same plot as the mesh command (Subsection 4.2.2) but without the column lines	`>>r=linspace(2.5,7.5,10);`
		`>>rc=linspace(7.5,12.5,10);`
		`>> [X,Y]=meshgrid(r,rc);`
		`>>e= Y-X.^2./(2*(Y-sqrt(Y.^2-X.^2)));`
		`>>waterfall(X,Y,e)`
		`>>xlabel({'Centroidal >>axis';' radius, cm'})`
		`>>ylabel({'Beam section';'radius, cm'})`
		`>>zlabel('Eccentricity,cm')` 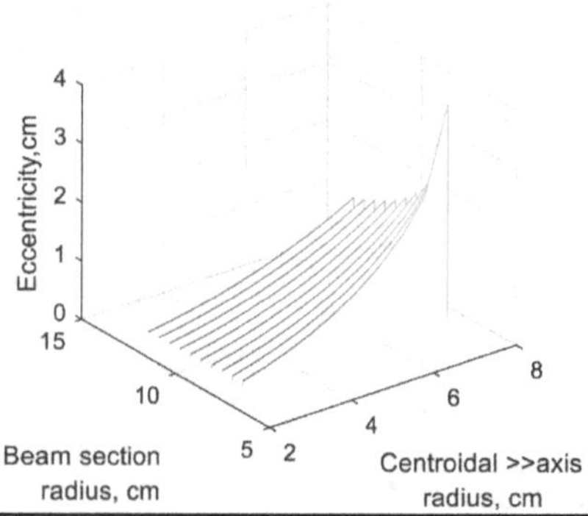
		Eccentricity,cm 4 3 2 1 0 15 10 5 2 4 6 8 Beam section radius, cm Centroidal >>axis radius, cm

4.4. APPLICATION EXAMPLES

4.4.1. Engine Piston Velocity

Velocity v of the engine piston with respect to crank angle θ can be calculated as

$$v = -r\omega \sin\theta + \frac{r^2\omega \sin\theta \cos\theta}{\sqrt{l^2 - r^2 \sin^2\theta}}$$

where *r* and *l* are the lengths of the crank arm and the connecting rod respectfully.

Problem: Calculate velocity at crank angles θ =ωt=0, …, 2π, ω=180 rad/sec and for crank and rod lengths of 0.12 and 0.25 m respectively. Generate *v*(θ) plot with the axes and plot captions. Write live script with name Ch_4_ApExample_1_live that contains explanatory text with *v*(θ) equation, codes, and generates plot using the "Output on right" options.

The Ch_4_ApExample_1_live program text, codes, and generated plot are (Figure 21).

Figure 21. The live script that generates plot of the engine piston velocity.

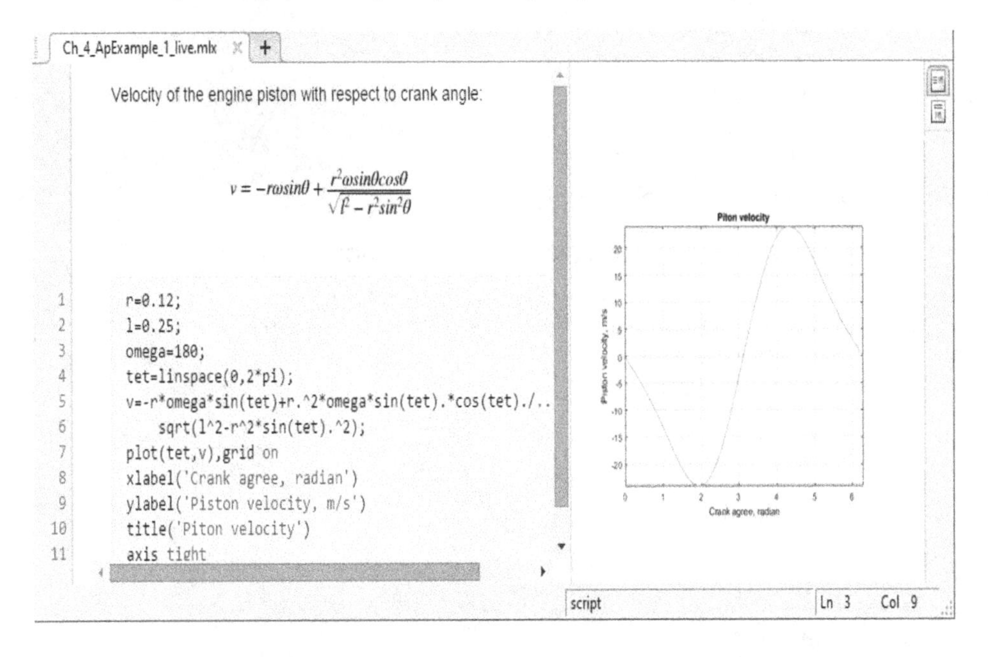

4.4.2. Power Screw Efficiency

Percent of a power screw efficiency *e*, when collar friction is ignored, is

$$e = \frac{\cos\alpha - \mu \tan\lambda}{\cos\alpha + \mu \cot\lambda} 100$$

Where μ is the coefficient of friction, λ is the screw leed angle, α is the thread angle.

Problem: Write script named Ch_4_ApExample_2 that calculates and generates plot for the power screw efficiency at $\lambda = 0, \ldots, \pi/4$ (radians), $\mu = 0.05$ and 0.025 (dimensionless), and $\alpha = 14.5$ (degrees). Add grid, axis labels, title, and legend to plot.

The Ch_4_ApExample_2 script pogram that solves this problem and its resulting plot (Figure 4.22) are:

```
alpha=14.5;                                              % degrees
mu=[0.05 0.15 0.025]';              % dimensionless,3x1 column vector
lambda=linspace(0,pi/4);                   % radians, 1x100 row vector
e=100*(cosd(alpha)-mu*tan(lambda))./...
  (cosd(alpha)+mu*cot(lambda));                % radians,[3x1]*[1x100]
l_degree=Lambda*180/pi;                 % converts radians to degrees
plot(l_degree,e')                     % e' to plot the curves by columns
grid on
legend('\mu=0.05','\mu=0.15',...
  '\mu=0.025','location','best')                % best legend location
xlabel('Lead Angle [degrees]')
ylabel('Efficiency, %')
title({'Power screw efficiency';...
  'thread angle \alpha=14.5^o'})                 % title with two lines
```

Figure 22. Efficiency of a power screw with the thread angle 14.5o.

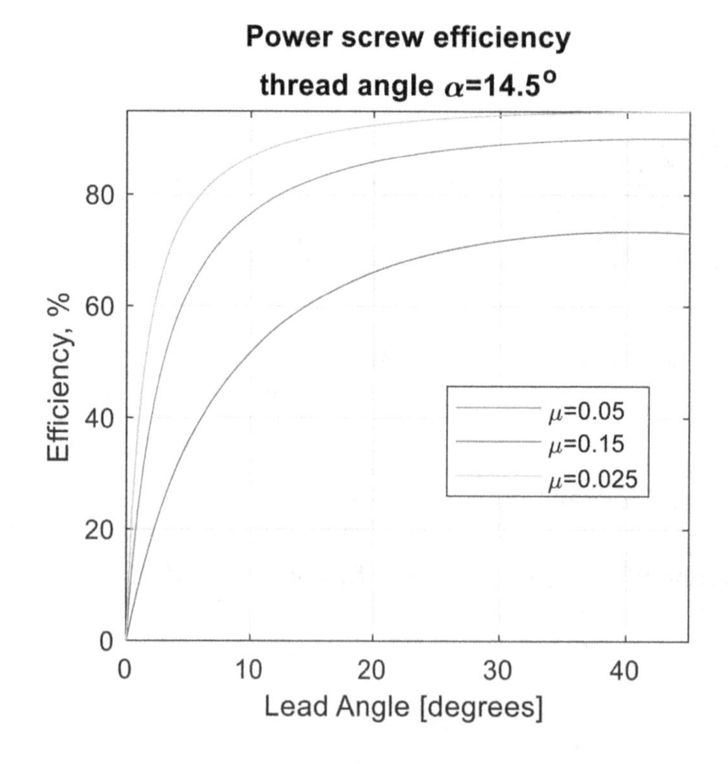

Note.

- The e expression incudes two products mu*tan(lambda)and mu*cot(lambda);each of this products contains two vectors that have dimensions [1x3] and [1x100].In this case the matrix multiplications are impossible. To avoid this inconvenience, the row vector mu was transposed into the column vector, so that the product [2x1]*[1x100] can be performed.
- The plot command builds lines according to the columns, for this the matrix e was transposed, so that each column contains the efficiencies along the lead angle.

4.4.3. Viscosity of a Motor Oil

The dependence of viscosity η, mm²/s, of a motor oil on temperature *T*, °C, was described by the equation.

$$\eta = Ae^{BT}$$

where A=906.8, B=-0.0504.

Th measured in tribology laboratory η−values are 658, 327, 178, 105, 66, and 45 31 mm²/s at *T*=10, 20, 30, 40, 50, and 60 °C respectively. The η-values were obtained with two-sided uncertainties (i.e. errors) 0.02·η mm²/s.

Problem: Write program containing an user-defined function named Ch_4_ApExample_3 that has not the input/output papameters and generates two plots in the same page (i.e. in the same Figure window), each of plot represemts both the calculated and measured viscosity values the left plot in the logarithmic *y*-coordinate (use the semilogy commands), and the right plot in ordinary *x,y*-coordinates; give for the second plot the bar of errors at each measured point and add legend; in the each plot designate the measured points with circle and add grid, axis labels, and captions.

```
function Ch_4_ApExample_3
                          % semilogaritmic and ordinal plots with oil viscosities
%                                                             to run: >>Ch_4_ApExample_3
eta=[658 327 178 105 66 45];
e=0.05*eta;
T=10:10:60;
A=906.8;B=-0.0504;
eta_fit=A*exp(B*T);
subplot(1,2,1)
semilogy(T,eta_fit,T,eta,'o'),grid on
xlabel('Temperature, ^oC')
ylabel('Kinematic viscosity, mm^2/s')
title('Logarithmic y-axis')
legend('Calculated','Measured')
axis tight
subplot(1,2,2)
```

```
plot(T,eta_fit),hold on
errorbar(T,eta,e,'o'),grid on
xlabel('Temperature, ^oC')
ylabel('Kinematic viscosity, mm^2/s')
title('Ordinal axes')
legend('Calculated','Measured')
axis tight
```

Entering Ch_4_ApExample_3 in the Command Window the following graph is generated (Figure 23).

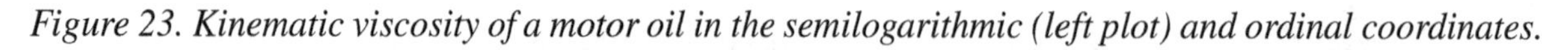
Figure 23. Kinematic viscosity of a motor oil in the semilogarithmic (left plot) and ordinal coordinates.

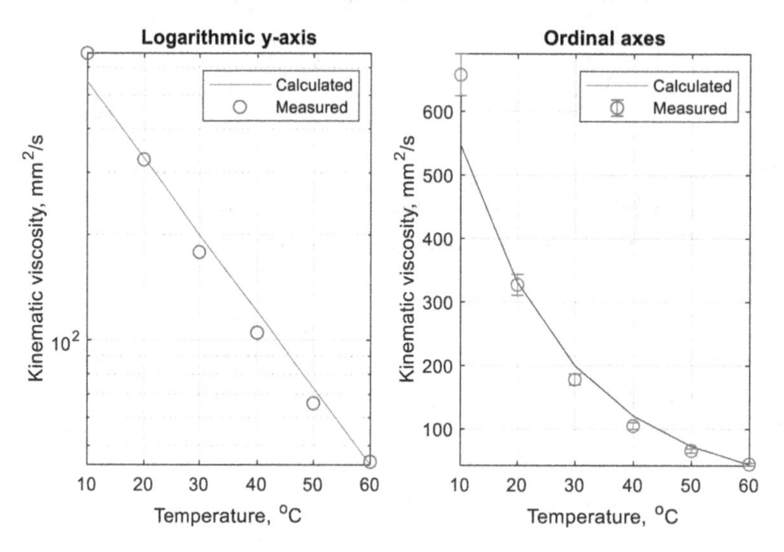

Note, the semilogy command can generate more than one curves.

4.4.4. Mode Shape of a Square Membrane

A one of possible mode shapes of a square membrane clamped on its outer boundaries can be calculated by the following expression

$$w_{32} = \sin\left(3\pi x\right)\sin\left(2\pi y\right)$$

where x and y is the membrane points coordinates.

Problem: Write script named Ch_4_ApExample_4 Calculate w and generate surface plot combined with the countour lines; use 30 values of the x and y from the range 0 and 1 each. Add axes labels and title to plot.

The following program, Ch_4_ApExample_4, after running, performs calculations and generates the $w_{32}(x,y)$ plot (Figure 24).

```
x=linspace(0,1,30);                                    % vector x
[X,Y]=meshgrid(x);                                     % X and Y matrix
w32=sin(3*pi*X).*sin(2*pi*Y);
surfc(X,Y,w32)                          % surface plot with contour lines
xlabel('x'),ylabel('y')
zlabel('Mode shape, w_{32}')
title('Square membrane mode shape')
```

Figure 24. Mode shape of a square membrane.

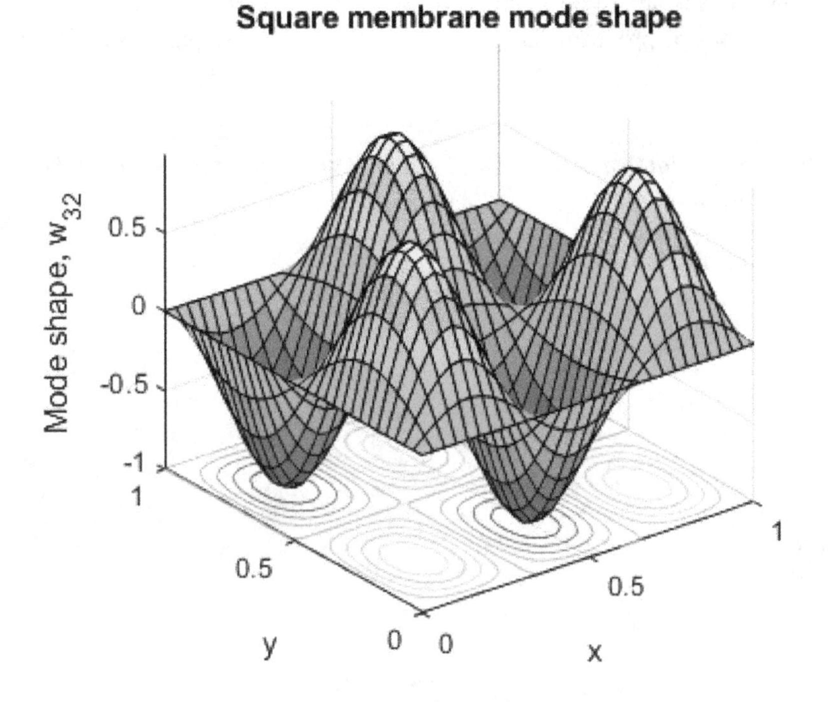

4.4.5. Rectangular Beam: Principal Shear Stress

In dimensionless coordinates ξ and η, the maximum principal shear stress τ' from the torsion of a beam of the square section is calculated as (Wang, 1953)

$$\tau' = \sqrt{\tau_{xz}'^2 + \tau_{yz}'^2}$$

where

$$\tau'_{xz} = -\frac{16}{\pi^2}\sum_{n=0}^{10}\frac{\left(-1^n\right)\sinh(k_n\xi)\cos\left(k_n\eta\right)}{\left(2n+1\right)^2\cosh\left(k_n\right)}$$

$$\tau'_{yz} = 2\eta - \frac{16}{\pi^2}\sum_{n=0}^{10}\frac{\left(-1^n\right)\sinh(k_n\xi)\sin\left(k_n\eta\right)}{\left(2n+1\right)^2\cosh\left(k_n\right)}$$

and k_n=(2n+1)π/2.

Problem: Write a live script with name Ch_4_ApExample_5_live that calculates and generates a plot τ'(ξ,η) when ξ and η are set in the range -1 … 1 each. Take 25 points for each of the mesh node coordinates. Add axes labels and title to plot. Use the "Output inline" view.

The following live script program, Ch_4_ApExample_5_live, performes the required solution (Figure 25).

Figure 25. The live script with generated plot of the maximum shear stress from the torsion of a cross section of the square beam.

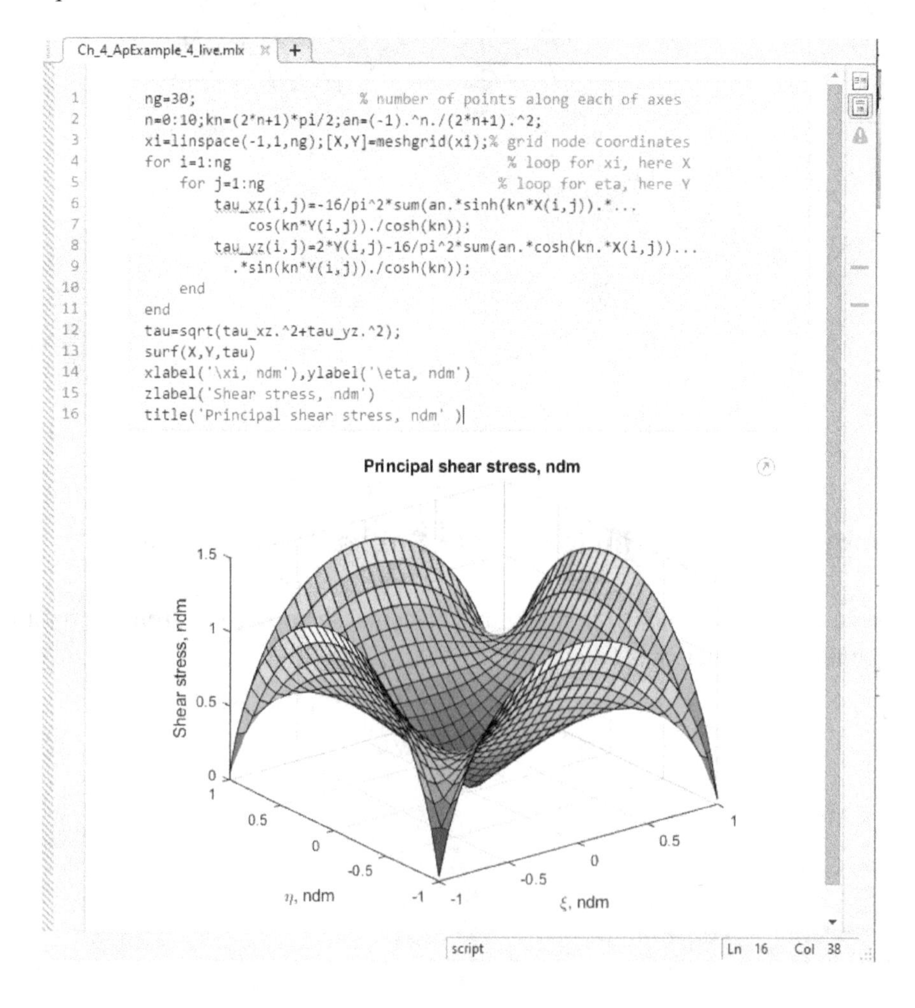

Note: In this live script program, according to the τ'_{xz} and τ'_{yz} expressions, the sum are calculated in respect to n values using the sum command for each point ξ,η of the greed node; two loops realize this: the external one for ξ and the inner one for η.

4.4.6. Steel Rainforced Concrete Beam With a Rectangular Cross-Section: Neutral Axis Location

The neutral axis location of a steel reinforced concrete beam with rectangular cross-section can be defined by the parameter *k* (Spiegal & Limbruner, 1992):

$$k = -\rho n + \sqrt{\left(\rho n\right)^2 + 2\rho n}$$

Where ρ and *n* are the dimensionless steel to concrete ratios of the cross section geometry parameters and the Young's modulus respectively.

Problem: Write program with user-defined function Ch_4_ApExample_6 having only the input parameters n_s, n_f, rho_s, rho_f, and n_grid, where the starting and finishing *n* values are n_s = 6 and n_f=12, and the starting and finishing ρ values are rho_s= 0.001 and rho_f= 0.009; the grid number n_grid=10. The program should generate surface graph $k(n, \rho)$ with axis labels and title.

The solution and the result is (Figure 26):

Figure 26. Parameter k for the steel reinforced concrete beam with a rectangular cross-section defined with the Ch_4_ApExample_6 *user-defined function.*

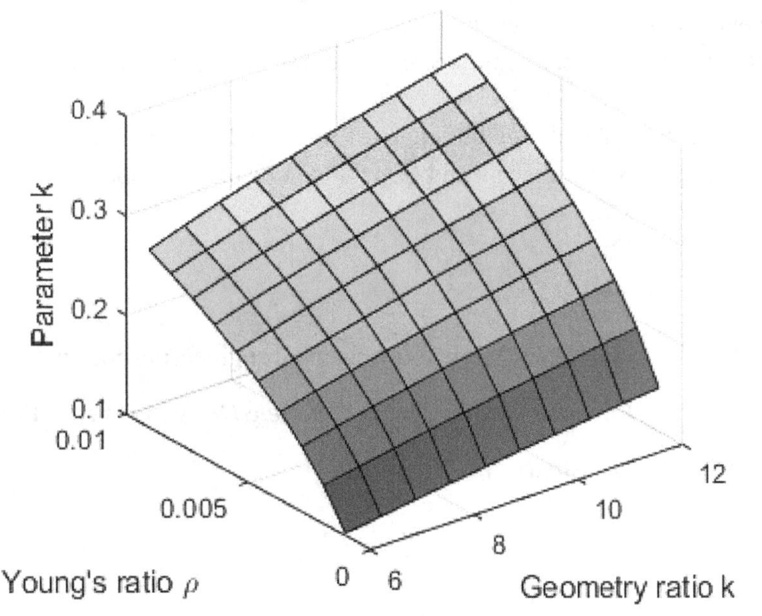

```
function Ch_4_ApExample_6(n_s,n_f,rho_s,rho_f,n_grid)
                         % neutral axis of the concrete beem, steel-reinforced
                           %   to run: >>Ch_4_ApExample_6(6,12,0.001,0.009,10)
n=linspace(n_s,n_f,n_grid);
ro=linspace(rho_s,rho_f,n_grid);
[X,Y]=meshgrid(n,ro);
rhon=X.*Y;
k=-rhon+sqrt(rhon.^2+2*rhon);
surf(X,Y,k)
xlabel('Geometry ratio k')
ylabel({'Young''s ratio \rho'})
zlabel('Parameter k')
title('Neutral axis location')
```

4.5. CONCLUSION

Basic and complementary commands for drawing and formatting 2D and 3D graphs were presented and examined in the chapter by various examples. The following M&T problems were solved and provided by plots generated with the discussed commands:

- Velocity of the engine piston,
- Efficiency of the power screw,
- Viscosity of a motor oil,
- Mode shape of a square membrane,
- Principal shear stress of a rectangular beam,
- Neutral axis location of a steel reinforced concrete beam with rectangular cross-section.

The materials in the current and previous chapters provide the necessary set of basic commands for writing the codes needed to solve, display and generate the resulting graphs and/or tables.

REFERENCES

Budynas, R. G., & Nisbett, J. K. (2011). Shigley's mechanical engineering design (9th ed.). McGraw-Hill.

Spiegal, L., & Limbrunner, G. F. (1992). *Reinforced Concrete Design* (3rd ed.). Prentice Hall.

Wang, C. T. (1953). *Applied Elasticity*. Mc-Graw-Hill.

ENDNOTES

[1] The plot generated with the stem3(x,y,z) and mesh(x,y,z) commands; used expression z=8+x+y

[2] Based on Table 3.3 in Burstein, 2015.

APPENDIX

Table 7. List of examples, problems, and applications discussed in the chapter

No	Example, Problem, or Application	Subsection
1	Generating a 2D graph of the wear of a micro-milling tool as function of the sliding distance.	4.1.1, 4.3.1
2	Friction coefficient – normal force plots with two and three curves for different material each.	4.1.2, 4.1.4.1
3	Three graphs in one figure window with a different number of friction coefficient-normal force curves.	4.1.3.
4	A formatted wear-sliding distance graph generated in the Live Editor.	4.1.4.
5	Generating a 3D line graph of the side helix.	4.2.1.
6	3D mesh graph for the distance from the centroidal to neutral axis of the circular beam section.	4.2.2.
7	3D surface graph for the distance from the centroidal to neutral axis of the circular beam section.	4.2.3.
8	Four surface graphs of various projections of the distance from the centroidal to neutral axis of the circular beam section.	4.2.4.
9	A graph with two y-axes for the kinematic viscosity and the specific volume of an engine oil represented as function of the temperature.	4.3.2.
10	Engine piston velocity.	4.4.1.
11	Power screw efficiency.	4.4.2.
12	Viscosity of a motor oil.	4.4.3.
13	Mode shape of a square membrane.	4.4.4.
14	Rectangular beam: principal shear stress.	4.4.5.
15	Steel rainforced concrete beam with a rectangular cross-section.	4.4.6.

Note, some small examples, mostly related to non-M&T issues, are not included in the list.

Chapter 5
MATLAB® Functions for Numerical Analysis and Their Applications in M&T

ABSTRACT

The chapter presents the MATLAB® commands that realize numerical methods for solving problems arising in science and engineering in general and in the field of mechanics and tribology (M&T) in particular. The most commonly used commands along with some information on numerical methods are explained. The topics of the chapter include interpolation and extrapolation, solving nonlinear equations with one or more unknowns, finding minimum and maximum, integration, and differentiation. All described actions are explained by examples from the field of M&T. At the end of the chapter, applications are presented; they illustrate how to interpolate the friction coefficient data, calculate elongation of a scale with two springs, determine the maxima and minima of the pressure-angle function, and solve some other M&T problems.

INTRODUCTION

Researches and laboratory practice in the field of mechanics and tribology involve a variety of wide-used math operations, such as interpolation, extrapolation, solving nonlinear algebraic equations, minimax, integration and differentiation. These operations frequently cannot be performed on the basis of a rigorous mathematical solution; in these cases, numerical methods are used. The available MATLAB® functions that can be implemented for the listed purposes are described below together with applications from the M & T area.

DOI: 10.4018/978-1-7998-7078-4.ch005

5.1. INTERPOLATION AND EXTRAPOLATION WITH THE INTERP1 COMMAND

Experimental or tabulate data are available at certain points and it is required sometimes to estimate values between the observed points, this can be done by interpolation. In case we need to evaluate values landing outside the data points, this can be done by extrapolation. For example, in some experimental technology, a preheated cylindrical part is cooled; the cooling temperatures at certain times were measured, the data is 690, 493, 392, and 355 C at times t=1, 3, 5, and 8 min respectively. The graph with measured, interpolated and extrapolated points is represented in Figure 1.

Figure 1. Original data (Δ), interpolation (*), and extrapolation (□) points

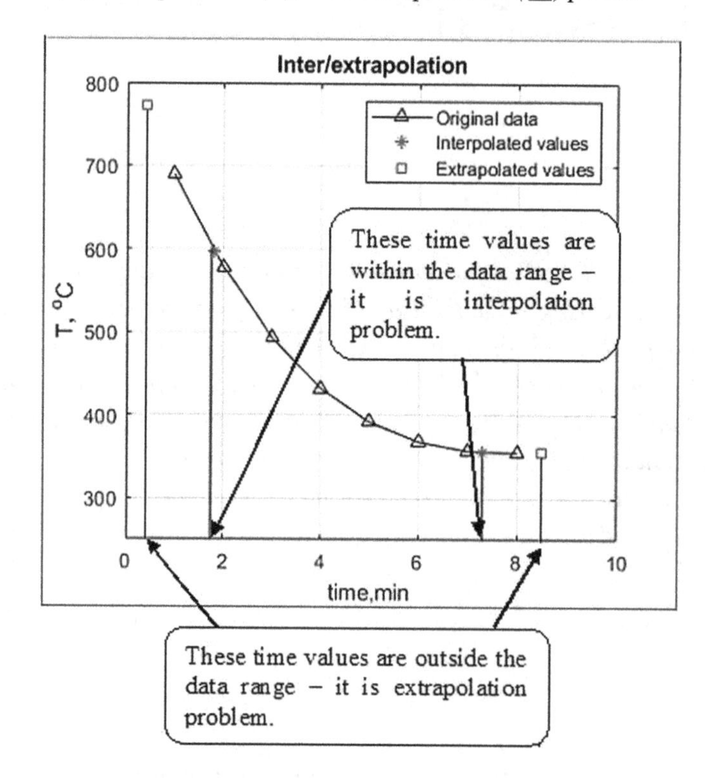

According to the data shown, the determination of T-values at times t=1.8 and 7.3 min (values within the range of the source data) is an interpolation problem, and the determination of T-values at t=0.4 and 8.5 min (values outside the data range) – extrapolation problem.

In MATLAB©, both actions: inter- and extrapolation can be performed with the interp1 function. The available forms are

```
yi= interp1(x, y, xi, 'method')
```

or

```
yi= interp1(x,y,xi,'method','extrap')
```

The first command form is used for interpolation data, and the second one is for extrapolation or for simultaneously inter- and extrapolation.

The input parameters of these commands are:

- x and y are respectively the argument and function vectors with data values;
- xi – a scalar or a vector with *x*-point values (arguments) for which the values *y*-points are sought;
- 'method' is a string specifying the name of the numerical method to be used for inter - or extrapolation; some available methods are – 'linear' (default), 'pchip' (called 'cubic' in earlier MATLAB© versions), and 'spline'; a default method does not need to be specified when the first command form is used;
- The 'extrap' string is required for extrapolation by the 'linear' method; for the 'pchip' and 'spline' methods the word 'extrap' can be omitted.
- The output parameter of these functions is yi – the value or vector of values defined (interpolated/extrapolated) at the xi points.

The linear method uses two adjacent points and a straight line between them to interpolate; the spline method uses cubic polynomials adjusted so that the fitting line is smooth, and the pchip method is the piecewise cubic Hermite interpolation using the polynomial method.

Demonstrate inter- and extrapolation by the following problem:

Problem: In the example in Figure 1 the time and temperature data are t=1:8 sec and T=690, 580, 493, 430, 392, 370, 360, 355 °C. Calculate and display the *T* values at the time points 1.8 and 7.3 min (interpolation), and 0.4 and 8.5 min (extrapolation) and generate the *T*(*t*) data graph along with the obtained inter- and extrapolated *T*-values. Use the spline method. Formate graph as in Figure 1.

The following commands, entered directly in the Command Window, solve the problem, display the defined T values, and generate the graph shown in Figure 1.

```
>>t=1:8;                                               % times (x axis)
>>T=[690 580 493 430 392 370 360 355];           % temperatures, (y axis)
>>ti=[1.8 7.3];                      %1.8 and 7.3 are the points of interpolation
>>te=[.4 8.5 ];                      % 0.4 and 8.5 are the points of extrapolation
>>T_i=interp1(t,T,ti,'spline');                         % spline interpolation
>>T_e=interp1(t,T,te,'spline','extrap');                % spline extrapolation
>>T_i,T_e                      % to display the inter- and extrapolated values
T_i =
 599.9881 358.0975
T_e =
 766.2666 354.6875
>>plot(t,T,'-^',ti, T_i,'*',te, T_e, 's'),grid on
>>axis([0 10 250 800])                                 % axes as in Figure 5.1
>>xlabel('time, min'),ylabel('T, ^oT')
>>title('Inter/extrapolation')
>>legend('Original data', 'Interpolated values', 'Extrapolated values')
```

5.2. NONLINEAR ALGEBRAIC EQUATION SOLUTION

Linear algebraic equations and their matrix solution, i.e. defining unknowns, was described in the second chapter (subsections 2.2.2, 2.4.1). Here we give a command that iteratively searches for a value x of a nonlinear equation with one variable f(x) =0 such that for this value the equation becomes zero. The command using for this is fzero and its general form is:

```
x=fzero('fun', x0)
```

where 'fun' is the name of the user-defined function containing the solving equation, or is the solving equation written as string; x0 is the initial or so-called guess value representing a priori given x-value, around which the command seeks the true x.

The x0 value can be pre-determined graphically on the $f(x)$ plot by checking the value of x at which the function is zero.

To demonstrate the command usage, consider the inverse problem for the pipe's coefficient of friction, fully developed turbulent flow (Colebrook formula, based on Hwang & Hita, 1987)

$$\lambda = \frac{1}{\left[2\log\left(3.7\frac{d}{k}\right)\right]^2}$$

where d is the pipe diameter, k is the surface roughness. If, for example, the required coefficient of friction λ is 0.024, what is the ratio d/k that meets the required value?

The equation should be rewritten in form $f(x)$=0:

here the d/k ratio is signed as x. The initial ratio value $x0$ must be within the actual d/k -range, for our case it can be $x0$=490.

The solution command is:

```
>> d_k=fzero('0.024-1./(2*log10(3.7*x)).^2',490)
d_k =
  456.3356
```

The string with the solved equation cannot include pre-assigned variables, e.g. it is not possible to define lambda=0.024 and then write fzero ('lambda-1./(2*log10(3.7*x)).^2',490). To use pre-assigned parameters the following fzero form is available:

```
x=fzero(@ (x) fun(x, preassigned_variables1, preassigned_variables2,…), x0)
```

Using the latter fzero form, we can write the above example as a life script file with the name ExFrictionCoeff, input parameters lambda and $x0$, and output parameter d_k. The live script program ExFrictionCoeff presents in Figure 2.

Figure 2. Live script with the fzero command defining ratio d/k for the specified friction coefficient λ

ExFrictionCoeff.mlx

Solution equation

$$\lambda - \frac{1}{[2\log(3.7x)]^2} = 0$$

where x is the d/k ratio

```
1  guess=490;
2  lambda=0.024;
3  fun=@(x)lambda-1./(2*log10(3.7*x)).^2;
4  d_k=fzero(fun,guess)

   d_k = 456.3356
```

Note: the form f=@ (x) fun(x, preassigned_variables1, preassigned_variables2,…) can be used not only with the fzero command, but also to calculate fun at any x (see example in the next subsection).

5.3. MINIMUM AND MAXIMUM OF A FUNCTION

In mechanical and tribological practice it is sometimes necessary to find the extremum (maximum or/and minimum) of a function to achieve the best its value. The maximum and minimum of the function corresponds to the point where the derivative is zero. The fminbnd command is destined for this purpose. This command searches in a specified local interval $x_1 \dots x_2$ for the such x and y values of an one-variable function $y=f(x)$ that correspond to a minima; the command has the following simplest forms

```
x=fminbnd('fun',x1,x2)
```

and

```
[x,f_x]=fminbnd('fun',x1,x2)
```

fun is the equation of function $f(x)$, for which the minimum is searched; equation is written as string between the single quotations; another possibility is to write the equation without quotations but with the @(x) symbols that should be written before the equation; note, that the 'fun' should be written using element-wise operators, e.g. '.*', '.^', and './'.

x1 and x2 are the input parameters denote the start and finish points of the interval where the x-value should be defined;

x and f_x are the output parameters containing the minimum point values x_{min} and f_{min} (the f-value at x_{min}).

The fminbnd command searches for the minimum; nevertheless, it can be used also to determine the maximum of the function by finding the minimum for which the function should first be multiplied by -1.

To illustrate the use of this command, we will find the minimum and maximum of the modified Buckingham pairwise interatomic potential which is used, for example, in physical mechanics and nano-tribology

$$v = Ae^{-\frac{r}{b}} - \frac{C}{r^6}$$

where r is interatomic distance; A, b, and C are constants.

Problem: Find values of the minimal and maximal distance r and potential v when $A = 5.5 \cdot 10^4$ eV, $b=0.16$, and $C=100$ eV/A^6; the search range of the r is 0.85 … 3.5 A for minimal point and 0.85 … 1.5 – for maximal point. Plot the $v(r)$ curve together with defined extremal points; add captions and legend. Display the minimal and maximal points with the fprintf command, use four digits points for each of the outputted values. Write the live script named ExBuckMinMax using the "Output on right" option.

The live script solves this problem are shown in the Figure 3.

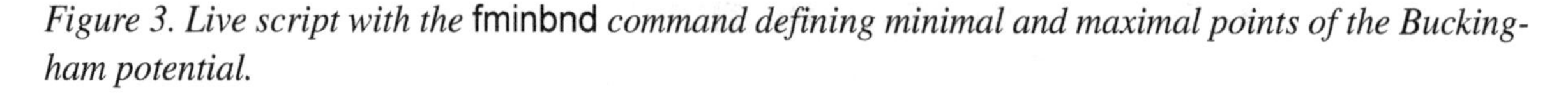

Figure 3. Live script with the fminbnd *command defining minimal and maximal points of the Buckingham potential.*

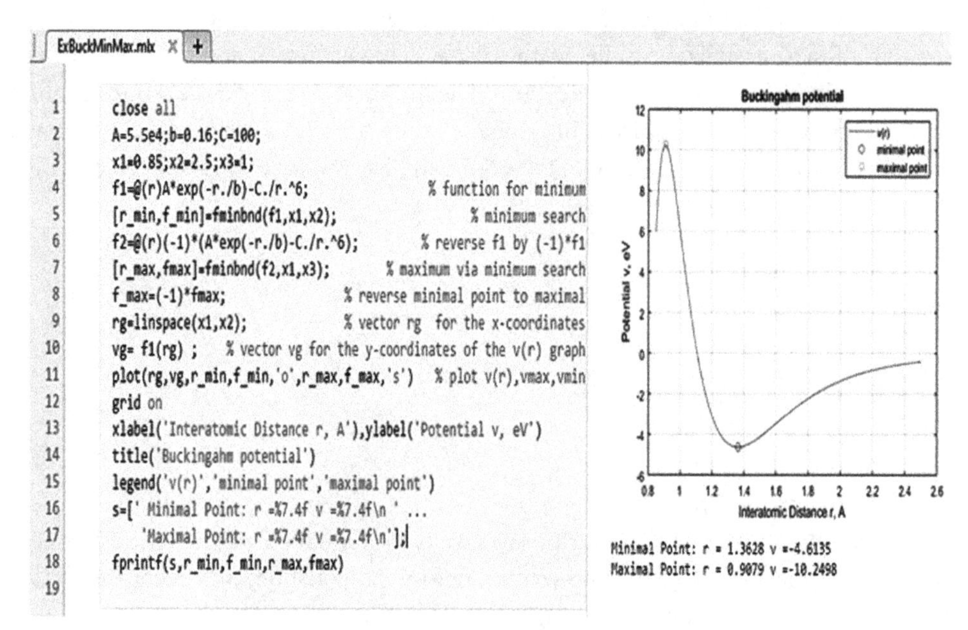

5.4. INTEGRATION

The engineers and scientists frequently need to define an integral that appear in a variety of practical situations. This problem can be solved with the MATLAB© commands, which calculate the integral numerically as the area beneath the function curve. The function can be represented by the mathematical expression $f(x)$, when the values of f can be obtained for any value of x, or by the tabular data, when the f values are given at the discrete x points – $f(x_1), f(x_2)$, …(Figure 5.4, a, b). To calculate an integral, the area beneath the f(x) curve is subdivided in small identical geometrical elements, e.g., rectangles, or trapezoids. With some uncertainty, the sum of areas of these elements gives obviously the value of the integral.

Figure 4. The areas (shaded) represent definite integrals when the f(x) function is given analytically (a) and by the data points (b).

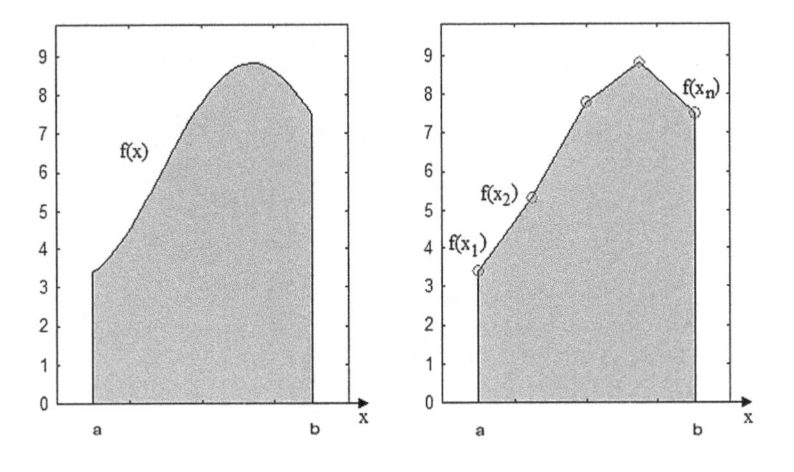

Below the **quad** and **trapz** functions of integration are described. The first is used when a integrating function is presented as an analytical expression, and the second – when it is presented as data points.

5.4.1. The Quad Command

This command calculates an integral approximating the function by the three-point parabolas; the areas beneath them are calculated and then summarized. The method is called the Simpson's rule. The quad comand uses a modified, so-called adaptive, Simpson's method and has the form:

```
I=quad('function', a, b, tolerance)
```

'function' denotes a string with the f(x) integrating expression, this expression can be written also not as string if the @(x) symbols is written before him. It is possible also to write here the name of a function file where the expression should be written, in this case the @ sign should be written before the function name;

a and b are the lower and upper limits of integration,

tolerance – is the desired maximal absolute error, this parameter is optional and can be omitted; the default tolerance is smaller than $1 \cdot 10^{-6}$;

I is the output variable containing the calculated value of the integral.

The function f(x) should be written using element-wise operators, e.g. '.*','.^', and './'.

For example, to calculate the flow rate of a fluid in the round pipe the following integral should be calculated:

$$Q = \int_0^R 2\pi v_{max} \left(1 - \frac{r}{R}\right)^{1/n} rdr$$

where R is the pipe radius, v_{max} – maximal flow profile velocity, n –coefficient, r – current position of the velocity profile. Calculate $Q, \frac{m3}{s}$, for R=0.06 m, v_{max} = 2 m/s, and n=7

The following command should be typed and entered from the Command Window

```
>> Q=quad('2*pi*2*(1-r/.06).^(1/7)*r',0,0.06)
Q =
  0.0185
```

```
>> R=0.06;vmax=2;n=7;
>> Q=quad(@(r)2*pi*vmax*(1-r/R).^(1/n)*r,0,0.06)
Q =
  0.0185
```

The latter form is suitable when the expression has parameters that should be inputted or pre-assigned, as shown above

Another option to use the quad command is to write in the Editor window the integrating expression in form of the user-defined function and then save it in the function file. For our example:

```
function y=Ch5_MyIntegral(x)
R=0.06;vmax=2;n=7;
y=2*pi*vmax*(1-x/R).^(1/n)*x;
```

To calculate the integral, previously saved in the Ch5_MyIntegral file, the following quad command should be entered from the Command Window:

```
>> Q=quad(@Ch4_Integr,-3,3)
Q =
  0.0185
```

5.4.2. The Trapz Command

The command has form

```
I=trapz(x,y)
```

For example, the dimensionless positive values of the hydrodynamic pressure distribution in thin lubricating film that separates two surfaces covered by semicircular pores are presented as function of the dimensionless coordinate (based on data from Burstein, 2016a):

Box 1.

***P*, nondim**	0	0.0283	0.0704	0.0980	0.1310	0.0655	0.000
***X*, nondim**	0.0	0.2	0.5	0.7	1.0	4.0	7.0

The film load support *W* along the pore cell is:

$$W = \int_0^7 P(X)\,dX$$

where the *P*(X) function under the integral is given numerically in the table.

The following commands should be typed and entered from the Command Window to find the integral *W*:

```
>>X=[0 .2 .5 .7 1 4 7];
>>P=[0 .0283 .0704 .0980 .1310 .0655 0];
>>W= trapz(X,P)
W =
     0.4618
```

5.5. DERIVATIVE CALCULATION

The derivative $\frac{dy}{dt}$ of a function of a single variable *y*(*t*) at a chosen *t* point can be approximately calculated as the ratio of infinitesimal changes of function $\triangle y$ to its argument $\triangle t$. In geometrical representation (Figure 5) it is the slope $\alpha \approx \frac{\Delta y}{\Delta t}$ of the tangent line to the curve at the *i*-th point.

Figure 5. Geometrical representation of the dy/dt derivative.

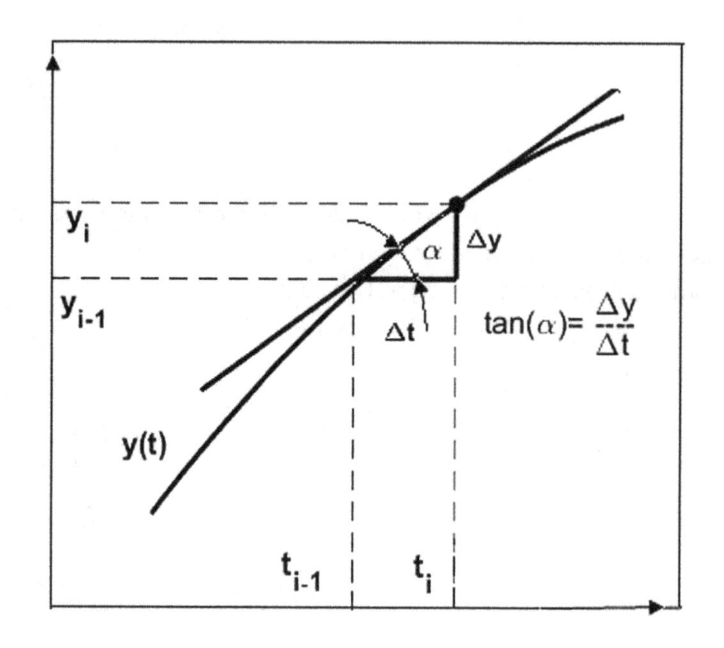

Following the above, the derivative at a point *i* is:

$$\left(\frac{dy}{dt}\right)_i = \lim_{\Delta t \to 0}\left(\frac{\Delta y}{\Delta t}\right)_i \approx \frac{y_i - y_{i-1}}{t_i - t_{i-1}}$$

According to this interpretation, the derivative can be calculated at each *i* -point as when we know the *y* and *t* values at the *i* and *i*-1 points.

To define the Δy and Δt differences, the diff command should be used. This command has the following forms:

```
dy=diff(y)
```

or

```
dy_n=diff(y,n)
```

y is the inputting vector with *y*-values that numbered corresponding to points *i* = 1, 2, …;

the n is the order of the particular difference that indicates how many times the diff should be applied, for example, if n=2 the diff should be applied twice;

the dy and dy_n are outputting vectors that contain the determined values of the 1st and n-th order differences respectfully.

Note: the dy vector with the defined differences is one element shorter than vector y, and correspondingly the dy_n vector is n elements shorter than y; e.g., if y has 8 elements thus dy has 7 elements, the second order difference vector dy_2 has consequently only 6 elements and so on.

As in the case of integration, the derivative can be defined for function presented analytically or in the form of a table. In the latter case, the number of calculation points is limited by the table, and the derivative value may be less accurate than in the case of an analytically given function, when you can specify a greater number of points to ensure greater accuracy. For example, coefficient of linear expansion L of a material due to temperature is $\alpha_L = \frac{1}{L}\frac{dL}{dT}$. The temperature-length data for a copper tube 35 m long at T=20 C shown in the table

Box 2.

T, °C	20	24.44	30	35.6	41.1	46.7	72.2
L, m	35.000000	35.0117	35.0175	35.0204	35.0233	35.0204	35.0233

Problem: Create live script named ExDeriv that uses these data to determine the linear thermal expansion coefficient at 30…72.2 C. Display results with the fprintf command showing the α_L at 30 … 46.7 C.

The Live Editor window with the ExDeriv live script program and its numerical results are shown in Figure 6.

Figure 6. Live script with the diff *commands defining coefficient of linear temperature expansion of a tube*

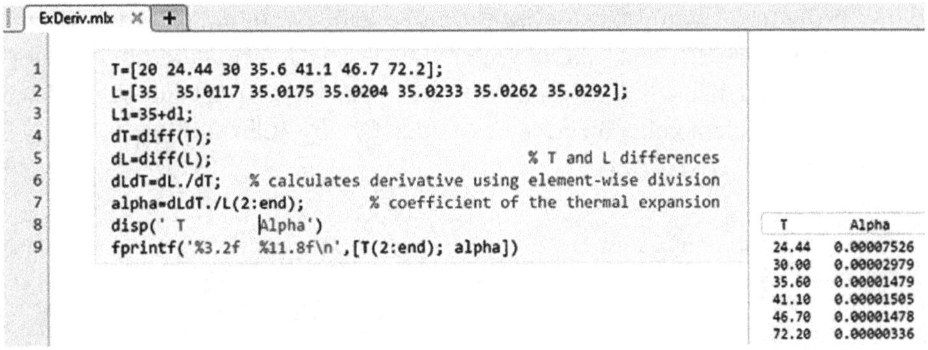

```
ExDeriv.mlx

1  T=[20 24.44 30 35.6 41.1 46.7 72.2];
2  L=[35  35.0117 35.0175 35.0204 35.0233 35.0262 35.0292];
3  L1=35+dl;
4  dT=diff(T);
5  dL=diff(L);                                  % T and L differences
6  dLdT=dL./dT;   % calculates derivative using element-wise division
7  alpha=dLdT./L(2:end);        % coefficient of the thermal expansion
8  disp(' T          Alpha')
9  fprintf('%3.2f  %11.8f\n',[T(2:end); alpha])
```

```
  T        Alpha
24.44   0.00007526
30.00   0.00002979
35.60   0.00001479
41.10   0.00001505
46.70   0.00001478
72.20   0.00000336
```

Note:

- If the step h of argument of the differentiating function $f(t)$ is the constant value, it can be used instead of diff(t);

- The derivatives of a function represented by an equation can be determined in the same way as from tabular data; the step in argument in such a case can be set smaller than in case tabulated data, that leads to more accurate values of the derivatives;
- If the derivative should to be determined for one specific point, for example, if the derivative value is needed at argument point t=1, thus two points are sufficient for the numerical differentiation - t and t-h, where h is the specified t-step;
- When the values of the derivatives are required at points other than those for which the derivatives were calculated (for example, by tabular points), then the interp1 command can be used to determine the derivatives at the required points from the calculated ones.

5.6. SUPPLEMENTARY COMMANDS FOR INTERPOLATION, EQUATION SOLUTION, INTEGRATION, AND DIFFERENTIATION

In addition to the commands discussed above, there are other commands that can be used for interpolation, solving nonlinear equations, integration, and derivative calculations; These commands are presented in Table 1.

5.7. APPLICATION EXAMPLES

5.7.1. Friction Coefficient Data Interpolation

Tribology laboratory measurements of the friction coefficients k as function of temperature T for cold sprayed commercially coating (pure Ni) are 0.8, 0.75, 0.64, and 0.37 at 273, 350, 400, and 450 K respectively.

Problem: Using the interp1 command, find the friction coefficient values for temperatures 300, 375, 425, and 460 K. Write for this program as user-defined function with the input vector containing the temperature for inter- extrapolation and output vector with defined temperatures. Try the linear, spline, and pchip (former 'cubic') methods and plot on the same page the original and interpolated points for each method. Save the program in the function file named Ch_5_ApExample_1.

The program solving this problem is

```
function k_interp= Ch_5_ApExample_1(T_interp)
                    %    interpolates coefficient of friction - temperatura data
                    %     To run:>> k_interp=Ch_5_ApExample_1([300 375 425 460])
T=[273 350 400 450];                                          % temperature vector
k=[0.8 0.75 0.64 0.37];                                % friction coefficient vector
k_interp(1, :)=interp1(T, k, T_interp, 'linear', 'extrap');                % linear
k_interp(2, :)=interp1(T, k, T_interp, 'spline', 'extrap');                % spline
k_interp(3, :)=interp1(T, k, T_interp, 'pchip', 'extrap');                  % cubic
subplot(1,3,1)
plot(T, k,'o-', T_interp, k_interp(1, :),'s'),grid on                    % 1st plot
xlabel('Temperature, K'), ylabel('Friction coefficient, nondim')
```

*Table 1. Some commands for inter- extrapolation, solution the nonlinear equations, integration, and derivative calculations**

Command	Purpose	Example
yi=spline(x,y,xi)	interpolates with the cubic splines. x,and y are the size-equal vectors with the given argument and function values respectively; xi is the x -point/s of interpolation; yi - defined interpolation values	>>x=(0:.2:1)*2*pi; >>y=cos(x);xi=4.6; >>yi=spline(x,y,xi) yi = -0.1154
yi=pchip(x,y,xi)	Piecewise Cubic Hermite Interpolating Polynomial (PCHIP). x, y, xi, and yi are the same as for the spline command	>>x=(0:.2:1)*2*pi; >>y=cos(x);xi=4.6; >>yi=pchip(x,y,xi) yi = -0.1165
x=roots(p)	returns the roots x of the polynomial; vector p contains the polynomial coefficients. The polynomial has form: $p_1x^n+...+p_nx+p_{n+1}=0$	%finds roots of %the polynomial %13.5x^2+3.2x-0.7=0 >>p=[13.5 3.2 -0.7]; >> r = roots(p) r = -0.3752 0.1382
x=fminsearch(fun,x0)	finds minimum x values of multivariable of the non-constrained function fun; the calculations begin from the x0 values specified for each variable.	>>f=@(x)(x(2) - x(1)^2)^2 + (1 - x(1))^2+x(1)+x(2); >>x0=[1 -1]; >>x=fminsearch(f,x0) x = 0.2500 -0.4375
I = integral(fun,a,b)	integrates expression written in the fun; a and b are the lower and upper limits of integration respectively; returns numerical value of the integral in the output variable I.	>> A=2.7;n=1.9;a=0;b=3; >>f=@(x)A*exp(-x.^n); >>I=integral(f,a,b) I = 2.3957
I=guadl(fun,a,b)	fun, a, b, and I are the same as for the integral command; fun can be written also as a string.	>> I=quadl('2.7*exp(-x.^1.9)',0,3) I = 2.3957
I=quadgk(fun,a,b)	All notations are the same as in the integral, but one of the limits can be infinite (noted inf).	>>f=@(x)2.7*exp(-x.^1.9); >> I = quadgk(f,0,Inf) I = 2.3959
I=quad2d(fun,a,b,c,d)	returns numerical value I of the double integral; fun is the same as for quadl; the integration limits a and b for the first variable (i.e. x) and c and d – for the second (i.e. y).	>>f=@(x,y)x.*sin(y); >>I = quad2d(f,pi,2*pi,0,pi) I = 29.6088
dcoeff=polyder(p)	calculates coefficients (returns in dcoeff) of the polynomial that is derivative of the polynomial giving by the p-coefficients; p has the same sense as for the root command.	%derivatives of %the polynomial %13.5x^2+3.2x-0.7=0 >>p=[13.5 3.2 -0.7]; >>dcoeff=polyder(p) dcoeff = 27.0000 3.2000 %dcoeff is the vector %with coefficients of %the polynomial: %27x+3.2=0

* The commands are presented here in their simplest form.

```
title('Linear Interpolation')
legend('Data','Interp.','location', 'southwest')                % left bottom
subplot(1,3,2)
plot(T,k,'o-', T_interp, k_interp(2,:),'s'),grid on              % 2nd plot
xlabel('Temperature, K'),ylabel('Friction coefficient, nondim')
title('Spline Interpolation')
legend('Data', 'Interp.', 'location', 'southwest')              % left bottom
subplot(1,3,3)                                                   % 3rd plot
plot(T,k,'o-', T_interp, k_interp(3,:),'s'),grid on
xlabel('Temperature, K'), ylabel('Friction coefficient, nondim')
title('Cubic Interpolation')
legend('Data','Interp.','location', 'southwest')                % left bottom
```

The Ch_5_ApExample_1 function is designed as follows:

- The first line of the function defines the function name and input and output variables – T_interp and k_interp respectively;
- Two following function lines present the help part of the function and provide its short purpose and command for running the function in the Command Window;
- In the subsequent function lines the temperatures and friction coefficient data is assigned to variables T and k respectively; then three interp1 commands calculate k_interp for each of the T_interp values and assign the interpolating values to the three-rows output matrix k_interp in such way that each row contains all k_interp values interpolated by the same method;
- Further function lines content commands for generating and formatting three plots in the same Figure window; the tabular data signed by circles and connected with the solid line and interpolating values signed by the red x; each plot is formatted with the axis labels, grid, a caption, and a legend.

After entering the running command in the Command Window, the calculated friction coefficients and plot for each of the interpolating methods are as follows:

```
>> k_interp= Ch_5_ApExample_1([300 375 425 460])
k_interp =
  0.7825  0.6950  0.5050  0.3160
  0.7839  0.7098  0.5302  0.2894
  0.7923  0.7080  0.5292  0.2984
```

5.7.2. Lengthening of a Two Springs Scale

Nonlinear springs (see image in the Figure 7) are constructed a scale in such way that force that acts of them are

$$F_s = K_1\left(L - L_0\right) + K_2\left(L - L_0\right)^3$$

with

$$L_0 = \sqrt{a^2 + b^2}$$

$$L = \sqrt{a^2 + \left(b + x\right)^2}$$

Figure 7. Live script for determining the lengthening of a two-spring scale; -- denotes the springs location before loading.

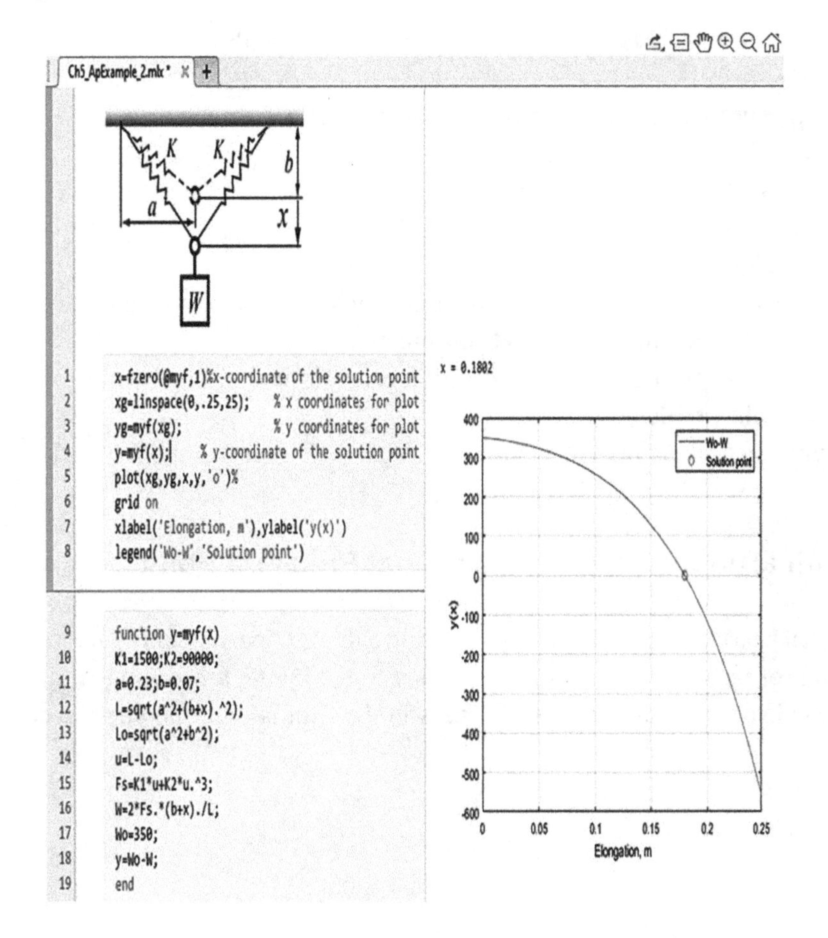

In these equations, K_1, K_2, a, and b are the spring constants, and L_0 and L are respectively the initial and current length of the joint point of the springs describing with the expressions

Different applied weights W cause the spring system elongation x according to the expression

$$W = 2F_s \frac{\left(b + x\right)}{L}$$

Problem: Write live script named Ch_5_ApExample_2 that calculates spring elongation x with the fzero function when the weight applied to the springs are W_0=350N, and spring constants K_1, K_2, a, and b are 1500, N/m, 90000 N/m^3, 0.23 m, and 0.07 m respectively; take x_0=1 (initial x-value). The live script should include an image of the system and generate a plot illustrating the graphical solution.

To use the fzero command, the W-equation should be rewritten as

$$W_0 - 2F_s \frac{\left(b + x\right)}{L} = 0$$

This expression is the solving equation and the x value of this equation is, apparently, the desired elongation value.

The Ch_5_ApExample_2 live script solving the problem with resulting x and generated plot are presented in the Figure 7.

The Ch_5_ApExample_2 live script is arranged as follows:

- The image at the beginning of the script was created and saved in a separate file with the name Ch5_Image, after which it was inserted into the script using the "Image" button on the "Insert" tab in the Live Editor menu;
- The first script command is the fzero function which finds the x- value representing the solution of the $y(x)$ equations written in the myf sub-function;
- Commands located above the myf sub-function generate and format the plot representing a graphical solution to the problem;
- The sub-function myf contains the above equations with their components – L_0, L, F_s, and spring constants.

5.7.3. Cylindrical Journal Bearing: Maximal Pressure Point

Cylindrical journal bearing comprises an inner rotating cylinder (journal) of radius R and outer immobile cylinder (bearing) of radius R_b. The cylinders are separated by the lubricating film. For long bearing, the dimensionless hydrodynamic pressure P arises in the film is described by the equation (based on Burstein, 2011)

$$P = \frac{\varepsilon\left(2 + \varepsilon\cos\theta\right)\sin\theta}{\left(2 + \varepsilon^2\right)\left(1 + \varepsilon\cos\theta\right)^2}$$

where $\varepsilon = e/(R-R_b)$ is the dimensionless clearance, and θ is the attitude angle changing in the range $0 \ldots 2\pi$.

Problem: Write a live script named Ch_5_ApExample_3 that uses the fminbnd command to find the maximum and minimum pressure points at ε=0.7 and generates a plot of the $P(\theta)$ curve with the marked extreme points. Use the following angle ranges for minimum and maximum point search – π …2π and 0… π respectively. Display numerical results with the fprintf command.

The live script solving the problem with resulting extreme points values and generated plot are presented in the Figure 8.

Figure 8. Live script for determining the maximal and minimal points: commands with numerical and graphical results.

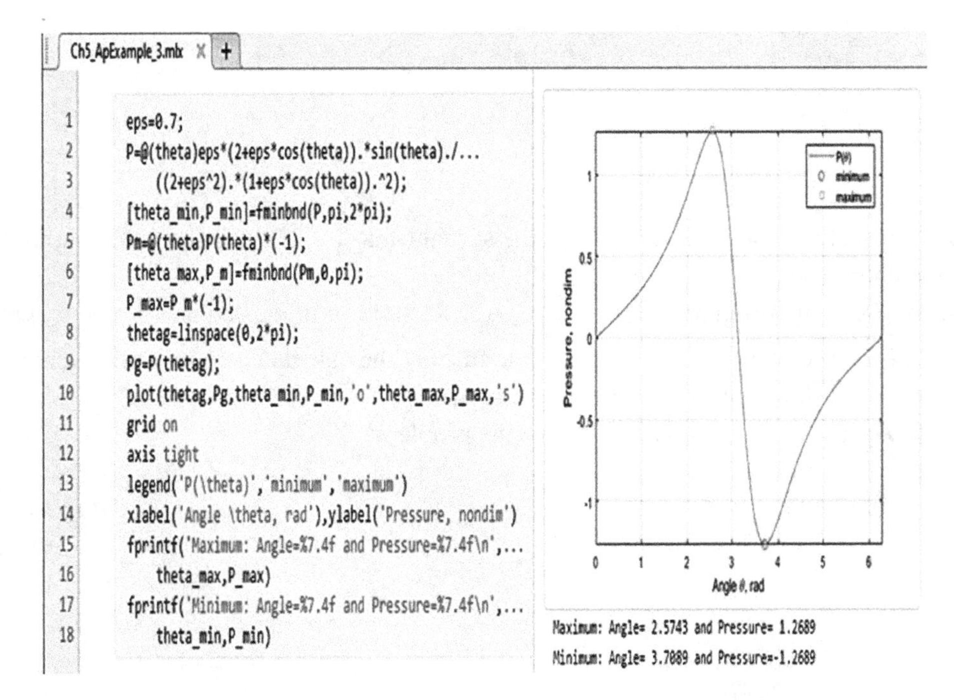

The script is arranged as follows:

- The ε value is assigned at the first lines of the script, and the pressure *P*-equation is presented in the P= @(theta)… form;
- The fminbnd function with angle range for the minimum point are used;
- To find the maximum point with the fminband command, the expression Pm are written; it represents the pressure equation multiplied by -1; then the new fminbnd function is used, and the resulting pressure is again multiplied by -1 to represent the maximum.
- Further commands generate and format plot, and display results as two lines with angle and pressure for the minimum and maximum points.

Note: a function defined in the form y= @(x)… can be used to calculate *y* for various *x* values with the command y (x); for example, the P=@(theta)… form above is used with the P(thetag) command to calculate pressure values at previously specified thetag values.

5.7.4. Centroid of the Cross Section of a Mechanical Part

The geometrical center location, called centroid, for the cross section of a some mechanical part can be calculated with the following expressions:

$$x_c = \frac{\int_{x_1}^{x_2} x\left(y_2 - y_1\right)dx}{\int_{x_1}^{x_2}\left(y_2 - y_1\right)dx}$$

$$y_c = \frac{\int_{x_1}^{x_2}\left(y_2^2 - y_1^2\right)dx}{2\int_{x_1}^{x_2}\left(y_2 - y_1\right)dx}$$

Consider the sectional shape described by the straight-line y_1=2*x and the cubic parabola y_2= x^3, where x is the coordinate.

Problem: Write script program named Ch_5_ApExample_4 that calculates x_c and y_c when x_1 and x_2 are 0 and $\sqrt{2}$ respectively. The program should display the calculated coordinates, generate plot of the cross section and show the defined centroid place.

The Ch_5_ApExample_4 program to solve the problem:

```
x1=0;x2=sqrt(2);                                        % integral limits
y1=@(x)2*x;                                          % cross section line1
y2=@(x)x.^3;                                        % cross section line 2
f_denom=@(x)y2(x)-y1(x);                        %denominator in the xc and yc
A=quad(f_denom,x1,x2);                          %integral in the denominators
f_nom_x=@(x)x.*f_denom(x);                              % numerator in xc
nom_x=quad(f_nom_x,x1,x2);                      %integral in the xc numerator
xc=nom_x/A                                                % calculates xc
f_nom_y=@(x)y2(x).^2-y1(x).^2;                          % numerator in yc
nom_y=quad(f_nom_y, x1, x2);                   % integral in the yc numerator
yc=1/2*nom_y/A                                            % calculates yc
xg=linspace(x1, x2);                                       % x for curves
plot(xg, y1(xg), xg, y2(xg), xc, yc, '*')      %cross section and centroid
axis tight
xlabel('x'),ylabel('y')
legend('y_1=2x','y_2=x^3','Centroid point',...
  'location', 'northwest')                        %left top angle location
```

The script is constructed as follows:

- At the first script lines the integral limits are assigned to the x1 and x2 variables;
- Next two lines represents the curves expressions written as @(x) functions;
- Two next lines are intended to calculate the integral in the denominators of the x_c and y_c expressions; the quad command is used for this;
- Three further commands calculate the integral in the numerator of the x_c -equation and x_c itself; the latter value is displayed;
- y_c is calculated with the next three commands similar to the x_c; the y_c value is displayed;
- After this the plot is generated and formatted.
- Now the results are plotted, formatted, and displayed:

After running in Command Window, the following results are displayed and plotted.

```
>> Ch_5_ApExample_4
xc =
  0.7542
yc =
  1.0775
```

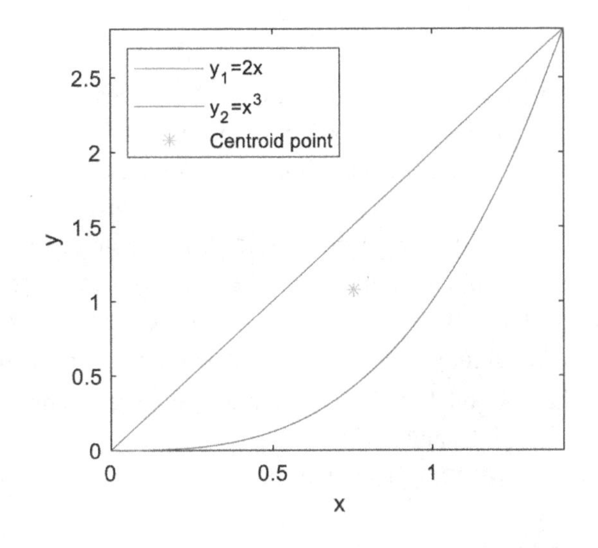

5.7.5. Roughness Parameter R_a for the Triangular and Sinusoidal Profiles

The R_a is the one of the roughness parameters that can be calculated by the following expressions (Bhushan, 2013):

$R_a = \frac{1}{\lambda}\int_0^{\frac{\lambda}{4}} z(x)\,dx$ - for a quarter of a triangular roughness profile, $0 \le x \le \lambda/4$

while

$$R_a = \frac{1}{\lambda}\int_0^{\frac{\lambda}{2}} z(x)\,dx$$ - for the half sinusoidal profile, $0 \le x \le \lambda/2$

The triangular profile line is described as

$$z = \begin{cases} \frac{4A_1}{\lambda}x, & x \le \frac{\lambda}{4} \\ 2A_1\left[1 - \frac{2}{\lambda}x\right], & \frac{\lambda}{4} \le x \le \frac{3\lambda}{4} \\ 4A_1\left[-1 + \frac{1}{\lambda}x\right], & \frac{3\lambda}{4} \le x \le \lambda \end{cases}$$

The sinusoidal profile line expression is

$$z = A_0 \sin\left(\frac{2\pi}{\lambda}x\right)$$

In these equations x and z are the Cartesian x and y coordinates respectively, λ is the wavelength, A_1 and A_0 is the maximal amplitude for the triangular and for the sinusoidal roughness profiles respectively. The equations are valid when the mean roughness line value is equal to zero.

Problem: Write the program named Ch_5_ApExample_5 that contains a user-defined life function called CLA_Ra and the preceding running command. The function should has the A_0, A_1, and λ as their input parameters and can be without output parameters. When accessing, the function should calculates R_a for the triangular and sinusoidal roughness profiles and generate a formatted plot containing the two profile lines and text with the defined R_a values. Assume A_0=6.3 µm (corresponds to internationally accepted roughness grade N9), A_1=A_0, and λ=7A_0.

The live program Ch_5_ApExample_5 solving the problem with the resulting plot is presented in the Figure 9.

The presented live program is arranged as follows:

- In the first four lines, the A_0, A_1, and λ values assign to the A0, A1, and l variables and the CLA_Ra command that access this live function and transmit to it the assigned values;
- The CLA_Ra function begins on the fifth line with the function definition command; and the next line contains a short comment about the purpose of the function;
- The subsequent four lines contain the equations under the R_a-integrals and the quad functions calculates the R_a parameters for the triangular and sinusoidal profiles;

Figure 9. Live function for determining the R_a parameter for triangular and sinusoidal roughness profiles: commands and generated plot containing numerical results.

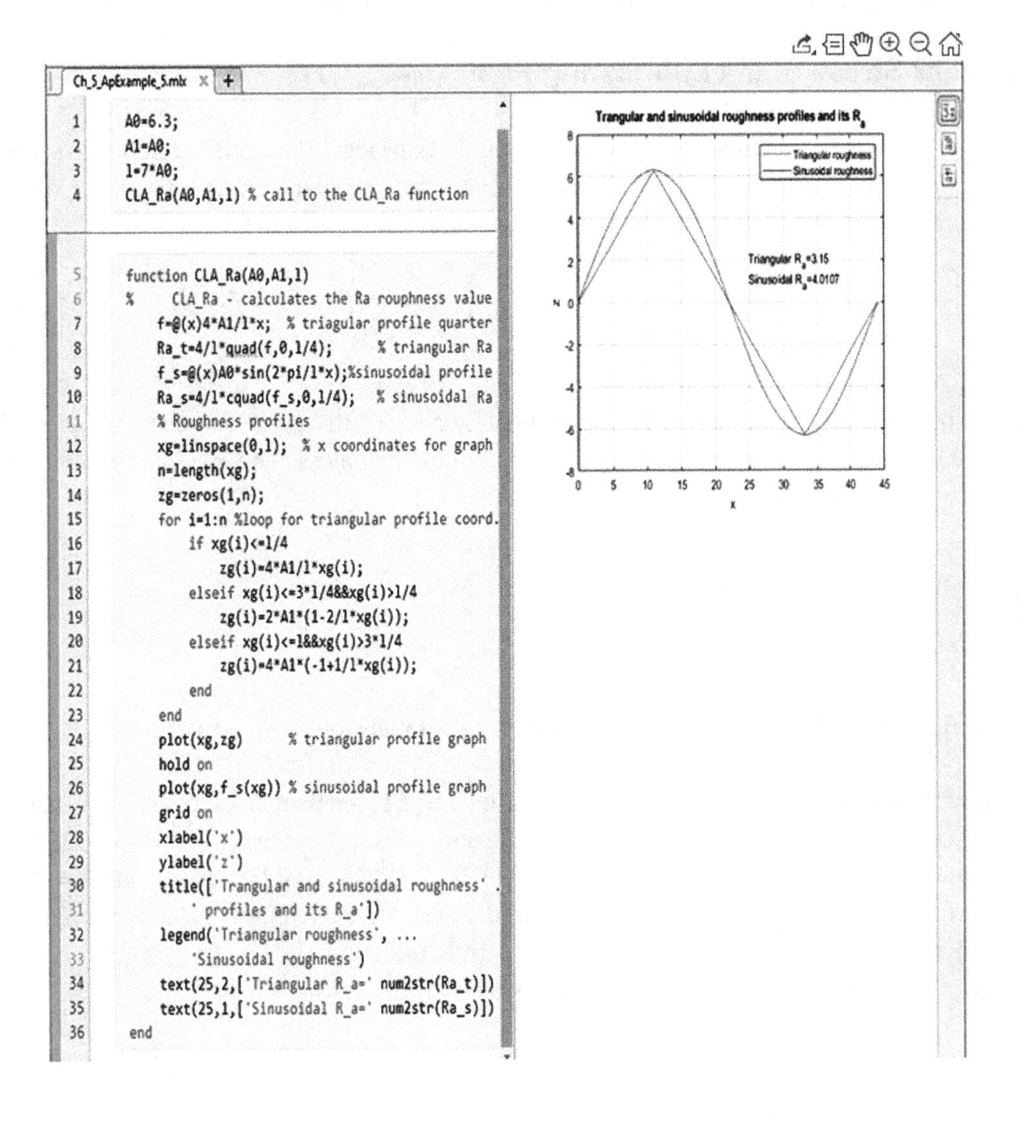

- Further commands generate the resulting graph; to calculate the triangular roughness profile represented by the piecewise expression $z(x)$; the for … end loop is used together with the if … end statement, which allows you to calculate zg coordinates at each of the hundred xg coordinates specified by the linspace command; two plot commands with the hold on command between them generate the two profile lines in the same graph that is formatted then with the grid on, xlabel, zlabel, title, and legend commands;
- The final two commands outputs R_a resulting values on the graph; the num2str command is used within each of the text commands to join in one vector the explanatory string and the numerical R_a value.

The calculations shows that the triangular and sinusoidal roughness profiles with the same amplitudes lead to significantly different values of the roughness height parameter R_a.

5.7.6. Shear Stress of the Lubricating Film

Shear stress τ arises in lubricating film between two parallel surfaces, unmoved covered with semicircular pores and moved smooth surface. For one pore cell (pore and adjacent surface), the shear stress on the top surface can be described with the following expression (Burstein, 2016b):

$$\tau = 3\frac{dP}{dX}H + \frac{1}{H}$$

where P is the hydrodynamic pressure, X- coordinate, and H is the gap between surfaces (called frequently as clearance); all variables in this equation are dimensionless. The gap can be calculated with the following equation

$$H = \begin{cases} 1 + \psi\sqrt{1 - X^2}, -1 < X < 1 \\ 1, 1 \leq X \leq \xi \end{cases}$$

where ψ is the pore-radius-to-gap ratio, and ξ – pore-cell-dimensions-to-pore-radius ratio.

Problem: write script program named Ch_5_ApExample_6 that calculates shear stress value at at X in the $-\xi \ldots \xi$ range when ξ=7 and ψ=0.5. The pressure and coordinate values are P=0, -0.0233, -0.0467, -0.07, -0.0933, -0.1167, -0.14, 0, 0.14, 0.1167, 0.0933, 0.07, 0.0467, 0.0233, and 0, and X=-7… 7 with step 1. Use the diff command for calculating the dP/dX derivative. Display τ values for point X=0 and ξ; use for this the interp1 command. Generate and format the τ (X) plot.

The Ch_5_ApExample_6 program that solves the problem are

```
psi=0.5;
xi=7;
dX=1;                                                    % step in X
X=-xi:xi;                                         % default step is 1
P=[0 -.0233 -.0467 -.07 -.0933 -.1167 -.14 0 ...
.14 .1167 .0933 .07 .0467 .0233 0];              %0s before '.' omitted
n=length(X);
H=zeros(1,n);                                  % H preassigned with zeros
for i=1:n
  if abs(X(i))<1
    H(i)=1+psi*sqrt(1-X(i).^2);                          % pore area
  elseif abs(X(i))>=1&abs(X(i)<=xi)
    H(i)=1;                                       % out of pore area
  end
```

```
end
dpdx=diff(P)./dX;                                   % dX=1 and can be omitted
tau=3*dpdx.*H(2:end)+1./H(2:end);                    % H is longer than dpdx
plot(X(2:end),tau), grid on
xlabel('Coordinate, nondim'),ylabel('shear stress, nondim')
title('Shear stress')
Xi=[0 xi];                          % vector with Xs for further interpolation
tau_i=interp1(X(2:end),tau,Xi);                      %finds tau at X=0 and xi
fprintf('Shear stress at X= 0 is %7.4f\n',tau_i(1))
fprintf('Shear stress at X=%2.0f is %7.4f\n',xi,tau_i(2))
```

The script is arranged as follows:

- At the first script lines the ψ, ξ, ξ-step, X and P values are assigned to the appropriate variables;
- The for … end loop organized then to calculated H for each of the X points; to calculate the piece-wise function X the if … end statement is used;
- Derivatives dP/dX are calculated as diff(P)/dX instead of diff(P)./diff(X) because the X values are given with a constant step dX; note, that when the step is 1 the dpdx derivative can be written simply as diff(P);
- After this step the tau values are calculated; for this, the H values are taken starting from the second point, as the H vector is one element longer than the vector dpdx;
- Now the results are plotted, formatted, and displayed; the interp1 command is used to obtain τ at the X= 0 and ξ points written in the tau_i vector; this vector is used in the fprintf command to display the calculated τ values with four digits after the point.

After running in Command Window, the following results are displayed and plotted.

```
>> Ch_5_ApExample_6
Shear stress at X= 0 is 1.2967
Shear stress at X= 7 is 0.9301
```

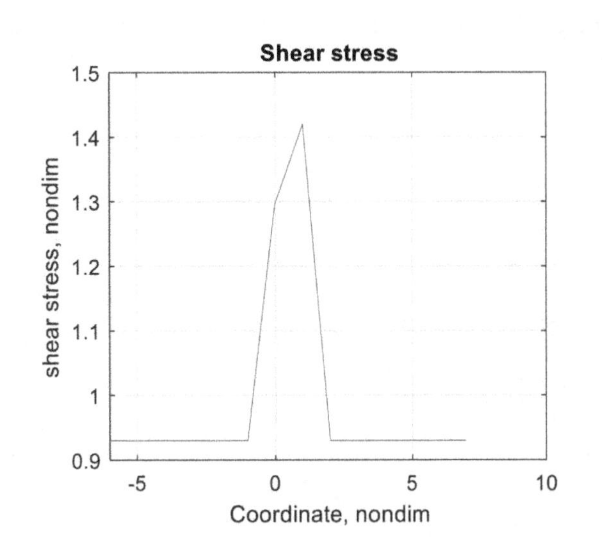

5.8. CONCLUSION

As follows from the considered material of the numerical analysis calculations, the presented command tools allow solving many actual engineering problems. Inter/extrapolation, equation set solution, minimax of function, integration and differentiation are widely used in M&T and can be implemented using the provided commands. This was demonstrated by solving the following problems:

- Friction coefficient data interpolation,
- Two-spring scale lengthening,
- Maximal pressure point of the cylindrical journal bearing,
- Centroid of the cross section of a mechanical part,
- R_a roughness parameter for triangular and sinusoidal profile model,
- Shear stress of the lubricating film.

Developed programs solve real problems and demonstrate that in the absence the rigorous mathematical solution, calculations can be performed with the commands of the numerical analysis.

REFERENCES

Bhushan, B. (2013). Introduction to Tribology (2nd ed.). John Wiley & Sons Inc. doi:10.1002/9781118403259

Burstein, L. (2011). Lubrication and Roughness. In J. P. Davim (Ed.), *Tribology for Engineers. A practical guide* (pp. 65–120). Woodhead Publishing. doi:10.1533/9780857091444.65

Burstein, L. (2016a). Hydrodynamic Behavior of the Sliding Surface with Semicircular Pores: Theoretical and MATLAB Considerations. *International Journal of Surface Engineering and Interdisciplinary Materials Science*, *4*(1), 45–68. doi:10.4018/IJSEIMS.2016010103

Burstein, L. (2016b). Friction Force of the Sliding Surface with Pores Having a Semicircular Cross Section Form. *International Journal of Surface Engineering and Interdisciplinary Materials Science*, *4*(2), 1–15. doi:10.4018/IJSEIMS.2016070101

Hwang, N. H. C., & Hita, C. E. (1987). Fundamentals of Hydraulic Engineering Systems (2nd ed.). Prentice-Hall.

APPENDIX

Table 2. List of examples, problems, and applications discussed in the chapter

No	Example, Problem, or Application	Subsection
1	Data and graph of a preheated cylindrical part cooling demonstrated inter- and extrapolation.	5.1.
2	Considering of the inverse problem for the pipe's coefficient of friction, fully developed turbulent flow.	5.2.
3	The minimum and maximum distance between two molecules according to the modified Buckingham potential.	5.3.
4	The fluid flow rate in a round pipe by numerical calculation of an integral.	5.4.1.
5	Load support of the lubricating film between surfaces when one of them is covered by semicircular pores.	5.4.2.
6	Calculations of the linear thermal expansion coefficient according to the material alongation-temperature data.	5.5.
7	Friction coefficient data interpolation.	5.7.1.
8	Lengthening of a two springs scale.	5.7.2.
9	Cylindrical journal bearing. Maximal pressure point.	5.7.3.
10	Centroid of the cross section of a mechanical part.	5.7.4.
11	Roughness parameter R_a for the triangular and sinusoidal profiles.	5.7.5.
12	Shear stress of the lubricating film.	5.7.6.

Note, some small examples, mostly related to non-M&T issues, are not included in the list.

Chapter 6
The ODE- and BVP-Solvers With M&T Applications

ABSTRACT

The chapter introduces solvers for solving initial and boundary value problems (IVP&BVP) of the ordinary differential equation (ODE). It begins with a description of the ODE solver commands applied to the initial value problem and presents the steps for solving the actual ODE. Further, the chapter presents the BVP-solver commands and steps for their usage. The solutions are presented through real examples. In the final part, the studied ODE and BVP commands are applied, mainly to problems oriented for mechanics and tribology (M&T). At the end of the chapter, applications to the M&T problems are presented; they illustrate how to solve IVP for the spring-mass system and particle falling, as well as BVP for a single clamped beam and hydrodynamic lubrication of a sliding surface covered with semicircular pores.

INTRODUCTION

Differential equations (DEs) occupy a central place in science and technology in general and in mechanical and tribological science and engineering in particular. In the natural sciences, various processes and phenomena are described with DEs, therefore they are studied and taught in colleges and universities and used by students, graduates, and professionals of M&T. However, actual DEs are often unsolvable analytically. In this case, the sole possibility is a numerical solution. Unfortunately, there is no universal numerical method. Thus, MATLAB® provides a special means, called *solvers*, for solving two groups of differential equations, specifically ordinary, ODE, and partial, PDE. The first group of solvers are described in brief below. ODE solvers subdivided by the problems they solve into two groups – initial value problem (IVP) solvers and boundary value problem (BVP) solvers. In the initial value problem, ODE is solved with specified value of the search function at given point from the range of solution (called domain). In the boundary value problem, ODE is solved with more than one, as a rule two, specified values of the function being solved from the domain. Sufficient familiarity with both categories of ODEs is assumed below.

DOI: 10.4018/978-1-7998-7078-4.ch006

6.1. INITIAL VALUE PROBLEM AND ODE SOLVER

The ordinary differential equation set, or sole equation, for which the ODE or BVP solvers are intended, are:

$$\frac{dy_1}{dt} = f_1\left(t, y_1, y_2, \ldots, y_n\right)$$

$$\frac{dy_2}{dt} = f_1\left(t, y_1, y_2, \ldots, y_n\right)$$

...

$$\frac{dy_n}{dt} = f_n\left(t, y_1, y_2, \ldots, y_n\right)$$

where $y_1,\ldots,y_n$ are the dependent variables while t is the independent that varies between the starting t_s and final t_f values; n is the number of first order ODEs that should be equal to the number of dependent variables; note, instead of t, the variable x or some other can be also used. To solve the high-order ODEs with the MATLAB® solver, this equation must be reduced to the first order by creating a set of first-order equations. It can be doing, for example, for the second order ODE in the following way: the first derivative dy_1/dt is denoted as y_2 after that the y_2 is substituted in the initial equation. Such derivative replacements can be used in case of the high-order ODEs up to lowering the DE to a set of the first order ODEs.

To illustrate the lowering of the order of the ODE, consider the second, third, and fourth order differential equations:

- The second order equation

$$\frac{d^2y}{dt^2} + \frac{y}{a}\left(\frac{dy}{dt}\right)^{1.3} = b\text{cost}$$

Denote, firstly, y as y_1 and, secondly, $\frac{dy_1}{dt}$ as y_2; substitute now y_2 into the initial equation, thus we obtain two first-order differential equations:

$$\frac{dy_1}{dt} = y_2, \frac{dy_2}{dt} = -\frac{y_1}{a}y_2^{1.3} + b\text{cos}t$$

To solve these equations, the $y_1(t)$ and $y_2(t)$ functions should be defined. The $y_1(t)$ is the solution of the initial second order equation while $y_2(t)$ is the derivative of the solution.

- The third order equation

$$\frac{d^3y}{dt^3} + a\frac{d^2y}{dt^2} - b\frac{t}{y} = d\tan t$$

Denote y as y_1, $\frac{dy_1}{dt}$ as y_2, $\frac{d^2y_2}{dt^2}$ as y_3 and substitute y_1, y_2, and y_3 in the equation above, this yield

$$\frac{dy_1}{dt} = y_2, \frac{dy_2}{dt} = y_3, \frac{dy_3}{dt} = -ay_3 + \frac{t}{y_1} + \mathrm{d}\tan t$$

In this case, the $y_1(t)$, $y_2(t)$ and $y_3(t)$ values should be defined to solve the initial equation; wherein, $y_1(t)$ is the solution of the original third order equation while $y_2(t)$ and $y_3(t)$ are, respectively, the first and second derivatives of the solution.

- The fourth order equation

$$\frac{d^4y}{dt^4} + b\frac{d^2y}{dt^2} + c\left[\left(\frac{dy}{dt}\right)^2 - \frac{y}{t}\right] = \mathrm{gy}$$

Analogously to the previous examples can be transformed to the set of the fourth first order equations

$$\frac{dy_1}{dt} = y_2, \frac{dy_2}{dt} = y_3, \frac{dy_3}{dt} = y_4, \frac{dy_4}{dt} = -by_3 - c\left(y_2^2 - \frac{y_1}{t}\right) + gy_1$$

For this case, the $y_1(t)$, $y_2(t)$, $y_3(t)$ and $y_4(t)$ values should be defined. The $y_1(t)$ is the solution of the initial fourth order equation while $y_2(t)$, $y_3(t)$ and $y_4(t)$ are, respectively, the first, second and third derivatives of the solution.

There is no universal numerical method for solving the IVP of any ODE. Therefore, the MATLAB® ODE solver includes a number of commands realizing different numerical methods for solving actual ODEs.

6.1.1. About Numerical Methods for Solving DEs

The main technique used to solve ODEs is to represent the derivative $\frac{dy}{dt}$ as the ratio of the finite differences, i.e.:

$$\frac{dy}{dt} \simeq \lim_{\Delta t \to 0} \frac{y}{t} = \frac{y_{i+1} - y_i}{t_{i+1} - t_i}$$

where Δt and Δy are the differences of the argument and function respectively, i is the point number in the $[t_0,t_n]$ -range of the argument t. The representation of the derivative as the ratio of the differences is apparently true for very small, but finite (nonzero) distances between points i+1 and i. The procedure for numerical solving the first-order differential equation, in accordance to the IVP, is to specify the initial, first-point value y_0 and calculate the value of the derivative at t_0 by the original ODE expression

$$\left.\frac{dy}{dt}\right|_0 = f(t_0, y_0)$$

assuming the constant argument difference Δt, the next-point function value y_1 can be determined as

$$y_1 = y_0 + \left.\frac{dy}{dt}\right]_0 t$$

Now, knowing y_1, we can calculate derivative at $t_1 = t_0 + \Delta t$

$$\left.\frac{dy}{dt}\right|_1 = f(t_1, y_1)$$

For the next point (argument $t_2 = t_1 + \Delta t$) we determine y_2

$$y_2 = y_1 + \left.\frac{dy}{dt}\right]_1 t$$

Such calculations are repeated until all function values y_i in the range $t_0 \ldots t_n$ are available (here n is the number of the final t-point).

This approach, first implemented by Euler, is used with various improvements and complications in advanced numerical methods such as the Runge-Kutta, variable order differences, Rosenbrock and Adams, etc. The following are MATLAB® commands that implement the finite difference approach and belong to the ODE solver.

6.1.2. The ode45 and ode15s Commands

The commands for the ODE solution have the following generalized form:

```
[t,y]=odeN(@fun_ode,tspan,y0)
```

- where N of the odeN is the identification number (sometimes with a letter) of the ODE solver command that denotes the numerical method realized for ODE solution, e.g. the ode45 command uses the Runge-Kutta method, and ode15s uses the variable order differences method.
- @fun_ode is the name of the user-defined function/file where the solving differential equations are written. The definition line of the user-defined function should have the form

```
function dy= fun_ode (t,y)
```

In the lines following the definition line, the differential equation/s, presented as first-order PDEs (see Subsection 6.1), should be written in the form

```
dy=[right side of the first ODE
right side of the second ODE
...];
```

- tspan - a row vector specifying the integration interval from the starting t_s and up to final t_f points, e.g. the [1 14] vector specifies a t-interval with the starting and final t-values equal to 1 and 14, respectively. The tspan vector can contain more than two values intended to display the solution at these values, e.g. [1:3:10 12 14] means that results of solution lays in the t-range of 1…14 and will be displayed at t-values of 1, 4, 7, 10, 12 and 14. The values given in tspan affect the output steps between the values but not the steps used in the solution. The solver command automatically chooses and changes the step to ensure the solution tolerance. The default tolerance is 0.000001 absolute units.
- y0 is a vector of initial conditions, e.g. for the set of two first-order differential equations with initial function values y_1=0 and y_2=4, this vector is written as y0=[0 4]; y0 can be given also as a column vector, e.g. y0=[0;4].
- y and t are the column vectors containing the defined y-values at the corresponding t-values; in case of the second or higher order ODEs y is the matrix with n-columns presenting the obtained y values to each of the n solved first-order ODEs.

The odeN comands can be used without the output arguments. In this case, the solver shows a solution graphically - in the automatically generated plot.

The ODE solver commands can solve so-termed non-stiff and stiff problems. The DE is called stiff when it contains the terms that lead to any singularities, e.g., jumps of the function, holes, gaps, ruptures or others. This leads to a divergence of numerical solutions, even at very small sizes of solution steps. Unlike stiff equations, non-stiff DEs are characterized by stable convergence of the solution.

The ode45 command is intended to solve non-stiff ODEs and ode15s to the stiff ODEs.

Unfortunately, it is impossible to know *a priori* if the equation is stiff or not and therefore impossible to determine in advance which ODE command should be used. When we solve the ODE that describes a real technology or phenomenon, we can select the certain ODE solver. For example, volatile processes, technologies using fast chemical reactions or explosions are apparently simulated with the stiff equations. In these cases the ode15s command should be used. Sometimes, a ratio of the maximal to minimal values of the ODE coefficients is used as a criterion of the stiffness. If this ratio is larger than 1000, the problem is categorized as stiff. However, this criterion is an empirical and not always correct. Therefore, if the type of the ODE is not known *a priori*, then the ode45 command is recommended to try first and after that - the ode15s.

6.1.2.1. ODE Solution Steps

1. 1. As stated above, the original differential equation should be lowered to the first order ODE (if it is higher than first order) and presented as set of the equations having the form

$$\frac{dy}{dt} = f(t,y), t_s \le t \le t_f$$

(t_s and t_f are the starting and final argument points) with initial condition

y=$y0$ at t_s=$t0$

given for the each of the first-order equations.

We demonstrate this and the next steps by an ODE for the so-termed second order system that describes many dynamic processes and phenomena in mechanics

$$a\frac{d^2y}{dt^2} + b\frac{dy}{dt} + cy = f(t)$$

Depending of inputted y and f(t) function, this equation describes many different phenomena, including piston-spring damper system, behavior of materials within some viscoelasticity simulations, RC circuits of the various test apparatuses, and other technological and physical processes. The y in this equation is the dependent variable denoting the seeking function at time t (independent variable) and a, b and c are the constants. Consider this equation for a=0.3, b=0.09, c= 3, with $f(t)$=sin(πt/90). The time values at which we want to display solution are 0, 3, 4, … 24 with initial values being $y(0)$=15 and $y'(0)$=0. All variables and coefficients are given as non-units because the equation and its solution can be used for different actual objects. For a numerical solution with the one of ODE solver commands, the equation should be rewritten as:

$$\frac{d^2y}{dt^2} = \sin(\pi t / 100) - \frac{b}{a}\frac{dy}{dt} - \frac{c}{a}y$$

Now denoting y as y_1 and $\frac{dy_1}{dt}$ as y_2, we obtain the following two ODEs of the first order:

$$\begin{cases} \dfrac{dy_1}{dt} = y_2 \\ \dfrac{d^2y_2}{dt^2} = \sin\left(\dfrac{\pi t}{100}\right) - \dfrac{b}{a}y_2 - \dfrac{c}{a}y_1 \end{cases}$$

2. In the second step, a file containing the user-defined function with solving equation should be created. The definition line of this function must include input arguments t and y and output argument dy that denotes the left side of the first order equations above. In the next lines, the right-hand part/s of the ODE/s should be written as vector in a separate line each or in the same line but divided with the semicolon (;). In our example, the function containing the differential equations is:

```
function dy=myODE(t,y)
a=0.3; b=0.09; c= 3;
dy=[y(2);sin(t*pi/100)-b/a*y(2)-c/a*y(1)];
```

This user-defined function should be saved in an m-file with the myODE name.

Note: If the argument t is absent in the right part of the differential equation, the tilde (~) character can be written instead. In this case the function definition line appears as follows: function dy=myODE(~,y); this can not be used for the discussed ODE set since the second equation has t in the right-hand part.

3. In the final step, the ODE solver command should be chosen and implemented. For our example, the ode45 solver should be selected because we have not any specific recommendations about the stiffness of the equation to be solved. In the Command Window, the following commands should be typed and entered:

```
>>[t,y]=ode45(@myODE,[0:3:24],[15 0])     %y0_1=15, y0_2=0
t =
   0
   3
   6
   9
  12
  15
  18
  21
  24
y =
  15.0000      0
  -9.5369   1.5966
```

```
6.1583   -2.0016
-3.7609   1.9299
2.5799   -1.6141
-1.3761   1.2989
1.1662   -0.9713
-0.3861   0.7335
0.6283   -0.5164
```

Calculations start with initial values of y_1=15 and y_2=0 at t=0; the starting and final times of the process are 0 and 24, respectively, with the time step being 3 (for reducing the output to nine points only). Therefore, the time span vector was inputted as [0:3:24].

The given commands solve the equation and display the one-column vector of the t values and two-column matrix with the resulting y values. To plot this solution, the plot command should be used but the achieved nine t,y-points are not sufficient to generate a smoothed plot. To produce the plot, the t_range vector should be inputted with the starting and final values only. In this case, the inner t-values will be automatically chosen. The number of points will be sufficient to generate a smoothed $y(t)$ plot. The commands that realize this purpose are:

```
>> [t,y]=ode45(@myODE,[0 24],[15 0]);                         % default t-step
>> plot(t,y(:,1),t,y(:,2),'--')                            % y1-solution, y2- y'
>> xlabel('Time'), ylabel('Y values')
>> title('Solution of the differential equation for the second order system')
>> legend('y','y''')                          %''-to display the ' derivative sign
>> grid on
>> axis tight
```

The generated plot is shown in Figure 1.

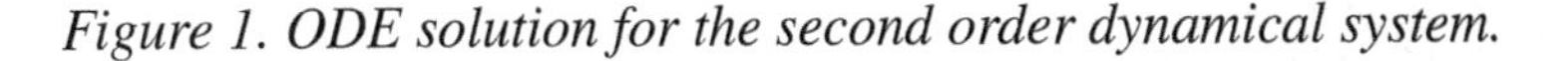

Figure 1. ODE solution for the second order dynamical system.

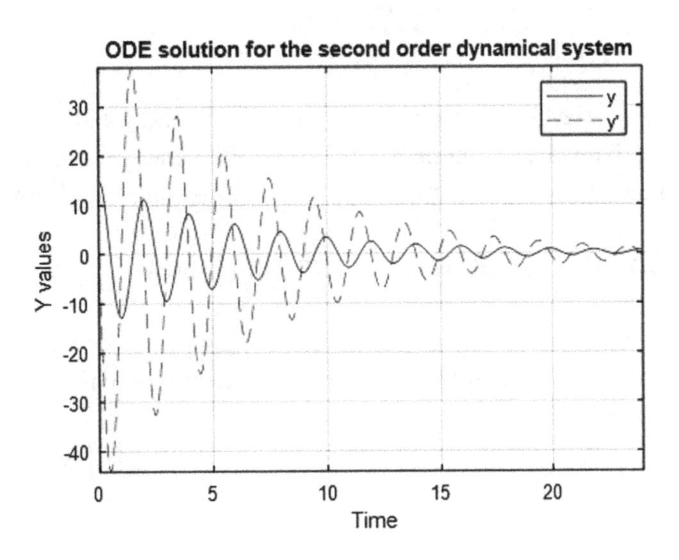

In the represented solution, the **ode45**, the plot, and plot formatting commands were written interactively in the Command Window while the ODE was written and saved as file. It is more reasonable to create a single program file that includes all these commands together. For this purpose, a user-defined function that includes a sub-function with the solving ODE/s should be written and save in a file with the same name. For the discussed example, this file - named **SOD_DE** (second order dynamic differential equation) - reads as follows:

```
function t_y=SOD_DE(t_span,y0)
                                    %              Solution of the second order ODE
                                 %              to run: >> t_y=SOD_DE([0 24],[15;0])
[t,y]=ode45(@myODE,t_span,y0);
plot(t,y(:,1),t,y(:,2),'--'),grid on
xlabel('Time'),ylabel('Y values')
title('Solution of the differential equation for the second order system')
legend('y(t)','y(t)''')
axis tight
t_y=[t y];                                %to output results as three-column table
function dy=myODE(t,y)                            % sub-function with solving ODEs
a=0.3; b=0.09; c= 3;
dy=[y(2);sin(t*pi/100)-b/a*y(2)-c/a*y(1)];
```

To run this program, the following command should be entered in the Command Window (see the second help line in the **SOD_DE** function):

```
>> t_y=SOD_DE([0 24],[15;0])
t_y =
        0  15.0000        0
   0.0000  15.0000  -0.0001
   0.0000  15.0000  -0.0001
   0.0000  15.0000  -0.0002
   ...
```

A three-column table displays the *t* values in its first column and *y* and y' values in the second and third two columns, respectively. Each column has 421 rows (shortened here to save the space). Additionally, the program generates the graph (Figure 1) with *y* and *y'* as function of time.

6.1.2.2. Extended Command Forms That Can Be Used With the ODE Solvers

When DEs include some parameters (e.g. in the discussed example the *a*, *b* and *c* variables are parameters), the ODE solver commands should be written in a more complex form to pass these parameters to a function containing the ODEs:

```
[t,y]=odeN(@fun_ode,tspan,y0,[ ], param_name1,param_name2,…)
```

- the empty brackets [ ] denotes an empty vector intended, in general case, for the so-termed options used to control the integration process[1]. In most cases, the default option values are used, which yield satisfactory solutions. Thus, we do not specyfy these options.
- param_name1,param_name2,... are the names of the arguments that we intend to use for transmitting their values into the fun_ode function. If the parameters are named in the odeN solver, they should also be written into the definition line of the fun_ode function.

For example, the SOD_DE file should be modified for introducing the a, b and c coefficients as arbitrary parameters:

```
function t_y=SOD_DE(a,b,c,t_span,y0)
                                      %               Solution of the second order ODE
                               %          to run: >> t_y=SOD_DE(.3,.09,3,[0 24],[15;0])
[t,y]=ode45(@myODE,t_span,y0,[],a,b,c);
plot(t,y(:,1),t,y(:,2),'--'),grid on
xlabel('Time'),ylabel('Y values')
title('Solution of the differential equation for the second order system')
legend('y(t)','y(t)''')
axis tight
t_y=[t y];                                  %to output results as three-column table
function dy=myODE(t,y,a,b,c)
                          %   sub-function with ODEs receiving a, b, and c parameters
dy=[y(2);sin(t*pi/100)-b/a*y(2)-c/a*y(1)];
```

In the Command Window, the following command should be typed and entered:

```
>> t_y=SOD_DE(.3,.09,.3,[0 24],[15;0])
```

The results are identical to those discussed in subsection 6.1.2.1.

The advantage of the ODE solver command form with the transfer parameters is its greater versality, e.g. the SOD_DE function with parameters can be used for any a, b and c without introducing their values into the myODE sub-function.

Another, possible more suitable, simpler form of the commands of ODE solver is

```
[t,y]=odeN(@(t,y)myODE,tspan,y0)
```

When this form is used, it is not necessary to create the sub-function with ODE/s. The live function program named AOD_DE_live are illustrate this form (Figure 2)

Figure 2. Live function with the @(t,y) *command form solving the second order dynamical differential equation with the* ode45 *command.*

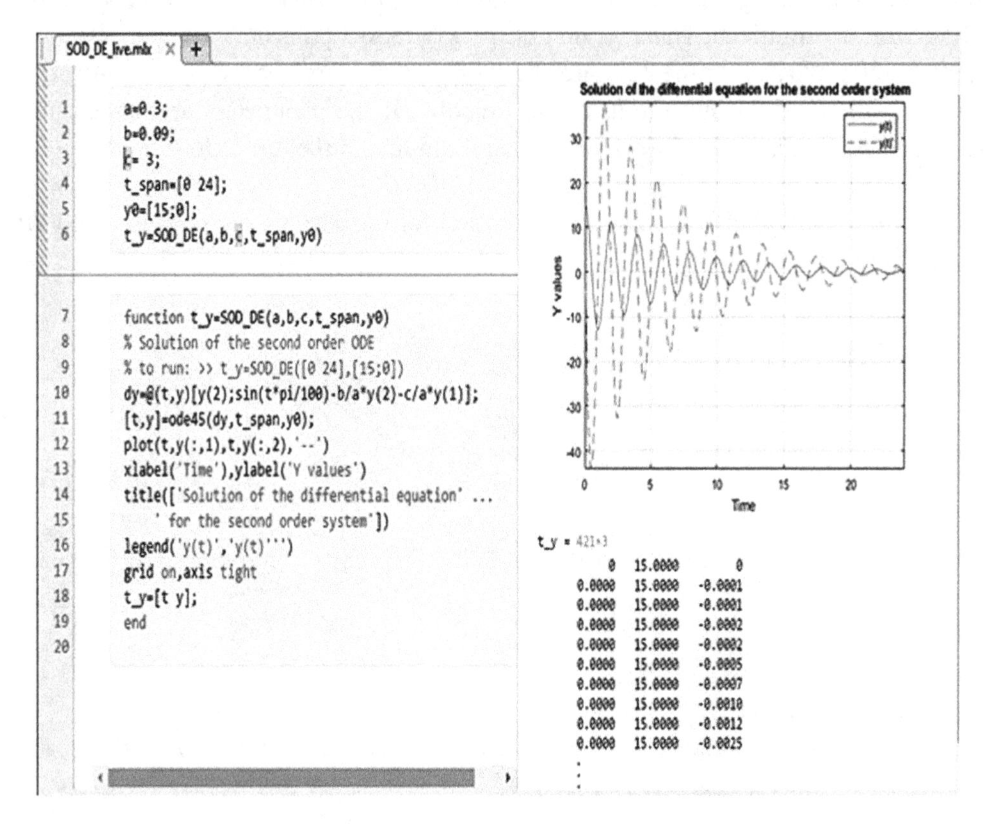

This @(t,y) form allows assign the parameters to the equation/s without writing them within the ode45 (or ode15s) or in the sub-function containing the solving equation with parameters, as it was made in the user-defined function myODE of the example above. The discussed form is preferable when we solve a single or two short differential equations.

6.1.3. Available Additional Commands of the ODE Solver

In addition to the two ODE-solver commands studied above, the solver has additional commands that can be used to solve the ODEs; some of these are listed in Table 6.1. The table represents the name of the available ODE solver commands, the numerical methods they use with the type of equations to be solved and examples that include the graphical solution each. The input and output variables of the represented commands is identical to those described previously and therefore not specified in the table.

Table 1. Initial value problem: the supplementary commands for solving ODEs[2]

ODE-command	Method and when to use	Example	
		Commands	**Graph**
ode23	Explicit Runge-Kutta method. The command is used for non-stiff and moderately stiff DEs. Quicker, but less precise than ode45	ODE: $y'=2t^2\sin t$, with y0=0, ts=0, tf=4. `>>[t,y]=ode23'(@(t,y) 2*t.^2.*sin(t)*,[0 4],0);` `>> plot(t,y,'o-')`	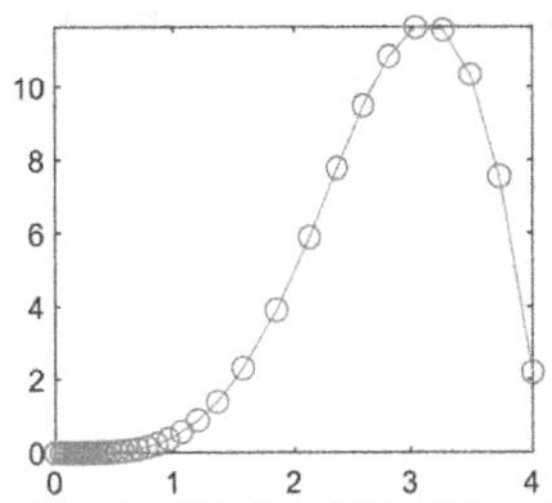
ode113	Adams' method. Non-stiff differential equations. The command is used for problems with stringent error tolerances or for solving computationally intensive problems.	ODE: the same example as for the ode23. `>>[t,y]=ode113(@(t,y) 2*t.^2.*sin(t)*,[0 2],0);` `>>plot(t,y,'o-')`	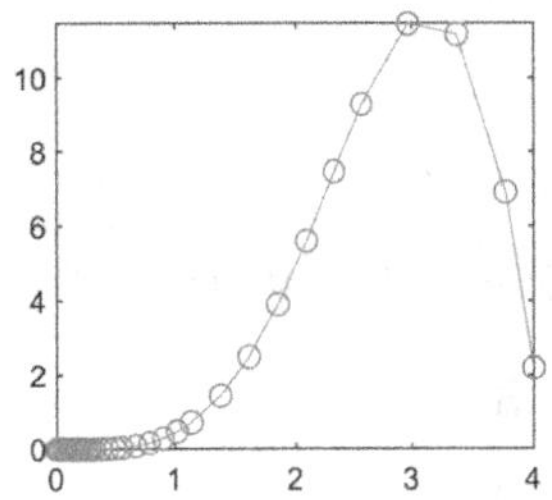
ode23s	Rosenbrock's method. The command is used for stiff differential equations. When ode15s is slow.	ODE: $y'=10t^2$ with y0=1, ts=0, tf=2. `>>[t,y]=ode23s(@(t,y) 10*t.^2,[0 2],1);` `>>plot(t,y,'o-')`	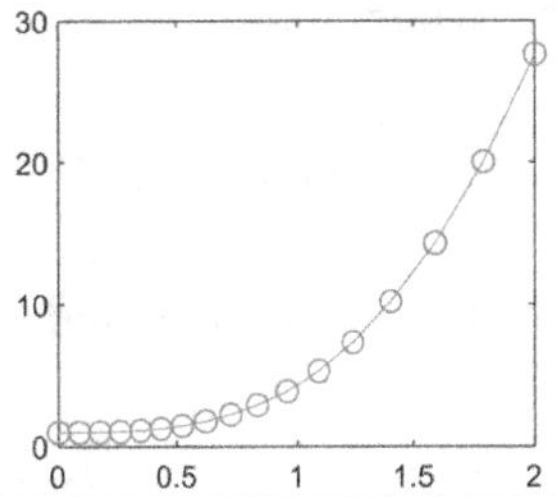
ode23t	Trapezoidal rule. The command is used for moderately stiff differential- and differential algebraic equations (DAEs).	ODE: the same example as for the ode23s. `>>[t,y]=ode23t(@(t,y)10*t.^2,[0 2],1);` `>>plot(t,y,'o-')`	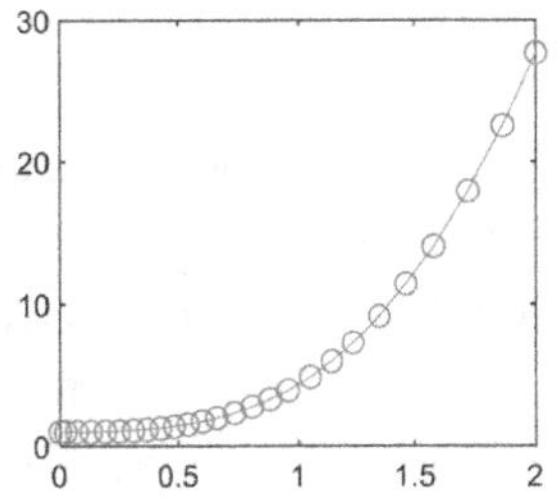
ode23tb	Trapezoidal rule/second order Backward differentiation formula,TR/BDF 2. The command is used for stiff differential equations; sometimes more effective than ode15s.	ODE: the same example as for the ode23s. `>>[t,y]=ode23tb(@(t,y)10*t.^2,[0 2],1);` `>>plot(t,y,'o-')`	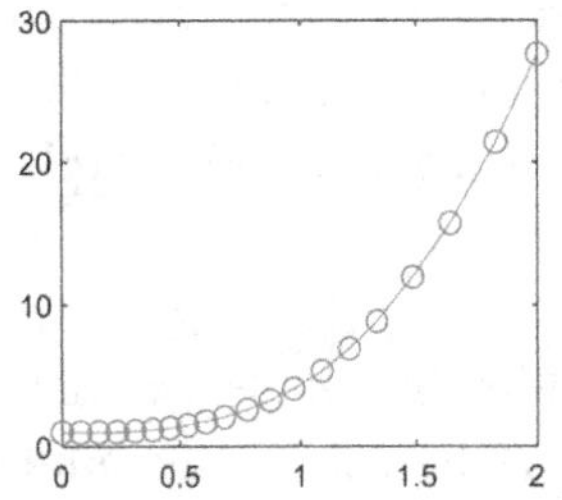

6.2. BOUNDARY VALUE PROBLEM AND BVP4C COMMAND FOR ODEs SOLUTION

MATLAB represents means to solve the two-point BVP when, unlike the IVP, at least two end points of the range should be specified to solve single or set of ODEs. His sort problem arises in the strength or mechanics of materials, in heat and mass transfer mechanics, physics, and various other applications. Mathematically the problem formulate so that ODE can be solved as an explicit first order system of ODEs

$$y_1'(x) = f\left(x, y_1(x), y_2(x), \ldots\right)$$

$$y_2'(x) = f\left(x, y_1(x), y_2(x), \ldots\right)$$

…

With boundary conditions *bc* specified at two points *a* and *b* of the *x* range [*a*,*b*]

bc(*y*(*a*),*y*(*b*))=0

In the DEs above *x* and *y* are the independent and dependent variable, respectively.
Below is the bvp4c command designed to solve a boundary value problem using the collocation method.

6.2.1. The bvp4c Command

The simplest form of this command is

```
sol = bvp4c(@fun_ode,@fun_bc,y_init)
```

- @fun_ode is the name of the user-defined function into which the single or set of ODEs should be written;
- fun_bc and y_init are, respectively, the function with *y* values at boundaries and structure with initial (guess) *y* values specified along the [*a*,*b*] interval;
- sol is a structure containing the *x* and *y* values defined in the solution.
- Structure is a more advanced concept than those used previously and further, we will give only the minimum information necessary to use the structure with the bvp4c command.
- The following description explains how to represent the input and output arguments in the bvp4c command.

The definition line of the user-defined fun_ode function should be look the same as it was explained for the fun_ode function of the odeN commands (sub-section 6.1.2):

```
function dy=fun_ode(x,y)
```

The extended command form @ (x,y) is also possible for the fun_ode function (sub-section 6.1.2.2). The definition line for user-defined fun_bc function should be as follows

```
function res= fun_bc(ya,yb)
```

- ya and yb are the column vectors that contains *y(a)* and *y(b)* values for each of the ODEs to be solved;
- res is an outputted column vector; this vector should be written so:

```
res=[ya(1)-y_at_point_a;yb(1)-y_at point_b]
```

- If the boundary values are y=0.1 at x=0 (point a) and y=4 at x=5 (point b), they should be rewritten as y-0.1=0 and y-5=0 and the res vector can be look as follows res=[ya(1)-0.1;yb(1)-5];
- If in addition to the y values at the boundaries, we have the derivatives y'=0 at point a thus the res vector is: res=[ya(1)-0.1;yb(1)-5; ya(2)].

The y_init variable can be specified using the bvpinit command specially developed for this purpose as follows:

```
y_init=bvpinit(x,y)
```

To evaluate the defined y values containing in the sol structure at the desired points x, the deval function are used y=deval(sol,x) where x must be specified in the solution range, and y is a matrix with rows each of which contains y obtained by x.

Demonstrate now using the bvp4c function by the solution of the equation of motion of a spring-mass harmonic oscillator:

$$\frac{d^2y}{dt} = -\frac{k}{M}y$$

where y is the displacement of a spring-mass system due to the stretch force; M – mass value; k- spring constant representing stiffness of a spring; t –time.

Problem: Solve this equation in the range t=0… 1.9154 sec with two-point boundaries y=0.1 at t=0 and y=0 at t=1.9154. Parameters in the ODE are k=300 N/m and M=10 kg.

Solution steps:

1. represent the original second order ODE as a set of two first order ODEs, for which denotes y as y_1 and y' as y_2, thus we have:

$$\begin{cases} \dfrac{dy_1}{dt} = y_2 \\ \dfrac{dy_2}{dt} = -\dfrac{k}{M} y_1 \end{cases}$$

2. Write now the fun_ode function of the bvp4c command

```
function dy=myODE(x,y)
k=300;M=10;
dy=[y(2);-k/M*y(1)];
```

3. In this step we should write the boundary conditions with the following user-defined function

```
function res=myBC(ya,yb)
res=[ya(1)-0.1;yb(1)];
```

4. Now the initial values of y at, for example, fife points along the x-range should be specified in the y_init structure using the bvpinit command:

```
y_init= bvpinit(linspace(0, 1.9154,5),[0.1 0])
```

5. In this step we obtain the defined y values from the sol structure at 100 x-values and represent results in graph using the following commands

```
x = linspace(0, 1.9154);
y = deval(sol,x);
plot(x,y(1,:),'^')
```

The ExBVP1 live script containing the above and graph formatting commands along with the generated graph is shown in Figure 3.

Figure 3. Live script with the bvp4c *function that solves two-point boundary value problem for the ODE describes the spring-mass oscillator.*

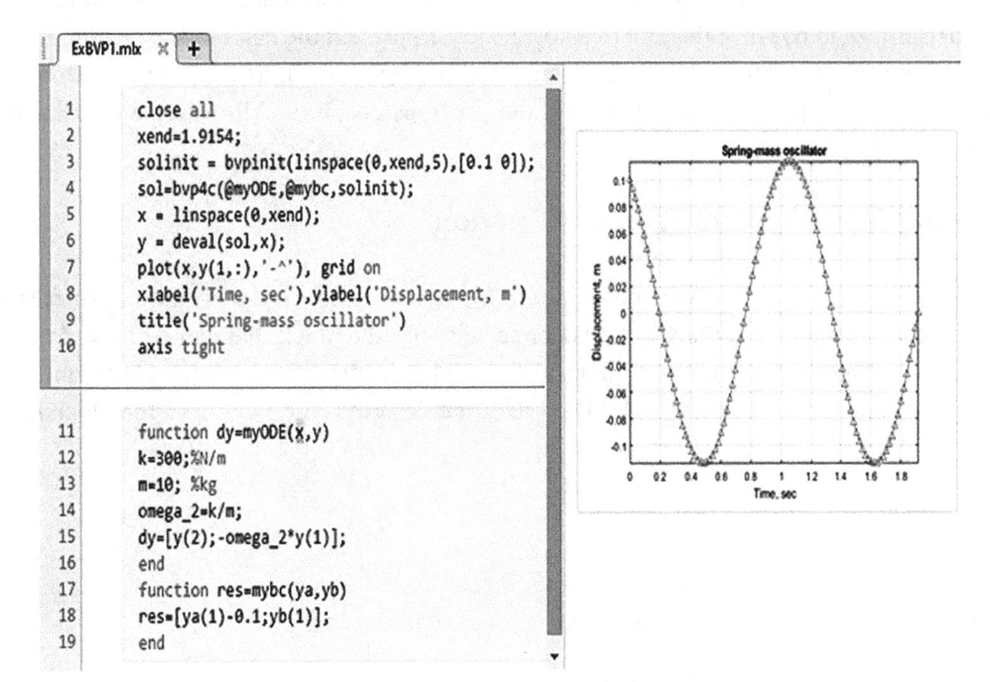

Note, the bvp4c command can be used to solve singular and multipoint boundary value problem, the more detailed information can be obtained by entering the doc bvp4c command in the Command Window.

6.2.2. About the bvp5c Command

This command has exactly the same simplest form as the bvp4c command

```
sol = bvp5c(@fun_ode,@fun_bc,y_init)
```

All comments given regarding the components of the bvp4c command are valid for bvp5c. For example, if in the ExBVP1 live script to change in the command name bvp4c only the digit 4 to the 5, the script works and shows the same results.

The bvp5c command solves the same class BVPs using the polynomial collocation method and provides the fifth-order accuracy that can be more suitable when the high accuracy and smooth solution are required.

6.3. APPLICATION EXAMPLES FOR INITIAL VALUE PROBLEMS

The commands in the examples below are written as live scripts or as user-defined functions with parameters, in the latter case they have a short help part, including a description line and a line with a command for running program in the Command Window. The codes of each program contain some differences, even if the program uses the identical IVP or BVP solution schemes; this is done to study the different possibilities to use the ode- and bvp solvers and to present the calculation results. Explanatory comments within the programs /functions are minimal or absent at all, as it is assumed that the reader has gained sufficient experience after studying the previous chapters. All necessary explanations are given directly in the text.

6.3.1. Spring-Mass System With Dry Friction

Consider the spring-mass system when block with M mass located on the horizontal surface and one of the spring end is fixed while the second is connected with the block that can be moved by the force applied to the block-spring system and then released. Dry friction between the block and surface is characterized by the friction coefficient μ. The ODE that describes the motion of the block is

$$\frac{d^2x}{dt^2} = -\mu g{\cdot}sign\left(\frac{dx}{dt}\right) - \frac{k}{M}x$$

where x is the block displacement, t – time, and g – acceleration of gravity.

Problem: Write a user-defined function with name Ch_6_ApExample_1 that solves this two-order ODE and obtains displacement x at times from the range 0 …2 sec when k=3000 N/m, M=6 kg, μ=0.4 (dimensionless), g=9.80665 m/s^2, x=0.1 m and $\frac{dy}{dt} = 0$ m/sec at t=0 sec. The function should include the input parameters only; take the k, M, μ, time range tspan, and initial values x_0 as input function arguments. Use the extended form of the ode-solver for transferring ODE-coefficients into the subfunction containing the ODEs. The function should generate the resulting $x(t)$ graph in the Live Editor window.

To solve the second order equation of motion, transform it to the two first order equations.

$$\begin{cases} \dfrac{dx_1}{dt} = x_2 \\ \dfrac{dx_2}{dt} = -\mu g{\cdot}sign\left(x_2\right) - \dfrac{k}{M}x_1 \end{cases}$$

To provide the solution, the appropriate ODE command should be chosen. Unless otherwise indicated, it should be the ode45 command.

The user-defined function Ch_6_ApExample_1 for the solution are

```
function Ch_6_ApExample_1(mu,k,M,t_span,x0)
%              Solution of the second order ODE
%  to run: >>Ch_6_ApExample_1(0.4,3000,6,[0 2],[0.1;0])
[t,x]=ode45(@myODE,t_span,x0,[],mu,k,M);
```

```
plot(t,x(:,1))
xlabel('Time, sec'),ylabel('Displacement, m')
title({'Solution of the ODE for the spring-mass '; ...
'system with dry friction'})
grid on
axis tight
function dx=myODE(~,x,mu,k,M)
g=9.80665;                         % m/s^2
dx=[x(2);-mu*g*sign(x(2))-k/M*x(1)];
```

The Ch_6_ApExample_1 function is designed as follows:

- The first function line defines the function name and the input variables – mu, k, M, t_span, and x0;
- Two following function lines present the help part of the function and provides a short purpose of the function and command to the function run;
- In the next line the ode45 command is written with the required input and output arguments; here the extended command form with parameters written as its input arguments (subsection 6.1.2.2) is used;
- Further function lines content commands for generating and formatting the resulting plot;
- Last two lines represent the myODE sub-function where the two solving ODEs are written.

To run the function should be saved into the same named file, after this the following command must be typed in the Command Window:

```
>> t_y=Ch_6_ApExample_6_1(0.4,3000,6,[0 2],[0.1;0])
```

After entering, the resulting graph appears:

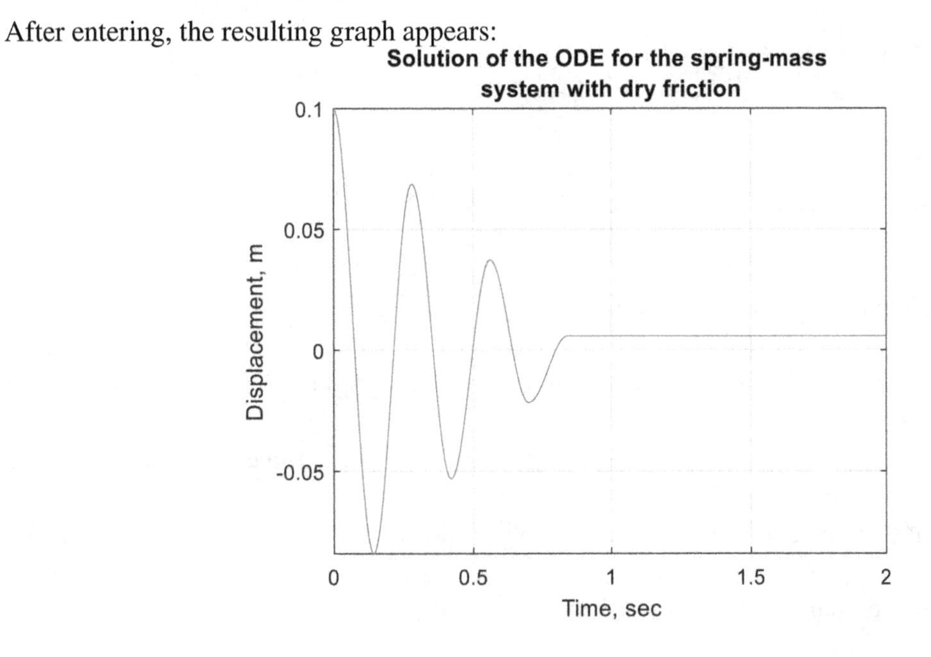

6.3.2. Particle Falling With Friction

The velocity $v(t)$ of the particle of mass m falling vertically satisfies the equation

$$m\frac{dv}{dt} = mg - p\left(v\right)$$

where t is the time and g - g – acceleration of gravity.

The particle is opposed by a frictional force; which affects on the particle velocity $p(v)=0.1v^2$.

Problem: Write a live script with name Ch_6_ApExample_2 that calculates v at t=0, 1, 2, 2.5, 3, 3.5, 4, 5, 8, and 10 s when m=5 kg, g=9.80665 m^2/s, and initial velocity v_0=0 m/s^2. The script should display the calculated t and v values and generate graph $v(t)$ with the axis labels, grid, and title.

For this equation solution, the ode45 command is suitable as there is no contrary reason. It is advisable to use the form with the ODE written strictly into the ode45 command.

The live script solving the problem, displaying the resulting t and v values and generated plot are presented in the figure 4.

Figure 4. Live script for solving IVP for ODE describing a particle falling with the opposite friction

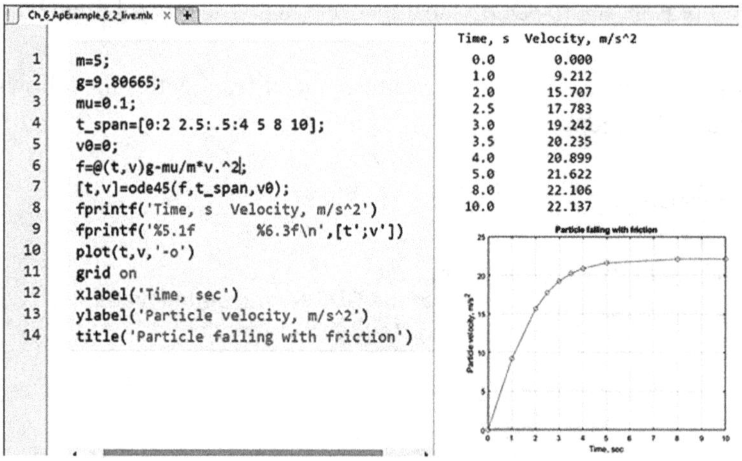

```
1   m=5;
2   g=9.80665;
3   mu=0.1;
4   t_span=[0:2 2.5:.5:4 5 8 10];
5   v0=0;
6   f=@(t,v)g-mu/m*v.^2;
7   [t,v]=ode45(f,t_span,v0);
8   fprintf('Time, s  Velocity, m/s^2')
9   fprintf('%5.1f          %6.3f\n',[t';v'])
10  plot(t,v,'-o')
11  grid on
12  xlabel('Time, sec')
13  ylabel('Particle velocity, m/s^2')
14  title('Particle falling with friction')
```

```
Time, s  Velocity, m/s^2
  0.0        0.000
  1.0        9.212
  2.0       15.707
  2.5       17.783
  3.0       19.242
  3.5       20.235
  4.0       20.899
  5.0       21.622
  8.0       22.106
 10.0       22.137
```

The Ch_6_ApExample_2 live script is designed as follows:

- The corresponding data values are assigned to the m, g, k, t_span, and v0 variables in the first four script lines;
- The ode45 command is implemented next with the first input argument containing the ODEs (here the @(t,x) form is used);
- Next two fprintf commands create the table caption and two column output with the obtained t and v values;
- Further commands generate the $v(t)$ graph and format it.

6.4. APPLICATION EXAMPLES FOR BOUNDARY VALUE PROBLEMS

6.4.1. Uniformly Loaded Single Clamped Euler Beam

The dimensionless uniformly loaded Euler beam equation has form (based on Magrab et al., 2015)

$$\frac{d^4 y}{\partial x^4} = q(x)$$

where y is the transverse displacement, $0 \le x \le 1$ – coordinate, q is the load that is a constant when the load is not change along the x (uniformly loaded beam) or change with x. All variables in this equation are considered dimensionless.

The displacement y and slope $\frac{dy}{dx}$ are equal to zero for the single clamped cantilever beam.

Problem: Write a live script with name Ch_6_ApExample_3 that uses the bvp4c command to calculate the transverse displacement of the uniformly loaded single clamped beam for $q(x)=1$, $y(0)=\frac{dy(0)}{dx}=0$ (clamped beam end) and at point $x=1$ (free beam end) the external negative bending moment $M_r = \frac{d^2 y}{\partial x^2}$ =- 0.6 is applied to the free beam end. Use the bvp4c command with the ODEs written in form @(x,y)… (see subsection 6.1.3); assume that the initial values for displacement y, slope $\frac{dy}{dx}$, moment $\frac{d^2 y}{dx^2}$, and shear force $\frac{d^3 y}{\partial x^3}$ are 0.1 each. The script should display the y-values at points x=0, 0.1, 0.2, …, 1 and generate the $y(x)$ graph with equally scaled axes and the displacement line four points width.

To solve the forth order beam equation, transform it to the four first order equations.

$$\begin{cases} \frac{dy_1}{dx} = y_2 \\ \frac{dy_2}{dx} = y_3 \\ \frac{dy_3}{dx} = y_4 \\ \frac{dy_4}{dx} = q \end{cases}$$

In accordance with the problem, the required boundary conditions are: y1(0)=y2(0)= y3(0)=0, and y3(1)=M_r.

The live script Ch_6_ApExample_3 that solves the problem, displays the resulting x and y values, and the generates the $y(x)$ graph is presented in the Figure 5.

Figure 5. Dimensionless displacement of the uniformly loaded beam clamped at the left end and with external moment applied to the right end of a beam.

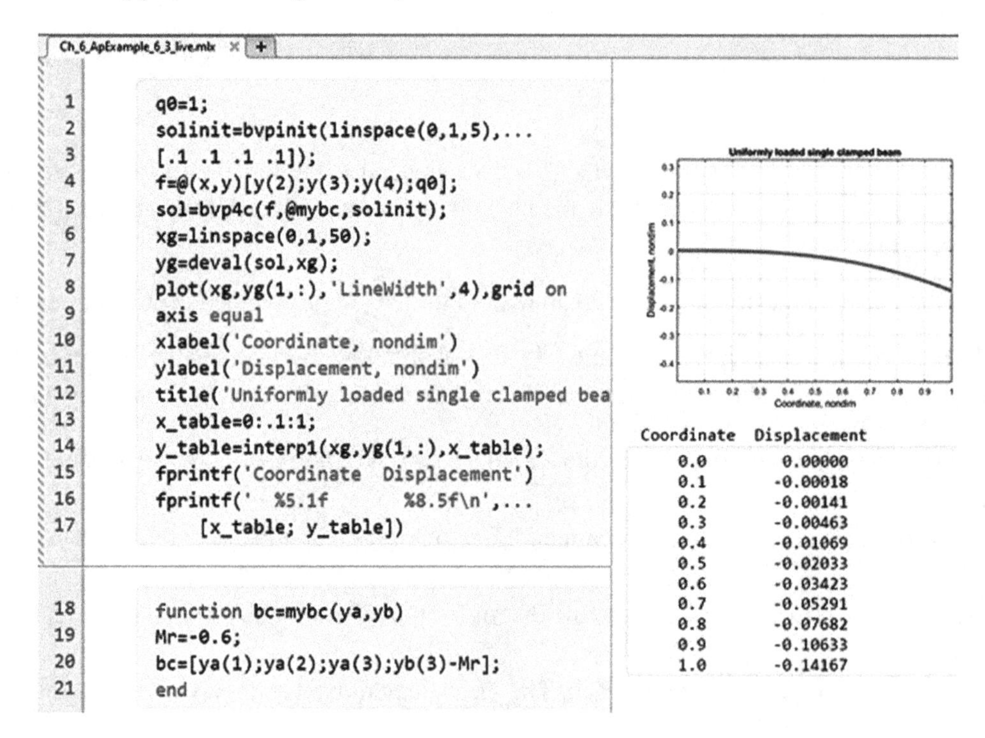

The Ch_6_ApExample_3 live script is designed as follows:

- The q values is assigned to the q0 variables in the first script line;
- In the next line the initial values of the y_1, y_2, y_3, and y_4 variables are assigned at the five equally spaced x points using the bvpinit command;
- Two following two commands are, respectively, the first input argument of the bvp4c command which contains the ODEs (given here in the @(y,x)… form) and the bvp4c command itself;
- Further 100 values of the x and y are obtained using the linspace and deval commands, respectively; these values are intended for the further graph generating;
- The follow five program lines include commands to graphical representing the obtained displacement values; in accordance with the problem requirements the 'linewidth' property is used in the plot command and then the axis equal command arranges the equally scaled plot axes;
- Next four commands are introduced to display the resulting table; two the fprintf commands create here the table caption and two columns with the obtained x and y values;
- Finally, the boundary condition function mybc is written (used by the bvp4c command). The function contains the applied moment value and four given boundary conditions at the a (left) and b (right) boundaries; the y_3 value at the b boundary is $y_b=M_r$ thus it can be rewritten as y_b-M_r=0 and therefore the left part of this expression should be written in the bc vector.

6.4.2. Hydrodynamic Pressure Distribution, Non-Contacting Surfaces With Pores

The one-dimensional Reynolds equation describing the hydrodynamic behavior of the surfaces separated by a thin lubricating film, when one of surfaces is covered by semicircular pores and moves, has the form (Burstein, 2016)

$$\frac{d^2P}{dX^2}+\frac{3}{H}\frac{dH}{dX}\frac{dP}{dX}=\frac{1}{H^3}\frac{dH}{dX}$$

where P is hydrodynamic pressure arising in the film, X – coordinate, H – gap between the surfaces; all variables in this equation are dimensionless.

The gap $H(X)$ and its derivative $\frac{dH(X)}{dX}$ for semicircular pore and adjacent surface parts (called pore cell) are given with the following equations

$$H=\begin{cases}1+\psi\sqrt{1-X^2}, |X|<1\\ 1, 1\le|X|\le\xi\end{cases}$$

$$\frac{dH}{dX}=\begin{cases}-\frac{\psi^2 X}{H-1}, |X|<1\\ 0, 1<|X|\le\xi\end{cases}$$

where ξ is the ratio of the cell dimension to the pore radius, and ψ – pore radii to gap ratio.

The hydrodynamic pressures at both ends of the pore cell range, $X\in[-\xi,\xi]$, are assumed to be zeros.

Problem: Write a live script with name Ch_6_ApExample_4 that uses the bvp4c command to calculate the pressure distribution P(X) at X in the range $-\xi\ldots\xi$ when $\xi=2$ and $\psi=8$. The script should generate a formatted $P(X)$ graph and display the P-values at points $X=-\xi$, $-\xi+0.2$, $-\xi+0.4$, …, ξ.

To solve the second-order Reynolds equation, convert it to two first-order equations.

$$\begin{cases}\frac{dP_1}{\partial X}=P_2\\ \frac{dP_2}{\partial X}=-\frac{3}{H}\frac{dH}{dX}P_2+\frac{1}{H^3}\frac{dH}{dX}\end{cases}$$

The boundary conditions are $P_1(-\xi)=P_1(\xi)=0$.

The live script Ch_6_ApExample_4 that solves the problem, displays the resulting X and P values, generates the $y(x)$ graph is presented in the Figure 6.

Figure 6. Pressure distribution in a lubricating film between surfaces, one of which is covered with semicircular pores and moves.

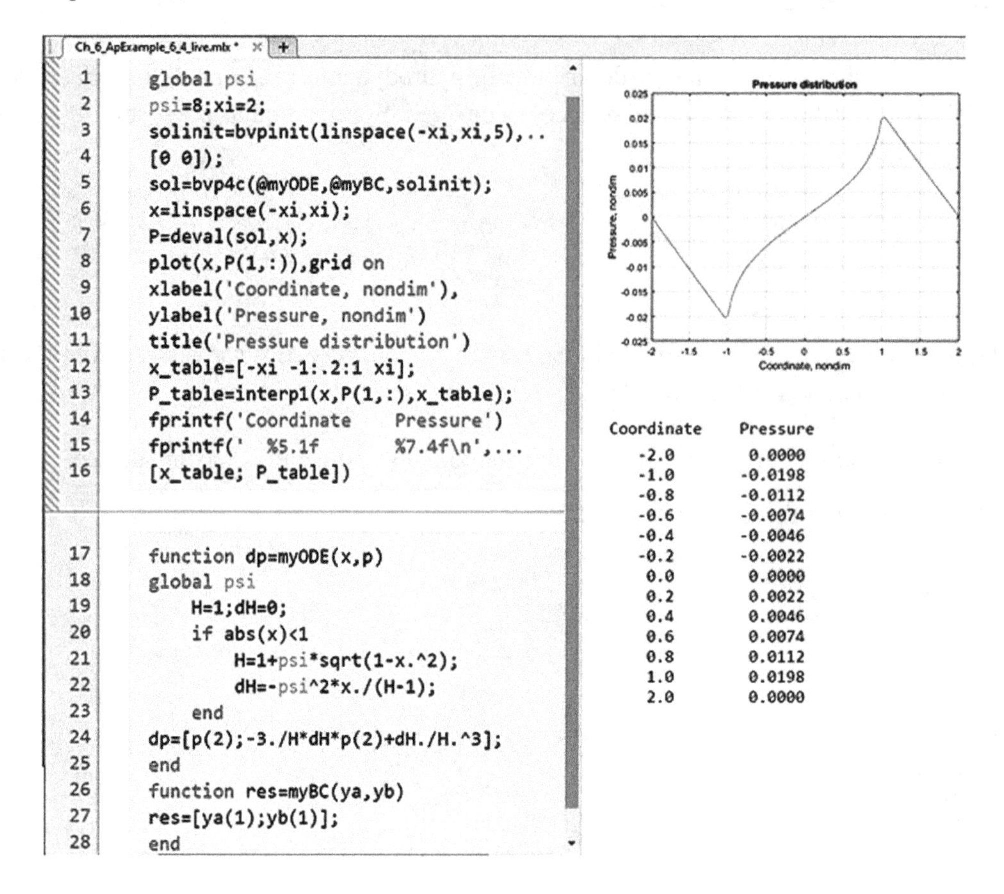

```
global psi
psi=8;xi=2;
solinit=bvpinit(linspace(-xi,xi,5),..
[0 0]);
sol=bvp4c(@myODE,@myBC,solinit);
x=linspace(-xi,xi);
P=deval(sol,x);
plot(x,P(1,:)),grid on
xlabel('Coordinate, nondim'),
ylabel('Pressure, nondim')
title('Pressure distribution')
x_table=[-xi -1:.2:1 xi];
P_table=interp1(x,P(1,:),x_table);
fprintf('Coordinate    Pressure')
fprintf('  %5.1f          %7.4f\n',...
[x_table; P_table])

function dp=myODE(x,p)
global psi
    H=1;dH=0;
    if abs(x)<1
        H=1+psi*sqrt(1-x.^2);
        dH=-psi^2*x./(H-1);
    end
dp=[p(2);-3./H*dH*p(2)+dH./H.^3];
end
function res=myBC(ya,yb)
res=[ya(1);yb(1)];
end
```


Coordinate	Pressure
-2.0	0.0000
-1.0	-0.0198
-0.8	-0.0112
-0.6	-0.0074
-0.4	-0.0046
-0.2	-0.0022
0.0	0.0000
0.2	0.0022
0.4	0.0046
0.6	0.0074
0.8	0.0112
1.0	0.0198
2.0	0.0000

The Ch_6_ApExample_3 live script is designed as follows:

- In the first script line, the xi variables is defined as global to allow using the assigned in the second line ξ values in the myODE function that designed further for the bvp4c function;
- In the next line the initial values of the y_1 and y_2 variables are assigned at the five equally spaced x points using the bvpinit command;
- Next line is the bvp4c command that should be used to solve the current problem;
- The following five program lines include commands for graphical representation of the obtained displacement values; for this, 100 points for *X* are used and the deval command to determine *P* form sol structure at these points..
- The next four commands are introduced to display the resulting table; among them, two fprintf commands create the table caption and two column output with the obtained *x* and *P* values;
- After that, the myODE function with the two solving equations is represented into the two-column vector dp; the function includes H and dH variables that are calculated at the *x*-values with the if…end command;
- Finally, the boundary condition function myBC is written (used by the bvp4c command). This function contains the P values for the two boundary points written within the bc vector.

6.5. CONCLUSION

Commands of the ODE and BVP solvers allow solving of the initial and boundary value problems, respectively. This was demonstrated step by step and implemented on the following M&T problems:

- Second order dynamic system,
- spring-mass harmonic oscillator,
- Spring-mass system with dry friction,
- Particle falling with friction,
- Uniformly loaded single clamped Euler beam,
- Hydrodynamic pressure distribution of the non-contacting surfaces with pores.

The programs that solve these problems perform the ODE solutions with the `ode45`, `ode15s`, and `bvp4c` commands that should be used for resolving the M&T problems described by single or a set of ODEs.

REFERENCES

Burstein, L. (2016). Hydrodynamic Behavior of the Sliding Surface with Semicircular pores: Theoretical and MATLAB Considerations. *International Journal of Surface Engineering and Interdisciplinary Materials Science*, *4*(1), 45–68. doi:10.4018/IJSEIMS.2016010103

Magrab, E. B., Azarm, S., Balachandral, B., Duncan, J. H., Herold, K. E., & Walsh, G. C. (2010). *An Engineer's Guide to MATLAB* (3rd ed.). Pearson.

ENDNOTES

1. Enter the help odeset command to review the possible options.
2. Based on Table 6.1 in Burstein, L. (2020). A *MATLAB® primer for technical programming in materials science and engineerings*. Cambridge: Elsevier-WP.

APPENDIX

Table 2. List of examples, problems, and applications discussed in the chapter

No	Example, Problem, or Application	Subsection
1	Second order dynamical system: a numerical solution.	6.1.2.1., 6.1.2.2.
2	Spring-mass harmonic oscillator boundary value problem solution.	6.2.1.
3	Spring-mass system with dry friction.	6.3.1.
4	Particle falling with friction.	6.3.2.
5	Uniformly loaded single clamped Euler beam.	6.4.1.
6	Hydrodynamic pressure distribution, non-contacting surfaces with pores.	6.4.2.

Note, some small examples, mostly related to non-M&T issues, are not included in the list.

Chapter 7
Solving One-Dimensional Partial Differential Equations

ABSTRACT

This chapter describes the pdepe command, which is used to solve spatially one-dimensional partial differential equations (PDEs). It begins with a description of the standard forms of PDEs and its initial and boundary conditions that the pdepe solver uses. It is shown how various PDEs and boundary conditions can be represented in standard forms. Applications to the mechanics are presented in the final part of the chapter. They illustrate how to solve: heat transfer PDE with temperature dependent material properties, startup velocities of the fluid flow in a pipe, Burger's PDE, and coupled FitzHugh-Nagumo PDE.

INTRODUCTION

In the ODEs discussed in the preceded chapter, an unknown function depends on one variable only. In partial differential equations (PDEs) the unknown function depends on several variables, like in the $P(x,t)$ function the pressure P depends on the location x and time t. Most real processes and phenomenon in technology and physics are described with PDEs. This is especially relevant and is applied in many disciplines of mechanical engineering and tribology (M&T), such as, for example, dynamics, heat and mass transfer, lubrication hydrodynamics, stress and strain analysis, engineering, machine parts, and many others. Actual engineering processes take place in three-dimensional space and at certain time moments. According to spatial dimensions the PDEs are divided into the one-dimensional, 1D, two-dimensional, 2D, or three-dimensional, 3D, equations. MATLAB® has various tools for solving the PDEs of each of these types. Here we present the pdepe solver with M&T application examples that provide solutions of relevant spatially one-dimensional partial differential equations.

DOI: 10.4018/978-1-7998-7078-4.ch007

7.1. STANDARD FORMS OF THE PDE, INITIAL AND BOUNDARY CONDITIONS REQUIRED BY THE PDE-SOLVER

7.1.1. PDE in Standard Form

To solve one-dimensional PDEs, the pdepe command is provided by MATLAB®. The command is designed to solve 1D PDE that can be presented in the following standard form:

$$c\left(x,t,u,\frac{\partial u}{\partial t}\right)\frac{\partial u}{\partial t} = x^{-m}\frac{\partial}{\partial x}\left(x^m f\left(x,t,u,\frac{\partial u}{\partial x}\right)\right) + s\left(x,t,u,\frac{\partial u}{\partial x}\right)$$

where *u* is the function that should be defined as a result of the PDE solution; *t* is the solution time range t0 (initial) ... tf (final) and the coordinate *x* is the coordinate range that is varied between *x*=*a* and *x*=*b*; m can be 0, 1 or 2 corresponding to Cartesian, cylindrical or spherical coordinates respectively;

$f\left(x,t,u,\frac{\partial u}{\partial x}\right)$ is called a flux term while $s\left(x,t,u,\frac{\partial u}{\partial x}\right)$ – a source term.

General form of this equation solution is *u* as function of coordinate and time - *u*(*x*,*t*).

In accordance with the presented standard equation form, the pdepe command is intended to solve PDEs of the first or second order with respect to coordinate *x*. The solving PDE can be an equation of the elliptical or parabolic type, which are usually solved in mechanics. Some examples of PDEs and their adaptation to the standard equation are presented in Table 1.

7.1.2. Initial Conditions for the PDE Solver

Solution of the PDE should be corresponds to certain initial conditions, that set the *u*-values at staring time *t*=0 for the all coordinate points. Initial values should be presented in the form:

u(*x*,*t0*)=*u0*(*x*)

where *u0* denotes the started values of the function *u* to be found; for example, if *u*(*x*,*t0*)=*2* (constant value at all *x*-points), thus *u0*=2; or if *u*(*x*,*t0*)=*5x*-1, thus *u0*=*5x*-1.

7.1.3. Boundary Conditions in Standard Form

The values of x for each of the both *x*=*a* and *x*=*b* (*a* and *b* – *x*-interval ends), boundaries should be set in the following general form

$$p\left(x,t,u\right) + q\left(x,t\right) f\left(x,t,u,\frac{\partial u}{\partial x}\right) = 0$$

Table 1. Some PDEs in traditional and pdepe*-adopted forms*

PDE	PDE –Adopted, Standard Form	Matching Terms of PDE and the pdepe Standard Form			
		m	c	f	s
$\frac{1}{x}\left(\frac{\partial u}{\partial t}\right)-\frac{\partial}{\partial x}\left(\frac{1}{t}u\right)=0$	$\frac{1}{x}\left(\frac{\partial u}{\partial t}\right)=\frac{\partial}{\partial x}\left(\frac{1}{t}u\right)$	0	$\frac{1}{x}$	$\frac{1}{t}u$	0
$\left(\frac{\partial u}{\partial t}\right)+u\frac{\partial u}{\partial x}=\nu\frac{\partial^2 u}{\partial x^2}$	$\left(\frac{\partial u}{\partial t}\right)=\frac{\partial u}{\partial x}\left(\nu\frac{\partial u}{\partial x}\right)-u\frac{\partial u}{\partial x}$	0	1	$\nu\frac{\partial u}{\partial x}$	$-u\frac{\partial u}{\partial x}$
$\frac{\partial T}{\partial t}=\frac{k}{\rho c_p}\frac{\partial^2 T}{\partial x^2}+\frac{q}{\rho c_p}$	$\frac{\partial u}{\partial t}=\frac{\partial}{\partial x}\left(\frac{k}{\rho c_p}\frac{\partial u}{\partial x}\right)+\frac{q}{\rho c_p}$	0	1	$\frac{k}{\rho c_p}\frac{\partial T}{\partial x}$	$\frac{q}{\rho c_p}$
$\rho C_p\frac{\partial T}{\partial t}=\frac{1}{r}\frac{\partial}{\partial r}\left(kr\frac{\partial T}{\partial r}\right)$	$\rho C_p\frac{\partial T}{\partial t}=\frac{1}{r}\frac{\partial}{\partial r}\left(kr\frac{\partial T}{\partial r}\right)$	1	ρC_p	$kr\frac{\partial T}{\partial r}$	0
$\frac{\partial \varphi}{\partial t}=D\frac{\partial^2 \varphi}{\partial x^2}$	$\frac{\partial u}{\partial t}=\frac{\partial}{\partial x}\left(D\frac{\partial \varphi}{\partial x}\right)$	0	1	$D\frac{\partial \varphi}{\partial x}$	0
$\pi^2\frac{\partial u}{\partial t}=\frac{\partial^2 u}{\partial x^2}$	$\pi^2\frac{\partial u}{\partial t}=\frac{\partial}{\partial x}\left(\frac{\partial u}{\partial x}\right)$	0	π2	$\frac{\partial u}{\partial x}$	0
$\frac{\partial u}{\partial t}=a\frac{\partial^2 u}{\partial x^2}-Fu$	$\frac{\partial u}{\partial t}=\frac{\partial}{\partial x}\left(a\frac{\partial u}{\partial x}\right)+\left(-Fu\right)$	0	1	$a\frac{\partial u}{\partial x}$	$-Fu$
$\frac{\partial u}{\partial t}=\frac{5}{r^2}\frac{\partial u}{\partial r}\left(r^2\frac{\partial u}{\partial r}\right)$ •	$\frac{\partial u}{\partial t}=x^{-2}\frac{\partial}{\partial x}\left(x^2 5\frac{\partial u}{\partial x}\right)$	2	1	$5\frac{\partial u}{\partial x}$	0

where *f* is the same function as in the standard form of PDE and does not be given additionally; *p* and *q* are the functions that should be written in concordance to the actual conditions at each of the *a* and *b* coordinate boundaries.

Thus, generally the boundary conditions should be rewritten to match the standard form $p+qf=0$. Therefore, to set the real boundary conditions we need to define two variables – *p* and *q*. For example, if $u=5$ at a boundary point thus matching this equation, $u-5=0$, and the standard boundary equation we need to require $p=u-5$ and $q=0$. In case $du/dx=0$ at a boundary point and *f* containing in PDE, say $f= D*du/dx$, the matching with the standard boundary form gives $p=0$ and $q=1$. These and some other possible boundary conditions with the *p* and *q* values corresponding to the standard form are presented in Table 2.

Table 2. Boundary conditions (BC) and their compliance with the standard boundary equation.

Boundary Conditions and *f* in the PDE	Rewritten in standard form	*p* and *q* matching the BC standard form
u=0	u=0	p= u, q=0
u=4.1	u-4.1=0	p = u-4.1, q=0
$\frac{\partial u}{\partial x}$=0 for PDE containing $f = k\frac{\partial u}{\partial x}$	$k\frac{\partial u}{\partial x}$=0	p =0, q=1
$\frac{\partial u}{\partial x}$=5.1 for PDE containing $f = k\frac{\partial u}{\partial x}$	$k\frac{\partial u}{\partial x} - 5.1k = 0$	p =-5.1k, q=1
$\frac{\partial u}{\partial x}$=2.5-$u$ for PDE containing $f = k\frac{\partial u}{\partial x}$	$k\frac{\partial u}{\partial x} - 2.5k + uk = 0$	p =-2.5k+uk, q=1
$g\cdot\frac{\partial u}{\partial x} = h$ for PDE containing $f = k\frac{\partial u}{\partial x}$	$k\frac{\partial u}{\partial x} - \frac{h}{g}k = 0$	p =- $\frac{h}{g}k$, q=1

The f function in all presented cases is the same as in the standard PDE and is taken equal to $k(du/dx)$ for example. The g in the last Table 7.2 line is constant or function of the x and t while the h is constant or function of the x, y, and u. If the boundary is $\frac{\partial u}{\partial x} = 0$, it is equivalent to $k\frac{\partial u}{\partial x} = 0$ (after multiplying both sides of the equation by k). The above conditions should be applied to boundaries a and b.

Note, in case of the cylindrical or spherical PDE (m=1 or 2), p and q must be set equal to zero at the a-boundary.

7.1.4. On the Numerical Methods use to Solve PDEs

The numerical solution of PDE is performed by replacing the derivatives with the finite difference and is similar to numerical solution of ODE described in the preceding chapter. Nevertheless, in the case of a PDE, apart from spatial differences, the time-dependent differences also appear. We illustrate this concept by the 1D second order ODE describing diffusion/adhesion-like processes that in its dimensionless has the form $\frac{\partial u}{\partial t} = D\frac{\partial^2 u}{\partial x^2}$, where D is a constant. Dividing evenly the space and time intervals into N+1 (from 0 to N) and M+1 (from 0 to M) points, respectively, this equation can be written as

$$\frac{u_i^{k+1} - u_i^k}{\Delta t} = D\frac{u_{i+1}^k - 2u_i^k + u_{i-1}^k}{\Delta x^2}$$

where i and k correspond to the current spatial and time point, respectively, while $\Delta x = x_{i+1}^k - x_i^k$ and $\Delta t = tk+1-tk$ are the x and t differences that are assumed to be constants.

Setting all the u-values at starting time ts (i.e. k=0) and at the boundary point x=0 (i.e. i=0), we can calculate u at time k=1 for each of the coordinate points, so for the i=1-point:

$$u_1^1 = u_1^0 + D\frac{u_2^0 - 2u_1^0 + u_0^0}{\Delta x^2}\Delta t$$

after this, for i=2,

$$u_2^1 = u_2^0 + D\frac{u_3^0 - 2u_2^0 + u_1^0}{\Delta x^2}\Delta t$$

and so on till the i=N-1

$$u_N^1 = u_N^0 + D\frac{u_N^0 - 2u_{N-1}^0 + u_{N-2}^0}{\Delta x^2}\Delta t$$

At this point, the u value at the boundary i=N is used, which should be given. When all u-values for k=1 time point are defined, the next time point k=2 can be calculated in the same way. This process can be continued up to the final given time point k=M.

Described scheme with some improvements and complexities applies to all finite differences methods used to solve the PDE. The discussed and more complex methods are studied in courses on numerical methods and their explanation is beyond the scope of this book.

7.2. THE pdepe COMMAND FOR SOLVING ONE-DIMENSIONAL PDEs

The command for solving the single or set of spatially one-dimensional PDEs is named pdepe and has the following form:

```
sol=pdepe(m,@fun_pde,@initial_cond,@boundary_cond,x_mesh,tspan)
```

- m is equal to 0, 1 or 2 as above (sub-section 7.1.1);
- fun_pde is the name of an user-defined function where the solving PDE must be written; definition line of this function reads

```
function [c,f,s]=fun_pde(x,t,u,DuDx)
```

$\frac{\partial u}{\partial x}$ $\frac{\partial u}{\partial x}$

- initial_cond is the name of an user-defined function with the initial conditions of the PDE being solved. The definition line of this function reads

```
function u0=initial_cond(x)
```

u0 being the vector $u(x)$ value at t=0;

- boundary_cond is the name of on user-defined function containing the boundary conditions for each of the both $x=a$ and $x=b$ (a and b – x-interval ends). The definition line of the boundary_cond function reads

```
function [pa,qa,pb,qb]=boundary_cond(xa,ua,xb,ub,t)
```

xa=a and xb=b being the coordinates of the boundary points, ua and ub are the u values at these points; pa, qa, pb, and qb represent the p and q values of the standard boundary condition form form (see subsection 7.1.3) given at a and b points.

- x_mesh is a vector of the x-coordinates at which the solution is sought for each time value contained in the tspan; xmesh values must be written in ascending order from a to b;
- tspan is a vector of the time points that must be written in ascending order;
- The output argument sol is a three-dimensional array comprising k (number of the solving PDEs) two-dimensional arrays with M (number of the time points) rows and N (number of the coordinate points) columns each. Elements in the three-dimensional arrays are numbered similarly to two-dimensional ones (see Section 2.2). For example, sol(2,4,2) denotes a term located in the second row and fourth column of the second array (termed sometimes as matrix page) while sol(2,:,1) denotes in the first array (page) all columns of the second row. In the pdepe solver, the sol(i,j,k) denotes defined u_k (u-values for k-th PDE, in the case of more than one PDE) at the ti time- and xj coordinate points, e.g. sol (:,:,1) is an array with each row containing u values calculated for each of the x-coordinates at a certain point in time. For a single PDE, there is one array only (k=1) for a set of two PDEs – two arrays (k=1 and 2), and so on. Note, for a single PDE the tree-dimensional array sol(:,:,1) is the resulting u and it can be used as the regular two-dimensional array.

For more information, type >>doc pdepe in the Command Window or use the MATLAB® Help window.

7.2.1. The Steps for Using the pdepe Command, by Example

For example, consider the simple thermo-diffusion equation with a heat source for temporal and spatial distributions of temperature *u* of a mechanical part (such as an insulated wire or thin rod):

$$\frac{\partial u}{\partial t} = k\frac{\partial^2 u}{\partial x^2} + S$$

where *t* and *x* are the time and coordinate, respectively, while *k* and *S* are the diffusivity coefficient and permanent heat source, respectively. All variables in this equation are given in dimensionless units; *S* and *k* are 1 each, *x* is in the range from 0 to 1 and *t* from 0 to 0.4.

This equation is taken with initial and boundary conditions as below.

The initial conditions:

$$u(x,0) = 0.4 + 0.6x$$

The boundary conditions:

$$u(0,t) = u_a = 0.4,\ u(1,t) = u_b = 0.4$$

The solution of this PDE goes through the following steps:

1. First, the equation must be presented in the standard form. The heat-diffusion equation rewritten in the required form reads

$$\frac{\partial u}{\partial t} = \frac{\partial}{\partial x}\left(k\frac{\partial u}{\partial x}\right) + S$$

Matching this and the standard equations, we obtain their identity when:

$$m = 0,$$

$$c\left(x,t,u,\frac{\partial u}{\partial t}\right) = 1,$$

$$f\left(x,t,u,\frac{\partial u}{\partial x}\right) = k\frac{\partial u}{\partial x},$$

$$s\left(x,t,u,\frac{\partial u}{\partial x}\right)=S$$

2. The initial and boundary conditions must be presented in the standard form. Accordingly, the initial condition:

$$u_0 = 0.4 + 0.6x$$

and the boundary conditions at the point boundaries *a* (*x*=*xa*=0) and *b* (*x*=*xb*=1) are the first-type boundary conditions (termed frequently Dirichlet boundary conditions):

$$u_a = 0.4, u_b = 0.4$$

Comparing this equations with the *p*+*qf*=0 standard form we obtain (Table 7.2, row 2):

$$p_a = u_a - 0.4, q_a = 0$$

$$p_b = u_b - 0.4, q_b = 0$$

3. At this stage, the user-defined function with pdepe command and with the three sub-functions containing the solving PDE/s, initial and boundary conditions should be written.

7.2.2. A Program That Solves PDE Using the pdepe Command

Now we represent a program that solves the above equation using the described steps and showing the results in two graphs on the same page (figure):

- the first – three-dimensional plot showing the obtained temperatures along the coordinate and at each of given time values (use for this the mesh command),
- the second – two-dimensional plot represented temperatures at three times – 0.001, 0.01, and 0.1 (use for this the contour command).

The following user-defined function named ExPDEPE and stored in the same name file solves the above PDE:

```
function ExPDEPE(n_x,n_t)
                    %solves an one-dimensional heat-diffusion PDE with Dirichlet BC
                                        %                          To run: >>ExPDEPE(25,25)
m=0;
xmesh=linspace(0,1,n_x);
```

```
tspan= linspace(0,0.4,n_t);
u=pdepe(m,@myPDE,@i_c,@b_c,xmesh,tspan);
[X,T]=meshgrid(xmesh,tspan);
subplot(1,2,1)
mesh(X,T,u)                                  % mesh plot with the X - T domain
xlabel('Coordinate'),ylabel('Time'), zlabel('Concentration')
title('1D diffusion with the pdepe')
subplot(1,2,2)
c=contour(X,u,T,[0.001 0.005 0.05]);                        % 3 iso-time lines
clabel(c)                          % labels for the iso-lines in the contour plot
xlabel('Coordinate, nondim'), ylabel('Temperature, nondim')
title('Isoth')
grid
function [c,f,s]=myPDE(x,t,u,DuDx)                          % PDE for solution
c=1;
k=1;
f=k*DuDx;
s=1;
function u0=i_c(x)                                        % initial conditions
u0=0.4+0.6*x;
function [pa,qa,pb,qb]=b_c(xa,ua,xb,ub,t)               % boundary conditions
pa=ua-0.4;pb=ub-0.4;
qa=0;qb=0;
```

The ExPDEPE function is written without output arguments. Its input arguments n_x and n_t are the serial numbers of the x- and t points given for the solution. The pdepe function calls here three sub-functions: myPDE comprising the heat-diffusion equation, i_c defining the initial condition and b_c defining the boundary conditions. The x_mesh and tspan vectors are created with two linspace commands in which the n_x and n_t must be inputted with the mydfsn run. The numerical results are stored in the u array, which in our case is two-dimensional, and used in the subsequent graphical commands. The mesh command is used to generate the 3D mesh sub-plot u(x,t) while the contour and clabel commands – to generate the 2D sub-plot u(x) showing the three labeled iso-time lines (0.001, 0.01, and 0.1) that were defined by the contour command as the vector of the selected t values.

After running this function in the Command Window, the following two plots are generated:

```
>> ExPDEPE(25,25)
```

Figure 1. Heat-diffusion equation solution obtained with the pdepe *command: temperature–coordinate – time 3D plot (a) and 2D contour plot with three iso-time lines (b)*

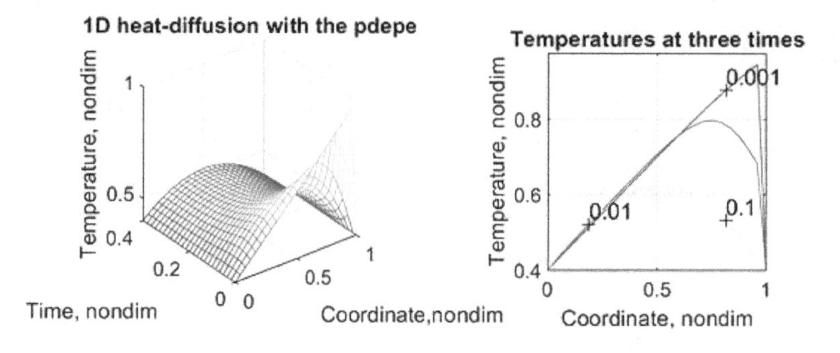

7.2.3. Extended Command Form That Can Be Used With the pdepe Command

When PDE includes some parameters (e.g. in the example above the k is the parameter), the pdepe command can be written in a more complex form to pass these parameters to functions containing the PDE, initial condition, and boundary conditions:

```
sol=pdepe(m,@fun_pde,@initial_cond,@boundary_cond,x_mesh,tspan,[ ], param_
name1,param_name2,…)
```

- The empty brackets [ ] denotes an empty vector intended, in general case, for the so-termed options used to control the integration process1. In most applications, the default option values are used and we do not specify these options here.
- Param_name1,param_name2,... are the names of the arguments that we intend to use for transmitting their values into the fun_pde function. If the parameters are named in the pdepe command, they should also be written in the definition line of each of the fun_pde, initial_cond, and boundary_cond functions.

For example, the ExPDEPE file should be modified for introducing the k coefficient as arbitrary parameters:

```
function ExPDEPE(n_x,n_t,k)
              % solves an one-dimensional heat-diffusion PDE with Dirichlet BC
                                  %                    To run: >>ExPDEPE(25,25,1)
m=0;
xmesh=linspace(0,1,n_x);
tspan= linspace(0,0.4,n_t);
u=pdepe(m,@myPDE,@i_c,@b_c,xmesh,tspan,[],k);
[X,T]=meshgrid(xmesh,tspan);
subplot(1,2,1)
mesh(X,T,u)                                  % mesh plot with the X - T domain
xlabel('Coordinate'),ylabel('Time'), zlabel('Concentration')
```

```
title('1D diffusion with the pdepe')
subplot(1,2,2)
c=contour(X,u,T,[0.001 0.005 0.05]);                   % 3 iso-time lines
clabel(c)                            % labels for the iso-lines in the contour plot
xlabel('Coordinate, nondim'), ylabel('Temperature, nondim')
title('Isoth')
grid
function [c,f,s]=myPDE(x,t,u,DuDx,k)                   % PDE for solution
c=1;
f=k*DuDx;
s=1;
function u0=i_c(x,k)                                  % initial conditions
u0=0.4+0.6*x;
function[pa,qa,pb,qb]=b_c(xa,ua,xb,ub,t,k)            %boundary conditions
pa=ua-0.4;pb=ub-0.4;
qa=0;qb=0;
```

In the Command Window, the following command should be typed and entered:

```
>> ExPDEPE(25,25,1)
```

The results are identical to those discussed in subsection 7.2.2.

The advantage of this form of the pdepe command is its greater versatility, e.g. the ExPDEPE function with parameter k can be used for any k-value without introducing it into the myPDE sub-function.

7.3. APPLICATION EXAMPLES

7.3.1. One-Dimensional Heat Transfer with Temperature Dependent Properties of a Material

The 1D heat transfer equation with temperature-dependent properties of a material has the form

$$\frac{\partial T}{\partial t} = \frac{\partial}{\partial x}\left(\alpha\left(T\right)\frac{\partial T}{\partial x}\right)$$

where T-temperature, x and t are the coordinate and time, respectively, and α - is the thermal diffusivity coefficient that varies with temperature, for example, linearly as $\alpha = 0.3T+0.4$. The x varies from 0 to 1, and t – from 0 to 0.3; all variables are dimensionless.

The initial and boundary conditions are as follows:

$T(x,0)=x$ (spatially linear temperature change from 0 to 1),

$T(0,t)=0.1,$

$T(1,t)=0.2.$

Problem: Write a user-defined function with name Ch_7_ApExample_7_1 that considers the heat equation with given initial and boundary conditions using the pdepe command. Design the function without the output parameters and with the following input parameters: starting times and coordinates, final time and coordinates, point numbers for the time and coordinate meshes; assume 20 points for each mesh. The problem is solved in Cartesian coordinates. The resulting temperatures as function of time and coordinate must be represented in the T(x,t) plot generated in the reverse y-view (use the ij option of the **axis** command for this).

Comparing the solving equation with the standard equation form (sub-section 7.1.1), we obtain that they are identical when:

$$m = 0$$

$$c\left(x, t, u, \frac{\partial u}{\partial t}\right) = 1$$

$$f\left(x, t, u, \frac{\partial u}{\partial x}\right) = \alpha\left(u\right)\frac{\partial u}{\partial x}$$

$$s\left(x, t, u, \frac{\partial u}{\partial x}\right) = 0$$

here u denotes the temperature T.

The initial and boundary conditions are identical to their standard form when

$$u_0 = x$$

x=xa=0:

$$p_a = u_a - 0.1$$

$$q_a = 0$$

x=xb=1:

$$p_b = u_b - 0.2$$

$$q_b = 0$$

The commands for solving this problem are:

```
function Ch_7_ApExample_7_1(ts,tf,xs,xf,nt,nx)
                  % solves heat equation,temperature-depending material property
                        %         To run: >>  Ch_7_ApExample_7_1(0,0.4,0,1,20,20)
m=0;
xmesh=linspace(xs,xf,nx);
tspan= linspace(ts,tf, nt);
u=pdepe(m,@mypde,@i_c,@b_c,xmesh,tspan);
surf(xmesh,tspan,u);
xlabel('Coordinate'),ylabel('Time'), zlabel('Temperature')
title('1D heat transfer, \alpha(T)')
axis square tight ij                                       %ij - reverses the y axis
function [c,f,s]=mypde(x,t,u,DuDx)                                            % PDE
c=1;
D=0.3*u+0.4;
f=D*DuDx;
s=0;
function u0=i_c (x)                                             % initial condition
u0=x;
function [pa,qa,pb,qb]=b_c(xa,ua,xb,ub,t)                     % boundary conditions
pa=ua-0.1;pb=ub-0.2;
qa=0;qb=0;
```

After saving this program in the ApExample6_3 file and entering the running command in the Command Window, the following plot appears:

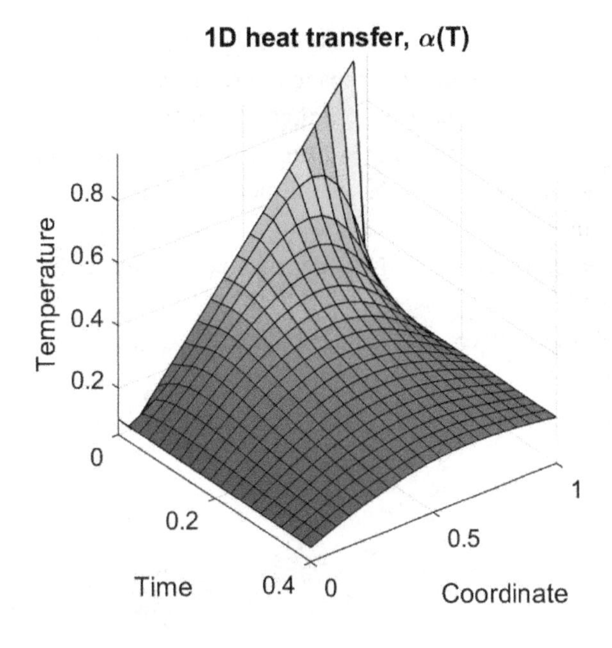

Presented Ch_7_ApExample_7_1 function is written without output arguments. Its input arguments ts, tf, xs, xf, nt, and nx are the start and final times, start and final coordinate points, and also the numbers of the x- and t points intended for the solution. The pdepe function invokes here three sub-functions: myPDE comprising the terms of the heat equation, i_c defining the initial condition, and b_c defining the boundary conditions. The xmesh and tspan vectors are created with two linspace commands. The numerical results are stored in the u array, which in our case is two-dimensional, and is used in the subsequent graphical commands. The surf command is used to generate the 3D sub-plot $u(x,t)$ while the axis command with the square, tight, and ij options is used to improve the presentation of the results.

7.3.2. Fluid Mechanics: Startup Flow Through a Pipe

In cylindrical coordinates, the dimensionless partial differential equation for velocity u of fluid flow through a pipe can be represented by the following equation (based on Batchelor, 2012):

$$\frac{\partial u}{\partial t} = 1 + \frac{1}{r}\frac{\partial}{\partial r}\left(r\frac{\partial u}{\partial r}\right)$$

here r is the radial coordinate that changes from 0 up to 1 (latter value corresponds to the dimensionless pipe radius), t –time.

Consider this equation with the following initial and boundary conditions:

$$u\left(x,0\right) = 0$$

$$u\left(1,t\right) = 0$$

Problem: Write a user-defined function with name Ch_7_ApExample_7_2 that calculates flow velocity u and generates the following three graphs in the same window: $u(r)$ at different t, $u(t)$ for center of the pipe (r=0), and $u(r,t)$. Take the r and t ranges from 0 to 1 each. The input parameters of the Ch_7_ApExample_7_2 function should be starting and final points for r and t and the amounts of the x and t points; and the output parameter – the two-dimensional matrix of each row of which is u-values along coordinate r at certain t. Take the r and t point numbers equal to 100 and 40 points respectively. To simplify the numerical output, display the u-matrix so that it has only six t-lines and five r-columns from the calculated range.

To solve the problem, the flow equation and initial and boundary conditions must be represented in its standard forms.

Comparing the solving equation with the standard form equation, we obtain that they are identical when:

$$m = 1\,(\text{cylindrical coordinates})$$

$$c\left(x,t,u,\frac{\partial u}{\partial t}\right) = 1$$

$$f\left(x,t,u,\frac{\partial u}{\partial x}\right)=\frac{\partial u}{\partial x}$$

$$s\left(x,t,u,\frac{\partial u}{\partial x}\right)=1$$

The standard forms of the initial and boundary conditions are

$$u_0=0$$

$p_a=0, q_a=0$ at r=ra=0 (since m=1)

$p_b=u_b, q_b=0$ at r=rb=1

The commands for solving this problem are:

```
function u_table=Ch_7_ApExample_7_2(xs,xf,ts,tf,nt,nx)
                        %  solves flow in pipe equation, in cylindrical coordinates
                          %     to run >> u_table=Ch_7_ApExample_7_2(0,1,0,1,20,100
m=1;
xmesh=linspace(xs,xf,nx);tspan=linspace(ts,tf,nt);
u=pdepe(m,@myPDE,@i_c,@b_c,xmesh,tspan);
t_table=round(linspace(1,nt,6));
r_table=round(linspace(1,nx,5));
u_table=u(t_table,r_table);
subplot(2,2,1)
plot(xmesh,u(round([3 nt/4 nt/2 nt]),:))
xlabel('Radial coordinate, nondim')
ylabel('Velocity, nondim')
title({'Velocity vs coordinate';'at different times'})
text(xmesh(round(nx/2+5)),u(nt,round(nx/2)),…
      ['time=',num2str(tspan(nt))])
axis square tight
grid on
subplot(2,2,2)
plot(tspan,u(:,1)), grid on
xlabel('Time, nondim'),ylabel('Velocity, nondim')
title({'Velocity vs time'; 'at pipe center'})
axis square tight
subplot(2,2,[3 4])
mesh(xmesh,tspan,u)
```

```
xlabel('Radial coordinate,nondim')
ylabel('Time, nondim'),zlabel('Velocity, nondim')
axis square tight
function [c,f,s]=myPDE(x,t,u,DuDx)                                    % PDE
c=1;
f=DuDx;
s=1;
function u0=i_c(x)                                      % initial condition
u0=0;
function [pa,qa,pb,qb]=b_c(xa,ua,xb,ub,t)             % boundary conditions
pa=0;qa=0;
pb=ub;qb=0;
```

After saving this program in the Ch_7_ApExample_7_2 file and entering the running command in the Command Window, the following table and plot appear.

```
>> u_table=Ch_7_ApExample_7_2(0,1,0,1,40,100)
u_table =
       0        0        0        0        0
  0.1655   0.1571   0.1301   0.0814        0
  0.2241   0.2105   0.1690   0.1015        0
  0.2407   0.2256   0.1801   0.1072        0
  0.2471   0.2315   0.1843   0.1093        0
  0.2491   0.2332   0.1856   0.1100        0
```

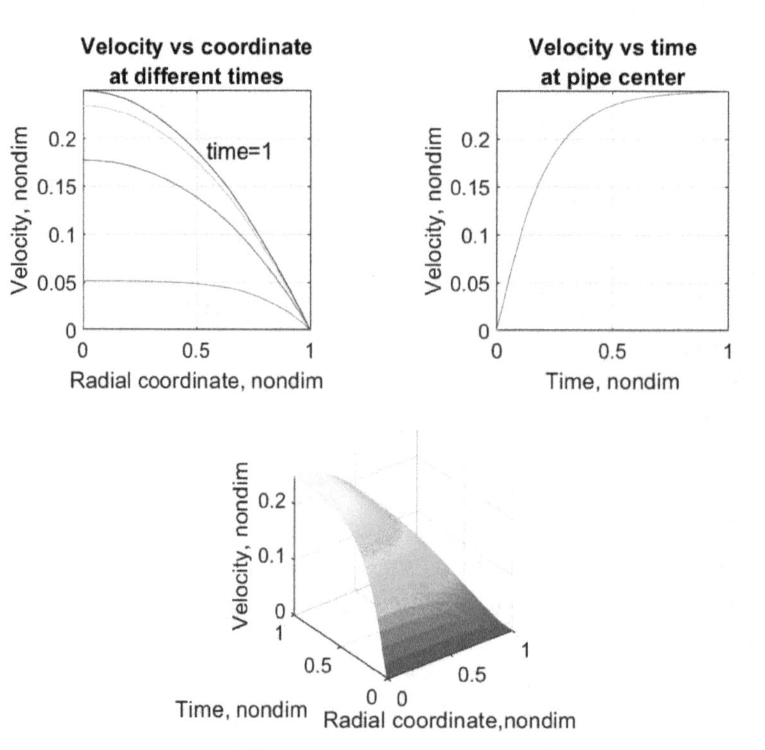

The presented function Ch_7_ApExample_7_2 includes:

In the definition line, the input arguments ts, tf, xs, xf, nt, and nx denoting the start- and final- times and coordinate points, and the numbers of the *t*- and *x* points intended for the solution; the output argument u_table is intended to display the calculated velocities in table with 6 time rows and 5 coordinate columns.

In the next lines, the xmesh and tspan vectors are created with two linspace commands and then the pdepe function is written that invokes three sub-functions: myPDE comprising the terms of the flow equation, i_c defining the initial condition and b_c defining the boundary conditions.

The calculated values are stored in the u array, which in our case is two-dimensional and used in the subsequent output and graphical commands. Two linspace commands are used to obtain six and five indices for subsequent creation the 6x5 u_table array with outputted u values; these indices must be integers so they are rounded with the round command.

The subplot commands are used to plot three graphs in the same Figure window; for this, two plot commands generate the *u(r)* and *u(t)* plots and the surf command generates graph *u(x,t)*. The axis commands use the square and tight options for better show the results.

7.3.3. The FitzHugh-Nagumo PDEs

Wave dynamics problems, simulations of the spike propagation in nerve, and other models are described by the FitzHugh-Nagumo -type PDE (abbreviated as the FHN equation) with a diffusion term:

$$\frac{\partial u}{\partial t} = D\frac{\partial^2 u}{\partial x^2} + u\left(a-u\right)\left(u-1\right) - w$$

$$\frac{\partial w}{\partial t} = bu - cw$$

where *u* and *w* are the action potential (e.g. electrical) and recovery processes variables respectively, *D* – diffusion coefficient; *a*, *b*, and *c* are parameters of the resting state and dynamics of the system.

The initial conditions depend on actual starting spike and are assumed here as

$$u\left(x,0\right) = e^{-x^2}$$

$$w\left(x,0\right) = e^{-\left(x+10\right)^2}$$

The boundary conditions at both ends, *l* and *r* (left and right), of the solving x-range are:

$$\frac{\partial u\left(l,t\right)}{\partial x} = \frac{\partial u\left(r,t\right)}{\partial x} = 0$$

$$\frac{\partial w(l,t)}{\partial x}=\frac{\partial w(r,t)}{\partial x}=0$$

Problem: Write a user-defined function with name Ch_7_ApExample_7_3 that calculates *u* and *w* and generates *u*(*x*,*t*) graph with coordinates *x* from 0 to 10 and at *t* from 0 to 400. Use the extended pdepe command form for transferring *D*, *a*, *b*, and *c* constants to sub-functions of this command. Assume *D*=0.05, *a*=0.1, *b*=0.01, *c*=0.01 and the numbers of the *x* and *t* points are 100 and 40 respectively. The Ch_7_ApExample_7_3 function should have one output parameter that displays the table of the *u* values. To shorten the output, display table with only five lines and five columns.

To solve the FHN equations, it is necessary to rewrite each of them in the standard form:

$$\frac{\partial u}{\partial t}=\frac{\partial u}{\partial x}\left(D\frac{\partial u}{\partial x}\right)+u(a-u)(u-1)-w$$

$$\frac{\partial w}{\partial t}=\frac{\partial w}{\partial x}\left(0\frac{\partial w}{\partial x}\right)+bu-cw$$

Comparing the solving equations with the standard form equation, we obtain their identity when *m*=1 (Cartesian coordinates) and:

$c\left(x,t,u,\frac{\partial u}{\partial t}\right)=1$ for each of the equations,

$f\left(x,t,u,\frac{\partial u}{\partial x}\right)=D\frac{\partial u}{\partial x}$ for the first equation,

$f\left(x,t,w,\frac{\partial w}{\partial x}\right)=0\frac{\partial w}{\partial x}$ for the second equation,

$s\left(x,t,u,\frac{\partial u}{\partial x}\right)=u(a-u)(u-1)-w$ for the first equations,

$s\left(x,t,w,\frac{\partial w}{\partial x}\right)=bu-cw$ for the second equation.

Alternatively, in the matrix form:

$$\begin{bmatrix}1\\1\end{bmatrix}.*\frac{\partial}{\partial t}\begin{bmatrix}u_1\\u_2\end{bmatrix}=\frac{\partial}{\partial x}\begin{bmatrix}D\dfrac{\partial u_1}{\partial x}\\0\dfrac{\partial u_2}{\partial x}\end{bmatrix}+\begin{bmatrix}u_1\left(a-u_1\right)\left(u_1-1\right)-u_2\\bu_1-cu_2\end{bmatrix}$$

where u1 and u2 denote the *u* and *w*, respectively.

The initial and boundaries conditions can be rewritten in the standard *p+qf* forms as:

$$\begin{bmatrix}u_1\left(x,0\right)\\u_2\left(x,0\right)\end{bmatrix}=\begin{bmatrix}e^{-x^2}\\e^{-\left(x+10\right)^2}\end{bmatrix}$$

$$\begin{bmatrix}0\\0\end{bmatrix}.*\begin{bmatrix}u_1\left(l,t\right)\\u_2\left(l,t\right)\end{bmatrix}+\begin{bmatrix}1\\1\end{bmatrix}.*\begin{bmatrix}D\dfrac{\partial u_1\left(l,t\right)}{\partial x}\\0\dfrac{\partial u_2\left(l,t\right)}{\partial x}\end{bmatrix}=\begin{bmatrix}0\\0\end{bmatrix}$$

$$\begin{bmatrix}0\\0\end{bmatrix}.*\begin{bmatrix}u_1\left(r,t\right)\\u_2\left(r,t\right)\end{bmatrix}+\begin{bmatrix}1\\1\end{bmatrix}.*\begin{bmatrix}D\dfrac{\partial u_1\left(r,t\right)}{\partial x}\\\alpha\dfrac{\partial u_2\left(r,t\right)}{\partial x}\end{bmatrix}=\begin{bmatrix}0\\0\end{bmatrix}$$

In the pdepe command notations:

$$c=\left[1;1\right];f=\left[0.01;0\right]*DuDx;s=\left[u\left(1\right)*\left(a-u\left(1\right)\right).*\left(u\left(1\right)-1\right)-u\left(2\right);b*u\left(1\right)-c*u\left(2\right)\right];$$
$$u_0=\left[\exp\left(-x.\hat{\ }2\right);0.2*\exp\left(-\left(x-2\right).\hat{\ }2\right)\right];pa=0;pb=0;qa=1;gb=1.$$

The commands for solving this problem are:

```
function u_table=Ch_7_ApExample_7_3(xs,xf,ts,tf,nt,nx,D,a,b,c1)
                                %                    solves two FitzHugh-Naguno PDEs
                   %>>u_table=Ch_7_ApExample_7_3(0,10,0,400,40,100,.05,.1,.01,.01)
  close all
  m=0;a=0.16;b=0.008;c1=0.008*2.54;
  xmesh=linspace(xs,xf,nx);tspan= linspace(ts,tf,nt);
  sol=pdepe(m,@myPDE,@i_c,@b_c,xmesh,tspan,[],D,a,b,c1);
  u=sol(:,:,1);w=sol(:,:,2);
  t_tab=round(linspace(1,nt,5));u_tab=round(linspace(1,nx,5));
  u_table=u(t_tab,u_tab);
  mesh(xmesh,tspan,u)
```

```
  xlabel('Coordinate,nondim')
  ylabel('Time, nondim'),zlabel('Potential, nondim')
  title('Acting potential')
  axis square tight
function [c,f,s]=myPDE(x,t,u,DuDx,D,a,b,c1)                          %PDEs
  c=[1;1];
  f=[D;0].*DuDx;
  s=[u(1)*(a-u(1))*(u(1)-1)-u(2);b*u(1)-c1*u(2)];
function u0=i_c(x,D,a,b,c1)                                % initial condition
  u0=[exp(-x.^2);exp(-(x+10).^2)];
function [pl,ql,pr,qr]=b_c(xl,ul,xr,ur,t,D,a,b,c1)
%boundary conditions
  pl=[0;0];ql=[1;1];
  pr=[0;0];qr=[1;1];
```

After saving this function in the Ch_7_ApExample_7_3 file and running it, the following resulting data and graph appear:

```
>>u_table=Ch_7_ApExample_7_3(0,10,0,200,20,400,.05,.1,.01,.01)
u_table =
  1.0000   0.0019   0.0000   0.0000   0.0000
 -0.2079  -0.1292   0.5578   0.0012   0.0000
 -0.0363  -0.0540  -0.1169  -0.0584   0.0459
 -0.0007  -0.0027  -0.0184  -0.0433  -0.0033
  0.0000   0.0001   0.0003  -0.0013   0.0006
```

The Ch_7_ApExample_7_2 function includes:

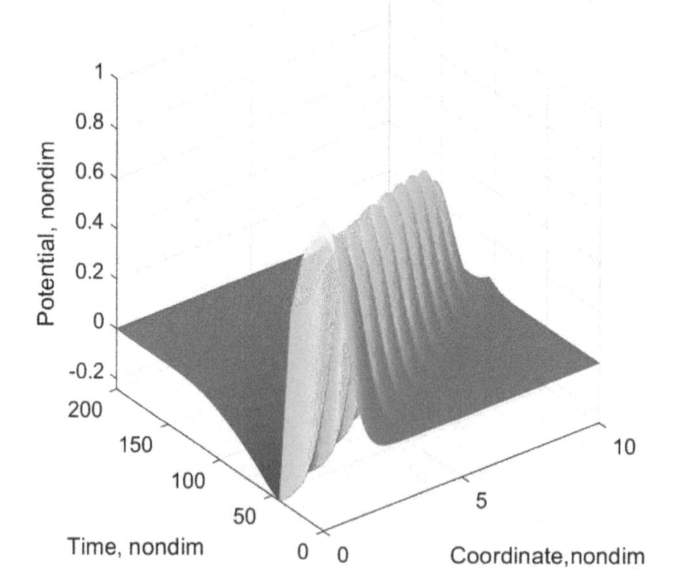

In the definition line, the input arguments ts, tf, xs, xf, nt, nx, D, a, b, c1 denote the start and final times and coordinate points, the numbers of the *x*- and *t* points intended for the solution, diffusion and other coefficients of the FHN equations (*c*1 designates here the *c*-coefficient of the FHN equations to avoid confusion with *c*-term in the myODE subfunction of the pdepe command); the output argument u_table is intended to display the calculated voltages in a table having 5 time rows and 5 coordinate columns.

In the next lines, the xmesh and tspan vectors are created with two linspace commands and this for using in the next pdepe function that invokes three sub-functions: myPDE comprising the terms of the flow equation, i_c defining the initial condition and b_c defining the boundary conditions. The numerical results are stored in the sol array, which in our case is three-dimensional, and contents matrix of *u*-values in the sol(:,:,1) page. Two linspace commands are used to obtain five indices for the subsequent creation of the 5x5 u_table matrix with *u* values for output; these indices must be integers, so they are rounded with the round command. The resulting 3D plot is generated by the mesh command.

7.3.4. The One-Dimensional Burgers' Equation

The Burgers' equation is a fundamental partial differential equation that describes phenomenon in various areas, such as fluid mechanics, nonlinear acoustics, gas dynamics, traffic flow, and others. Its one-dimensional form for the case of fluid flow are as follows:

$$\frac{\partial u}{\partial t} = \nu \frac{\partial^2 u}{\partial x^2} - u\frac{\partial u}{\partial x},\ 0 \le x \le 1, t \ge 0$$

where *u* is the fluid velocity, *x* – coordinate, *t* - time, and ν – fluid viscosity.

Consider this equation with the following initial and boundary conditions

$$u(x,0) = \sin(\pi x)$$

$$x(0,t) = x(1,t) = 0$$

Problem: Write a live program with name Ch_7_ApExample_7_4 that considers the Burger's equation with given initial and boundary conditions using the pdepe command. The user-defined function named Burgers should generate the following three graphs in the same window: *u*(*x*) at four different *t*, *u*(*t*) for the four different *x* (*x*=1/2), and *u*(*x*,*t*). Assume *v*=0.1 and the *x* and *t* ranges from 0 to 1 each. The input parameters of the Ch_7_ApExample_7_4 function should be the starting and final points for *x* and *t* and the amounts of the *x* and *t* points. Take the *x* and *t* point numbers equal to 100 and 40 points respectively.

To solve the problem, the flow equation and initial and boundary conditions must be represented in its standard forms.

Comparing the solving equation with the standard form of the PDE equation, we obtain that they are identical when:

$m = 0$ (Cartesian coordinates)

$$c\left(x,t,u,\frac{\partial u}{\partial t}\right)=1$$

$$f\left(x,t,u,\frac{\partial u}{\partial x}\right)=u\frac{\partial u}{\partial x}$$

$$s\left(x,t,u,\frac{\partial u}{\partial x}\right)=-\nu\frac{\partial u}{\partial x}$$

The standard forms of the initial and boundary conditions are

$$u_0=0$$

$$p_a=u_a, q_a=0 \text{ at xa=0}$$

$$p_b=u_b,\ q_b=0 \text{ at xb=1}$$

The live program solving this problem are shown in Figure 2.
The Ch_7_ApExample_7_4 live program includes:

- The first line represents the command that runs the live function Burger;
- The Burger function definition line has the input arguments xs, xf, ts, tf, nt, and nx denoting the start- and final- coordinate and time points, and the numbers of the t- and x points intended for the solution;
- The function help line contains a short function destination, the run command is not included in help because it is presented in the first line of the program;
- In the next line, the xmesh and tspan vectors are created with two linspace commands and then the pdepe function is written that invokes three sub-functions (which are presented in lines 33 … 41 at the end of the program): myPDE comprising the terms of the Burgers' equation, i_c containing an initial condition and b_c presenting the boundary conditions; the '~'symbol replaces a variable not used in these sub-functions; the calculated values are stored in the u array, which in the considering case is two-dimensional and is used in the subsequent graphic commands;
- The commands in the 8 … 17 lines generate a formatted graph *u*(*x*) with four iso-time curves taken at time instants *t*=3, *nt*/4, *nt*/2, and *nt*; to display the string 't=' with subsequent *t*-value for each of the four curves, a four-row string matrix str1 is generated with *t* values rounded to two decimal digits;
- The commands in the 18 … 27 lines generate a formatted graph *u*(*t*) with four iso-coordinate curves taken at the coordinates *x*=3, *nx*/4, *nx*/2, and *nx*; to display the string with the 'x=' characters and subsequent *x*-value for each of the four curves, a four-row string matrix str1 is generated with the *x* values rounded to two decimal digits;

Figure 2. Live program Ch_7_ApExample_7_4 *with the Burger function and the resulting plots*

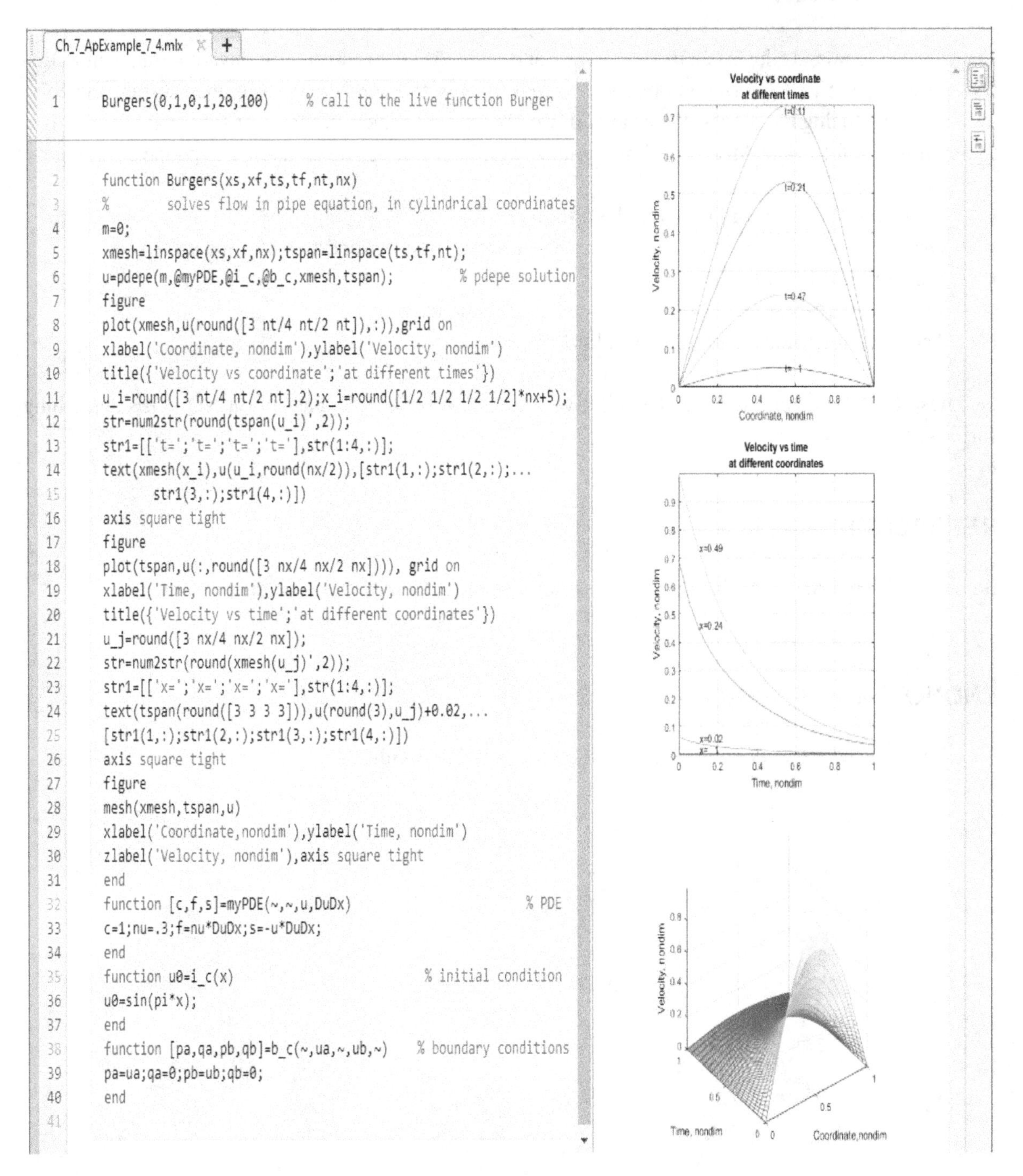

- The commands in the 28…31 lines generate a three-dimensional $u(x,t)$ graph using the mesh and formatting commands.

7.4. CONCLUSION

The pdepe command allows you to solve single or a set of the spatially one-dimensional transient PDEs. By presenting the PDE, initial and boundary conditions in the appropriate reference forms, and solving tthe PDE/s using the discussed command, we obtain a numerical solution of the actual PDE/s. It is shown how the following M&T problems containing PDEs can be solved using the pdepe command:

- Thermo-diffusion 1D PDE with a heat source,
- One-dimensional heat transfer with temperature dependent properties of a material,
- Sturt-up flow in pipe,
- The FitzHugh-Nagumo - type PDEs with a diffusion term,
- The one-dimensional Burger's PDE for fluid flow.

Thus, the pdepe command can be effectively used to solve a wide class of the 1D PDEs describing various M&T problems.

REFERENCES

Batchelor, G. K. (2012). *An Introduction to fluid dynamics*. Cambridge University Press.

ENDNOTE

1 Enter the help odeset command to review the possible options.

APPENDIX

Table 3. List of examples, problems, and applications discussed in the chapter

No	Example, Problem, or Application	Subsection
1	The solution of one-dimenssional diffusion/adhesion PDE.	7.1.4.
2	Thesolution of one-dimensional heat-diffusion PDE with a heat source.	7.2.1.,7.2.2.
3	One-dimensional heat transfer with temperature dependent properties of a material.	7.3.1.
4	Fluid mechanics: startup flow through a pipe.	7.3.2.
5	The FitzHugh-Nagumo PDEs.	7.3.3.
6	One-dimensional Burgers' equation.	7.3.4.

Note, some small examples, mostly related to non-M&T issues, are not included in the list.

Chapter 8
Solving Two-Dimensional Partial Differential Equations

ABSTRACT

This chapter describes the PDE Modeler tool, which is used to solve spatially two-dimensional partial differential equations (PDE). It begins with a description of the standard forms of PDEs and its initial and boundary conditions that the tool uses. It is shown how various PDEs and boundary conditions can be represented in standard forms. Applications to the mechanics and tribology are presented in the final part of the chapter. They illustrate the use of PDE Modeler to solve the Reynolds equation describing the hydrodynamic lubrication, to implement the mechanical stress modeler application for a plate with an elliptical hole, to solve the transient heat equation with temperature-dependent material properties, and to study vibration of a rectangular membrane.

INTRODUCTION

The pdepe command described in the preceding chapter solves one-dimensional PDEs only. To solve two-dimensional and some simplified object geometries of three-dimensional PDEs, the special PDE Modeler tool (formerly known as the PDE Tool) is designed. The modeler is part of the Partial Differential Equation Toolbox™. Further, we study two-dimensional PDEs only. The chapter described the steps that should be used to solve 2D PDEs with various boundary and initial conditions and presents solutions for some examples and applications from the M&T fields. Among them are

- Reynolds equation describing the hydrodynamic lubrication of surfaces with hemispherical pores;
- The "Mechanical Stress" PDE Modeler option applied to a thin plate with an elliptical hole;
- The transient heat equation with temperature-dependent coefficients describing the material properties;
- The wave equation adopted to the problem of vibration of a rectangular membrane.

DOI: 10.4018/978-1-7998-7078-4.ch008

8.1. ABOUT NUMERICAL METHOD USED FOR SOLUTION OF THE PDEs

The method that the PDE toolbox uses for PDE solution is called the finite element method, FEM. This method is more complicated than the finite difference method but more suitable for the real object geometry and for the inclusion of heterogeneous properties of materials. The simplified PDE solution collocation scheme used by FEM is as follows. The body shape in the x,y plane, referred to as the domain, is divided on small triangles. For each of the triangles, the trial second order polynomial solution is assumed; the polynomial coefficients determined from residuals between the trial solution and true solution, for the latter the polynomial solution of the differential equation is taken. The residuals are searched so to be zero for each of triangle while the coefficients are determined from the general set of the residual equations. Polynomials with defined coefficients are used to calculate the solution values at the triangle nodes. This is a rough description given for a first-order 1D differential equation, more detailed explanations for 1D, 2D and 3D PDEs can be obtained at the courses of numerical analysis and are beyond the scope of this book.

8.2. PDE TOOLBOX INTERFACE

The PDE toolbox provides the PDE Modeler which represents an interface for solving elliptic, parabolic, hyperbolic and eigenvalue types of scalar PDEs with Dirichlet or Neumann boundary conditions. The general standard forms of the PDE-types and boundary conditions (BC) that can be solved with this modeler, as well as examples of PDEs with its coefficients matching the standard form PDE or BC are given in Table 1.

Table 1. The standard forms of PDEs and BCs, designed to be solved using PDE Modeler

Standard Form of the PDE or BC	PDE or BC Name	Example	
		PDE and His Name	**Variables and Coefficients to Match Standard Form**
$m\frac{\partial^2 u}{\partial t_2}+d\frac{\partial u}{\partial t}-\nabla\left(c\nabla u\right)+au=f$	Elliptic equation, when m=0, d=0. Parabolic equation, when m=0. Hyperbolic equation, when d=0.	$\frac{\partial T}{\partial t}=\nabla\left(k\nabla T\right)$ +s - parabolic PDE	u=T, m=0, a=0, f=s, c=k, d=1.
$-\nabla\left(c\nabla u\right)+au=\lambda du$ or $-\nabla\left(c\nabla u\right)+au=\lambda^2 mu$	Eigenvalue PDE.	$-\nabla(\nabla u)=\lambda u$,	m=0, c=1, a=λ, d=0.
$hu=r$	Dirichlet boundary: the u value is given.	T=0	T=u, h=1, r=0.
$\vec{n}\left(c\nabla u\right)+qu=g$	Neumann boundary: given the du/dx, du/dy values with/without the u values.	$\frac{dT}{dx}=0$	T=u, c=1, q=0, g=0

In the equations presented within the table:

u is the name of a dependent variables that can be, for example, the temperature *T*, as in the table examples, or any other dependent PDE variable;

x, *y*, and *t* are the independent variables coordinates and time respectively;

$\nabla = \frac{\partial}{\partial x} + \frac{\partial}{\partial y}$ - denotes an operator termed nabla; note, $\nabla(\nabla) = \nabla^2 = \frac{\partial^2}{\partial x^2} + \frac{\partial^2}{\partial y^2} =$ is called Laplacian, in vector form Δu =div(grad u);

m, *d*, *c*, *f*, *a*, λ, *h*, *r*, *q* and *g* are the coefficients that can be constant or vary depending on *x*, *y*, *t* and *u*;

$\vec{n}$ is the outward unit normal vector used to indicate the normal direction of the d*u*/dx and d*u*/dy in respect to the boundary surface;

div - divergence, differential operator that can be defined as dif(**F**)= $\nabla \cdot \boldsymbol{F}$, where **F** is a vector field with Cartesian components $\mathbf{F}_x$, $\mathbf{F}_y$, $\mathbf{F}_z$;

grad – gradient, operator that is some equivalent to nabla.

Note, the system of equations can be also solved with the PDE Modeler tool; for this, each of the PDEs and accompanying BCs should be presented in standard form.

These notations will be further clarified when the actual PDEs are solved using the PDE Modeler.

To launch the PDE Modeler interface, the pdeModeler command should be entered in the Command Window:

```
>> pdeModeler
```

Figure 1. The PDE Modeler window.

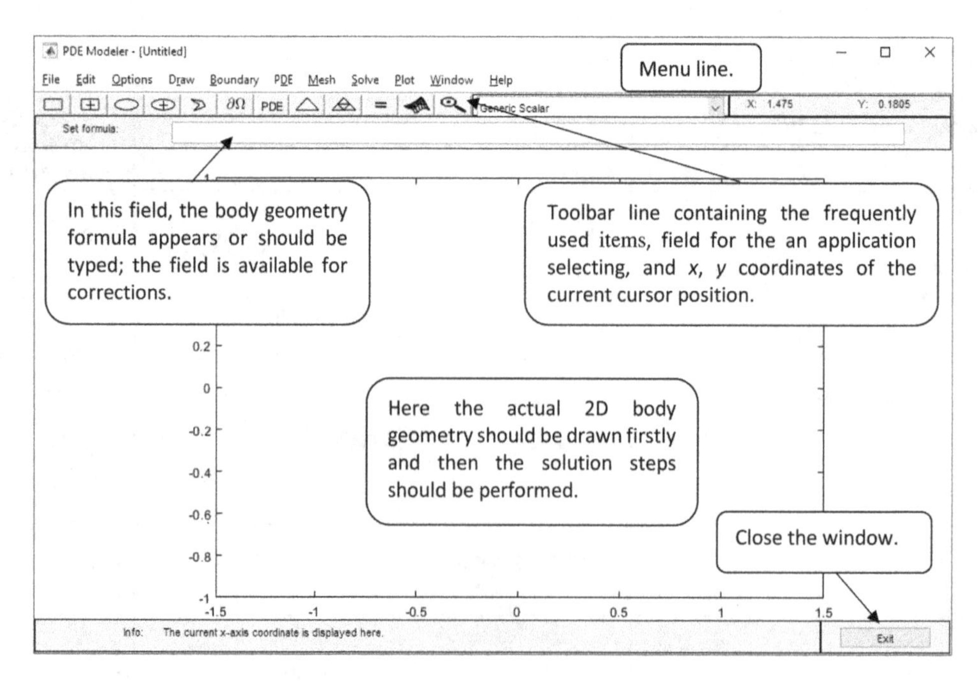

After this, the PDE Modeler window appears - Figure 1. The toolstrip of this window contains the menu, line of the buttons presenting the frequently used options, and the field line to insert the formula described the required body shape. The large empty field with *x* and *y* coordinate axes and ticks is the place where the 2D geometry of the actual body must be drawn to further PDE solution. The menu contains a line with the File, Edit, Options, Draw, Boundary, PDE, Mesh, Solve, Plot, Window, and Help options, each of which has a popup menu with list of additional options. We will describe all necessary options below along with the steps for solving the PDE.

8.2.1. Solution Steps in the PDE Modeler

To describe the steps of a solution with the PDE Modeler tool, we use the example of a laminar flow in a horizontal pipe. The equation describes the axial flow field is

$$-\Delta u = f$$

and Dirichlet BCs

$$u=0$$

at the inner wall of the pipe $Y^2+Z^2=1$.

In these equations: $\Delta = \frac{\partial^2 u}{\partial Y^2} + \frac{\partial^2 u}{\partial Z^2}$ is the Laplacian; *u* - axial fluid velocity in the *Y* and *Z* –directions ; $f = \frac{r}{\mu}\frac{dP}{dX}$ with the constant pressure gradient $\frac{dP}{dx}$ and the dynamic viscosity of the fluid μ; *Y*=y/*r*, *Z*=*z*/*r* dimensionless coordinates; *y* and *z* are dimensional Cartesian coordinates of the pipe cross-section: *r* - pipe radius. Assume μ=0.4 N·s/m^2, $\frac{dP}{dX}$=3·10^8 Pa/m, and *r*=7.5 mm, accordingly *f*=(0.0075^2/0.4)· 3·10^8=4.2188e+04 m/s. The center (0,0) of the coordinates placed at the pipe center, and *Y* and *Z* are in the range -1 … 1.

Step 1. Firstly, the applied model matching the solving problem should be chosen. This can be done by selecting the appropriate Application line in the popup menu of the of the Option button located on the main menu – Figure 2.

Figure 2. First step in PDE solution: selecting the application mode

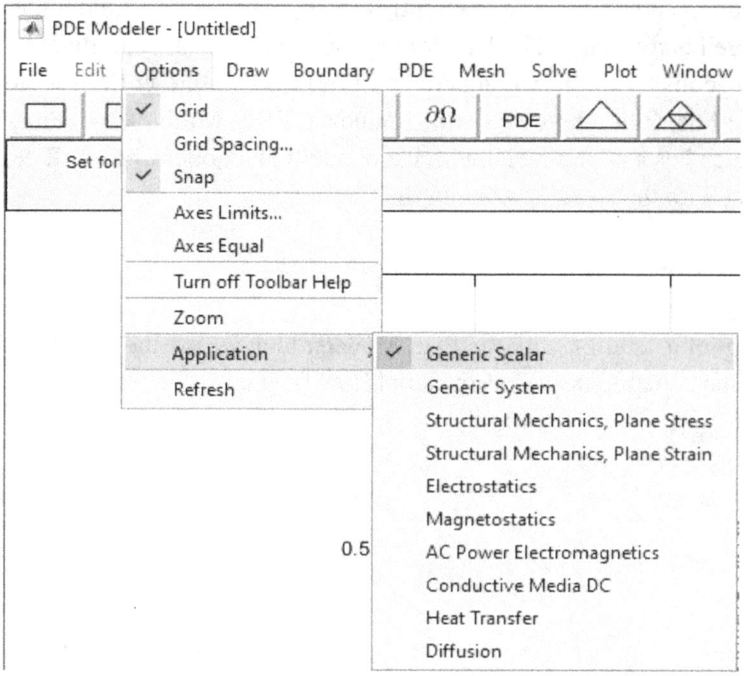

As can be seen, there is not the option connecting to the flow dynamics. In this case the Generic Scalar options (default) is chosen for the further solution.

Step 2. At this stage, the 2D body geometry should be drawn. To do this, the Grid and Snap options (see Figure 8.3) must be marked, and the limits of the axes must be settled correspondingly to the sizes of the body being studied. After selecting the Axes Limits or Grid Spacing -option, small panels appears with fields requiring the appropriate values. For our pipe cross-section dimensions, the x-axis and y-axis limits -1.25 and 1.25 were entered – Figure 8.3a. The default grid spacing was selected for x-axis and 0.5 for y-axis– Figure 8.3b. In addition the Axes equal line was clicked to

Figure 3. Axes Limits (a) and Grid Spacing (b) panels, filled in accordance with the equation being solved

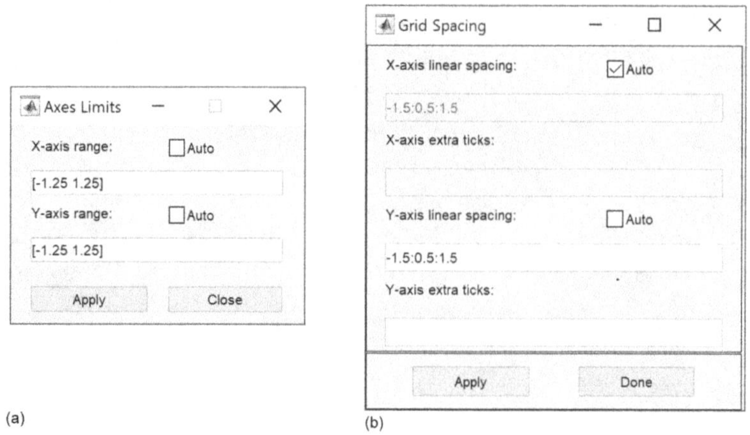

the non-deformed circle view at the drawing and other steps; choosing this option automatically changes X-axis limits to set equal axis scaling. The selected limits and grid lines appear in the graph. The general case the drawing is done by combining shapes, which can be selected from the popup menu of the Draw menu button – see Figure 4.

Figure 4. Second step of the PDE solution; selecting the shape of a body

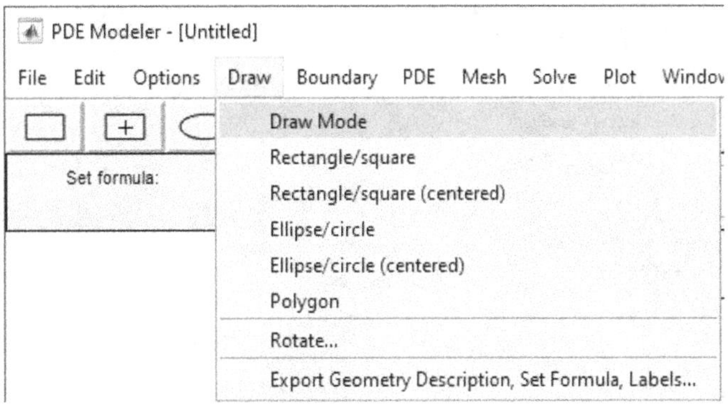

The same shapes – rectangle, square, ellipse, circle or polygon - can be selected with a click on the appropriate button of the frequently used button line. To draw the pipe cross-section, click the Ellipse/Circle(centered) button, press then the mouse + pointer at the (0,0) point and drag the mouse pointer with the pressed button to the (1,0) and then to the (0,1) grid points; after releasing the mouse button, the circle appears – Figure 5

Figure 5. The PDE Modeler window with a drawn circled pipe cross-section

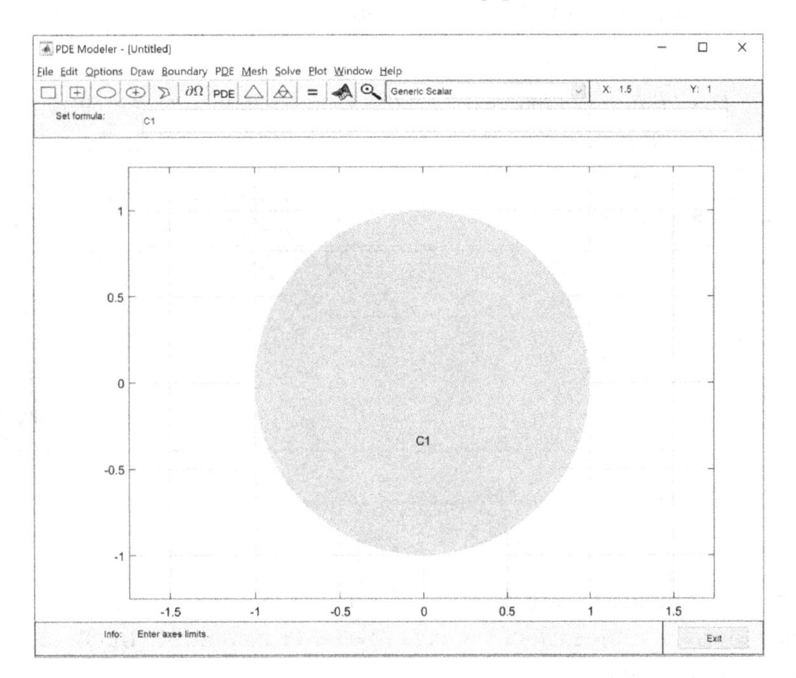

Instructions for drawing bodies with more complicated geometries using the rectangle, ellipse or polygon options are presented in subsection 8.3.4.

Step 3. BCs should be specified at this step. To do this, select the Boundary button in the main menu and click the Show Edge Labels line to see the segments of the circle where the BC should be inputted. BC can be entered for the selected circle segment after selecting the "Specify Boundary Conditions …" line of this menu – Figure 6.

Figure 6. Third step of the PDE solution: inputting the BCs.

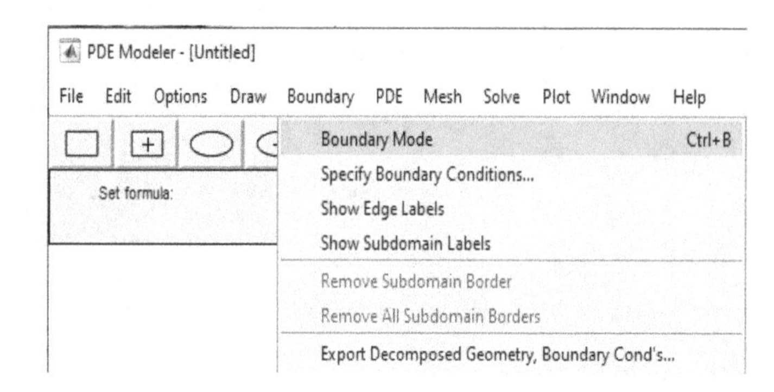

Nevertheless, the more suitable way to specify BC is to click the real boundary line/segment of the drawn body. The corresponding type of the BC should be marked with the necessary boundary equation parameters specified in the appropriate fields. In our case, the default NC is the Dirichlet condition.

For our problem, for each of four segments of the circle, the following coefficients should be entered: h=1 and r=1 – Figure 7. Note, that r here is the standard notation adopted by the PDE Modeler and not r used above to designate the pipe radius.

Figure 7. The Boundary Condition panels with the inputted Dirichlet boundaries

Boundary Condition

Boundary condition equation: h*u=r

Condition type: Neumann / Dirichlet

Coeffici	Value	Description
g	0	
q	0	
h	1	
r	0	

OK Cancel

Note, the Dirichlet boundaries are default and it is not required to input them, nonetheless it is recommended to check it for each of segment in case of complicate figure to avoid confusion as results of man actions that is realized during the solution steps.

Step 4. Now the PDE should be specified. Select the PDE Mode line of the PDE menu button and then the PDE Specification line (Figure 8) or click within the drawn shape (circle, in our case).

Figure 8. The PDE mode selection

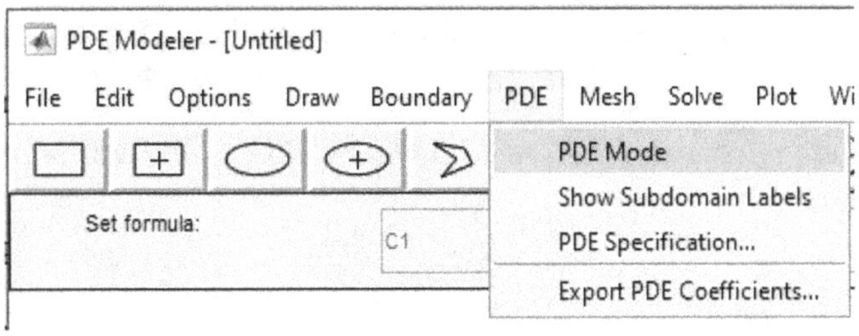

After the PDE Specification panel appears, we should mark the PDE type option required for solving the equation. Matching the solving equation and standard equation (Table 8.1) we can determine that in our case we have an elliptic equation and the Elliptic option (default) should be marked. Further, it needs to input coefficient into the appropriate fields of this panel. These coefficients can be determined by comparing our equation and the equation written at the top of the panel. The equation appearing in the panel is written in a general vector form via the divergence, *div*, and gradient, *grad*, as $-div\left(c * grad\left(u\right)\right) + au = f$; for scalar case it can be rewritten as $-\nabla\left(c\nabla\left(u\right)\right) + au = f$. Therefore, the coefficients *c*, *a* and *f* should be inputted as *c*=1, *a*=0 and *f*=4.2188e+04 (Figure 9) to match our equation.

Figure 9. The PDE Specification panel filled in for the studied example

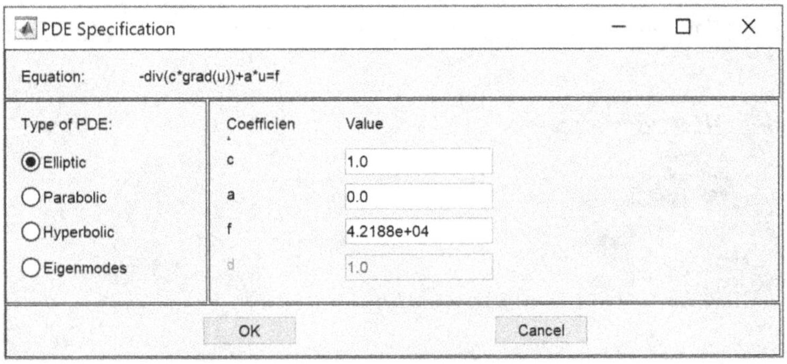

Note for Figure 9:

- the set of parameters of the PDE Specification panel varies for different PDE types; these parameters may include initial conditions (when there is not the time term in PDE).
- in case of complex body geometry, the "Show Subdomain Labels" line should be clicked to see the number of each of the body subdomains and to specify the PDE for each of them.

- in the *c*, *a*, *f*, or other (depending on the PDE type) the panel fields, expressions can be written as via coordinates *x*, y and time *t* (these notations are mandatory), for example, if $c = 0.5\sqrt{y^2 + z^2}$ in the c-field we should write 0.5*sqrt(x.^2+y.^2) .

Step 5. At this stage, a triangular mesh should be generated to obtain a solution in their nodes. To achieve this, the Mesh Mode line in the popup menu of the Mesh button should be selected. To see the node numbering, the Show Node Labels line should be also marked – Figure 10.

Figure 10. The Mesh Mode selection; the Show Node Labels line is marked; triangle nodes is numbered as for the studied example

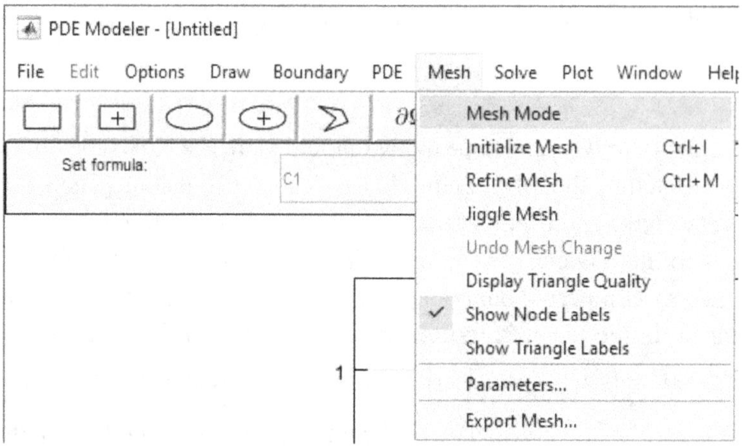

The mesh can be refined by selecting the Refine Mesh or Jiggle Mesh line and returned to the starting view by selecting the Initialize Mesh line. In our case we click twice the Refine Mesh line to better circle matching and the higher result accuracy.

Step 6. Now the Solve PDE line of the popup menu of the PDE button should be selected – Figure 11.

Figure 11. The Solve PDE mode selection

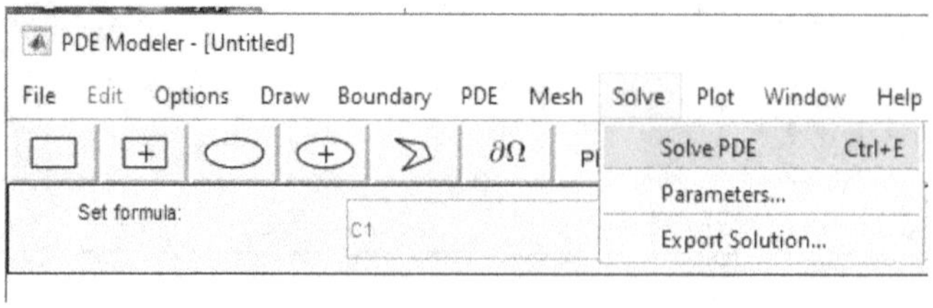

When non-default solver parameters are required, the Parameters option can be selected. In such case, the Solve Parameters panel is opened. The default view of this window for a parabolic PDE is presented in Figure 12.

Figure 12. The Solve Parameters specification panel; view for an elliptic equation

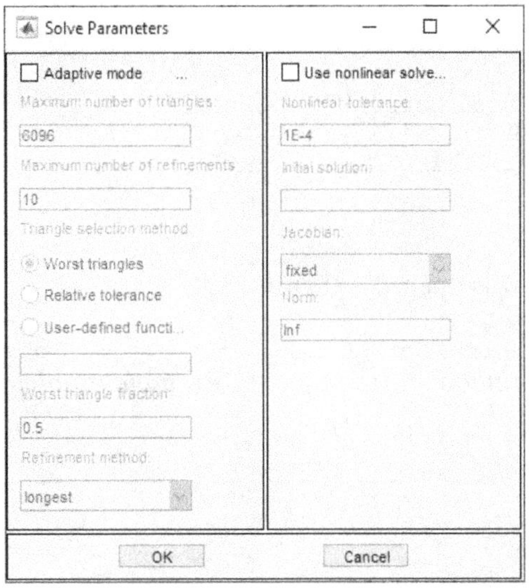

Note, in case of parabolic, hyperbolic and eigenvalue equations has another view adopted to the actual PDE type. For example, for an elliptic equation with a non-default initial condition (u(t0)=0, default), it should be typed in the u(t0) field as a string line - in single quotes, e.g. 'sqrt(x.^2+y.^2)'. For the discussed example, it is not necessary to introduce any changes in the Solve Parameters panel. Therefore, after selecting the Solve PDE mode, the resulting graphical solution presenting the flow velocities appears automatically. For our example it looks as per Figure 13.

Figure 13. View of the PDE solution in the PDE Modeler

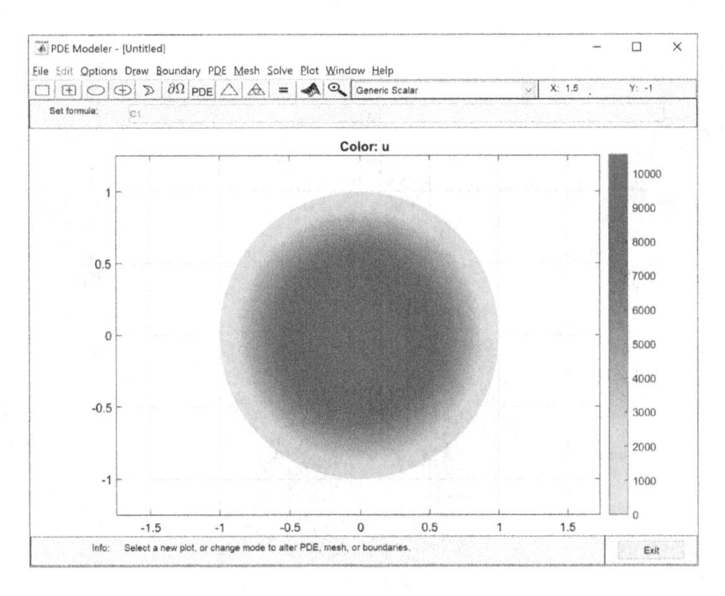

Step 7. To provide some graphical changes or generate 3D plot the Parameters line in the popup menu of the Plot button should be selected – Figure 14

Figure 14. Plot mode, the Parameters line selection

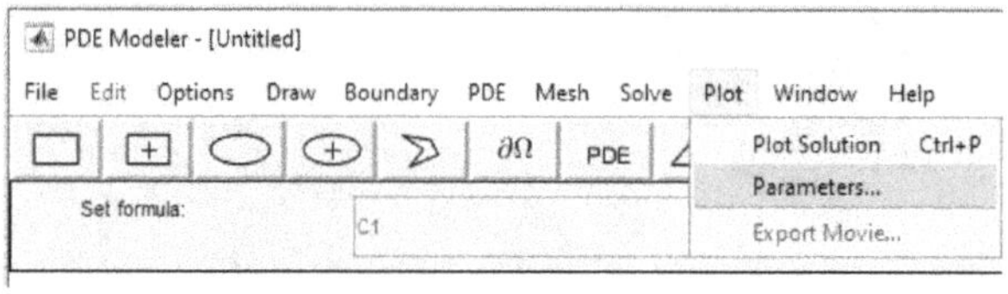

The Plot Selection window is opened – Figure 15

Figure 15. The Plot Selection window; filled in for the example problem

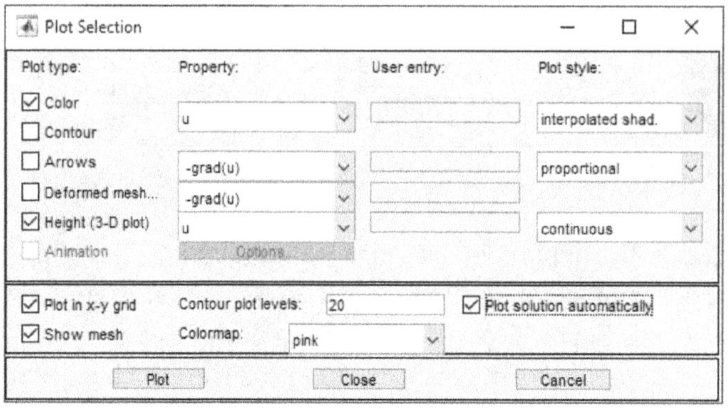

To add contour lines (iso-velocity lines, in our case) to the early produced plot the Contour box should be marked; results shown in the Figure 16.

Figure 16. Resulted 2D plot with the contour line added with the Plot mode "Parameters ..." option

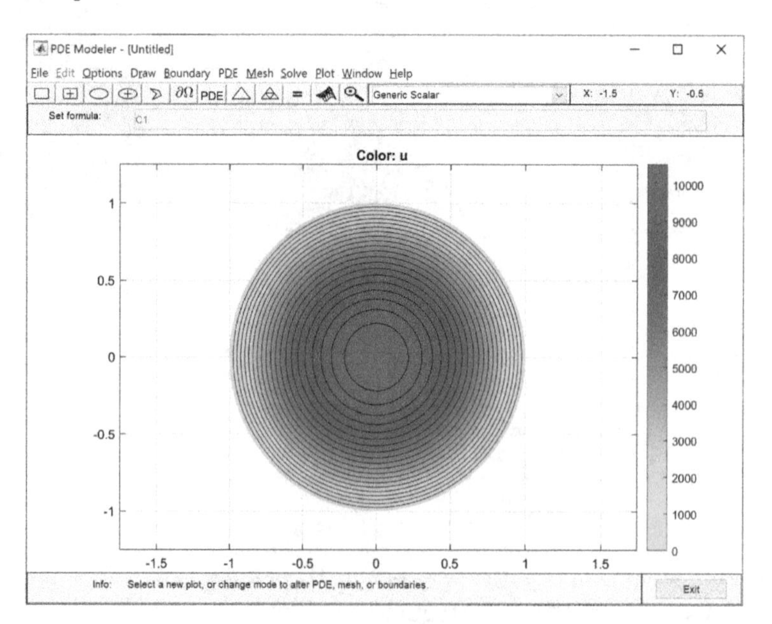

By default, 20 iso-lines are generated. To change this number for example to the five iso-lines, type the digit 5 in the 'Contour plot levels' field.

To generate a 3D graph, the Height (3D) box should be marked. The rectangular mesh can be shown by marking the "Show mesh plot" and "Plot in x-y grid" boxes. If it is desirable to change the look-up colors, the needed map colors can be selected with the "Colormap:" popup menu.

Finally, after selecting all the desired parameters of the graph (as shown in the Figure 15), click the Plot button of the Plot Selection window. The resulting plot is presented in Figure 17.

Figure 17. The 3D plot with solution representation by the PDE Modeler

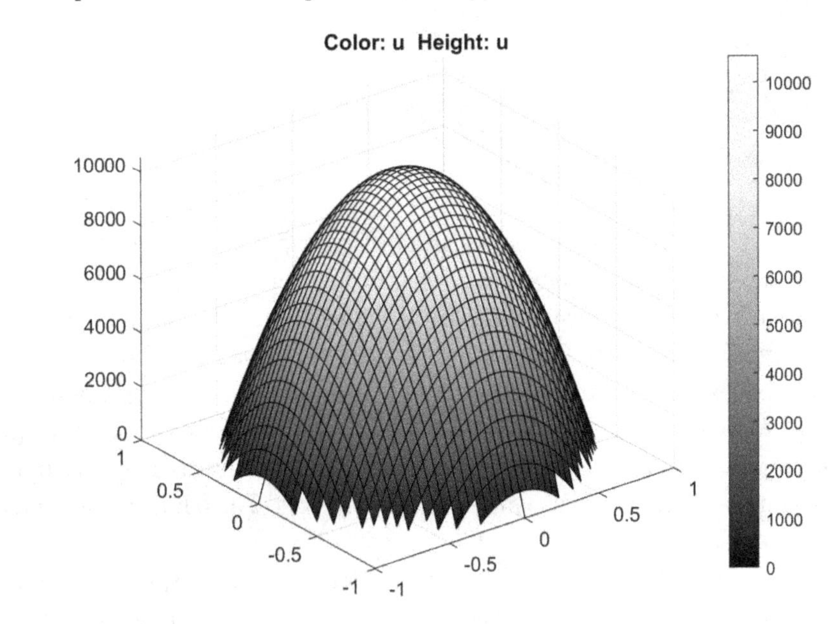

Step 8 (optional). The PDE Modeler generates a program with commands, which carry out all of the steps described above. To save this program, select the 'Save As' line of the popup File options of the main menu and type the desired file name in the File name field of the 'Save As' panel that appears.

The PDE Modeler generates the file with the commands that carry out all of the PDE solution steps, and it can be opened and run for any future calculations. To save the program, select the Save As line of the popup menu of the main menu File option and type the name ExPipeFlow in the File name field of the panel that appears.

8.2.2. Export Solution and Mesh Values to the MATLAB® Workspace

The PDE Modeler presents the results of the PDE solution in graphical form with a color bar showing data with the color scale. Nevertheless, the numerical values are frequently required in technical calculations. This can be achieved by exporting the defined u-values and coordinates of the mesh nodes at which these values were obtained to the MATLAB® workspace. To export the solution, select the Export Solution line in the popup menu of the Solution main menu button. In the small panel, that appears (Figure 8.18a), choose to change/not change the default variable name and click the OK button. In the Workspace MATLAB® window, the variable with the selecting name appears. *u* is the matrix in general case with columns corresponding to the current time each, but in stationary cases (the studied example), *u* is the one-column vector containing the 2097 obtained *u*-values.

Figure 18. Panels intended to export the PDE Modeler solution in the MATLAB® workspace; a) export obtained u-values, b) export x,y-node coordinates (p-matrix) and some other mesh parameters (e and t-matrices)

To export the mesh data, select the Export Mesh line in the popup menu of the Mesh main menu button. In the small panel that appears (Figure 16b), change/not change the default mesh variable names and click the OK button. In the Workspace MATLAB® window, three variables with the selecting names appear. In a two-row matrix *p* (points). the first row is the *x*-coordinates of the mesh points while the second row – y-coordinates; the six-row matrix e (edges) contains ending point indices, starting and ending parameter values, edge segment and subdomain numbers; the four-row matrix *t* (triangle) contains the indices of the corner points and the subdomain numbers. For practical needs, only the *p* data is important but some commands may also use the *e* and *t* parameters. Thus, it is recommended to transfer all parameters to the workspace.

After exporting solution and node coordinates, their values can be displayed. In our example, to display in the shortg format the coordinates *y* and *z* (correspond to the *x* and *y* coordinates in the PDE Modeler notations) together with u-values at the 1, 201, 401, ..., 2001 points and at the final point, the following commands can be entered in the Command Window:

```
>>pos=[1:200:2001 length(u)];format shortg
>>disp('           y           z              u'),disp([p(:,pos)' u(pos)])
          y            z          u
         -1  -1.2246e-16          0
    0.65376     -0.50638     3332.1
    0.22928      0.12644     9822.6
   -0.67156      0.74095          0
   0.073497     -0.65106     6018.7
   -0.92568     -0.24895     855.39
    0.61163      0.49238     4043.3
   -0.19285    -0.041658      10136
  -0.010092      0.36331       9153
     0.2487      0.25474     9207.8
    0.77148     -0.14849     4036.5
    0.44292      0.55504       5225
```

8.2.3. Conversion of the Solution From the Triangular to Rectangular Grid

Exported u-values are obtained for the node coordinates of a triangular mesh but, practically, the rectangular grid coordinates is of the greatest engineering interest. To transform the data from the triangle mesh nodes to rectangular grid nodes, the tri2grid command can be used. The simplest form of the command is

```
U= tri2grid(p,t,u,x,y)
```

For our example:

```
U =
      NaN        NaN          NaN       -6.1132e-15     NaN          NaN           NaN
      NaN        1165.9       4680.1        5854.8      4681.5       1167.3        NaN
      NaN        4681.1       8196.8        9369.9      8196.3       4684.2        NaN
-8.7197e-15      5856.1       9373.3        10538       9371.6       5852.5          0
      NaN        4681.9       8198          9368.1      8196.3       4682.1        NaN
      NaN        1163.8       4681.2        5852.6      4679.5       1164.2        NaN
      NaN           NaN       NaN       -3.1522e-14     NaN          NaN           NaN
```

Note, in the case of a circle or more complicated part geometry that includes non-rectangular shapes, cavities, scale-different elements, etc., the node points of a rectangular grid may appear outside the body, in this case these points are designated as NaN (not a number) in the outputted u-values, as presented above.

8.3. DRAWING TWO-DIMENSIONAL OBJECTS

The geometry of the real object should be drawn in the first steps of the PDE solution. For this purpose, the PDE Modeler provides some basic shapes and appropriate buttons on the toolbar line or in the Draw menu. These shapes are an ellipse/circle, rectangle/square and polygon. In addition, these shapes can be drawn using MATLAB® commands by entering them in the Command Window. Button icons, their purposes and example shapes with formula fields, and also the alternative command forms and its examples for the given shapes are presented in the Table 2.

Shapes of more complicated geometry can be drawn with a combination of the shapes presented in Table 3 with additional correction (if necessary) of the expression appearing in the Set formula field of the toolbar line. Some examples of drawing various complicated objects are presented in Table 3.

Table 2. Draw buttons and alternative commands with examples[1]

Icon, button name, and description	Example of the button usage	Alternatively	
		Command form	Example
Rectangle or Square button. Draws a rectangle or square starting at a corner and dragging the + (mouse pointer) to the diagonal corner; the + shown at the end point.	Rectangle:	pderect([xmin xmax ymin ymax]) where xmin, xmax, ymin, and ymax are the coordinates of the diagonal corners.	>>pderect([-.5 .5 -.5 2.5])
Rectangle or square (centered) button. Draws a rectangle or square starting at the center.	Square:	The same command as in previous case.	>>pderect([-0.5 0.5 -0.5 0.5])
Ellipse or circle button Draws an ellipse or circle starting at the perimeter. The + (mouse pointer) are shown in the example at the end point.	Ellipse:	pdeellip(xc,yc,a, b, phi) where xc, yc, are the ellipse center coordinates, a and b are the ellipse semi-axes, phi – rotation angle in radian.	>> pdeellip(0.25,0,1,0.5,0)

continued on following page

Table 2. Continued

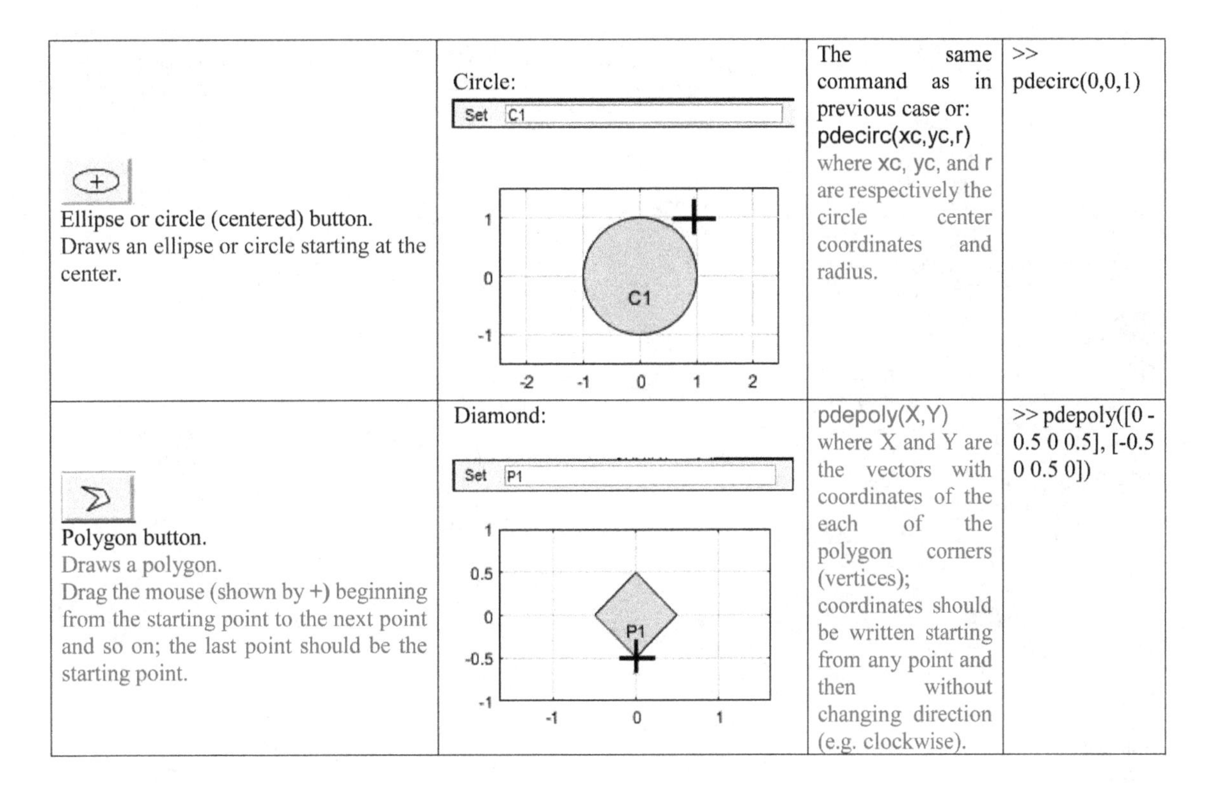

Ellipse or circle (centered) button. Draws an ellipse or circle starting at the center.	Circle:	The same command as in previous case or: pdecirc(xc,yc,r) where xc, yc, and r are respectively the circle center coordinates and radius.	>> pdecirc(0,0,1)
Polygon button. Draws a polygon. Drag the mouse (shown by +) beginning from the starting point to the next point and so on; the last point should be the starting point.	Diamond:	pdepoly(X,Y) where X and Y are the vectors with coordinates of the each of the polygon corners (vertices); coordinates should be written starting from any point and then without changing direction (e.g. clockwise).	>> pdepoly([0 -0.5 0 0.5], [-0.5 0 0.5 0])

Table 3. Examples of some complicated geometric shapes drawing using the PDE Modeler[2]

What to draw	Steps to draw with the PDE Modler tool	Figure and Set formula, view in Draw mode	Draw with commands
A shape with two sub-domains: rectangle 0.3x4 and circle with r=0.5 with center at the (0,-1) point.	• Mark the Grid and Snap options; use the default parameters for the axes limits and grid spacing. • Click the Rectangle (centered) button. Place the mouse pointer on point (-0.15,-2) and drag to (0.3, 2). • Click the Circle (centered) button. Place the mouse pointer on point (0,-1) and drag to (0,-0.5) with the right mouse button pressed. • Check the expression in the Set Formula field - it should be R1+C1.	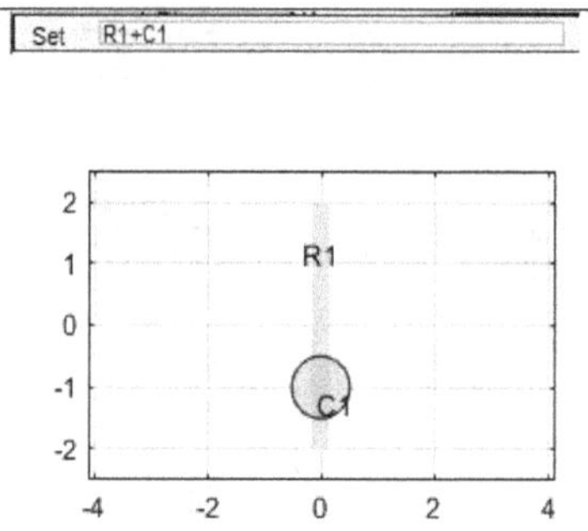 Set R1+C1	`>>pderect([-.15 .15 -2 2]) >>pdecirc(0,-1,.5)`
A 1x2 rectangular plate centered at (0,0) and with a radius of 0.2, rounded corners.	• Mark the Grid and Snap options and adjust the Axes Limits, and Grid spacing as follows: **x**-axis to `-1.5:0.1:1.5` and **y**-axis to `-1.5:0.1:1.5`. • Click the Rectangle button. Place the mouse pointer on point (–1,-0.5) and drag to (1,0.5). • Draw four circles with the radius 0.2 and the centers at (–0.8,–0.3), (–0.8,0.3), (0.8,–0.3), and (0.8,0.3) – with the ellipse (circled) button each. • Add four squares with the side 0.2, one in each corner. • Enter the following expression R1-(SQ1+SQ2+SQ3+SQ4) +C1+C2+C3+C4 in the Set Formula field. • Select Boundary mode line in the Boundary menu option. • Select the 'Remove All Subdomain Borders' line.	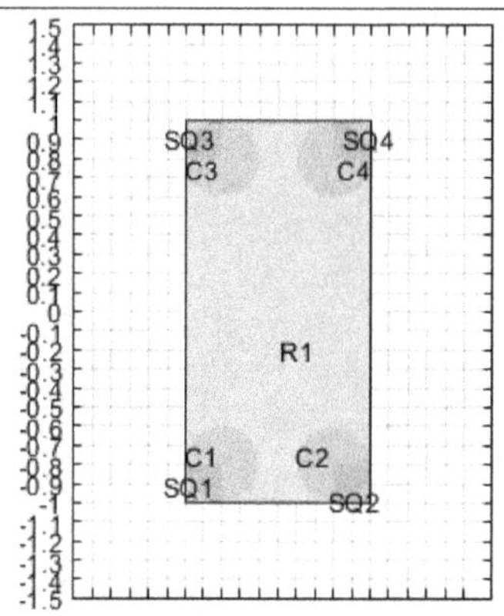 Set formula: R1-(SQ1+SQ2+SQ3+SQ4)+C1+C2+C3+C4 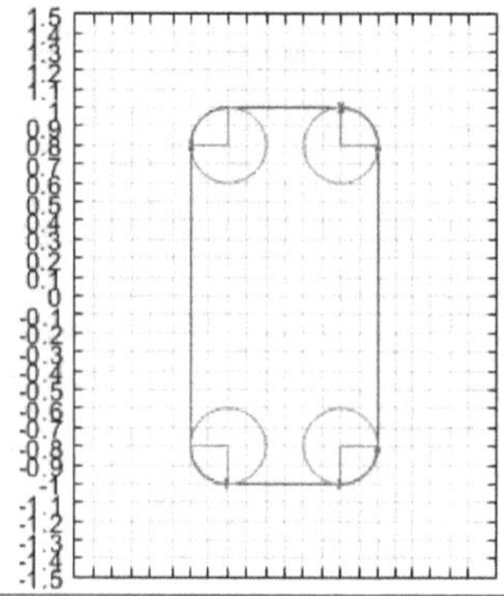	`>>pderect([-.5 .5 -1 1])` `>>pdecirc(-.3,-.8,.2)` `>>pdecirc(.3,-.8,.2)` `>>pdecirc(-.3,.8,.2)` `>>pdecirc(.3,.8,.2)` `>>pderect([-.5 -.3 -1 -.8])` `>>pderect([.3 .5 -1 -.8])` `>>pderect([-.5 -.3 .8 1])` `>>pderect([.3 .5 .8 1])` After this enter R1-(SQ1+SQ2+SQ3+SQ4) +C1+C2+C3+C4 in the Set formula field.

continued on following page

Table 3. Continued

		The two first plots show the plate view before and after activating the Boundary Mode. The lower plot represents the plate with rounded corners after marking the Remove All Subdomain Borders Note, in the PDE Modeler window this body scale/grid is some different.	

A centered squared 2x2 plate with centered rectangular 0.2x1.2 cavity	• Mark the Grid and Snap options and adjust the Axes Limits, and Grid spacing as follows: each **of the x-** and y-axis to -`1.5…1.5` with spacing 0.1
	• Click the Rectangle (centered) button. Place the mouse pointer to point (–0,0) and drag to (1,0.5).
	• Click the Rectangle (centered) button again. Place the mouse pointer on point (–0,0) and drag to (0.6,0.1).
	• Enter the following expression in the Set Formula field: SQ1-R1
	• Select the Boundary mode line in the Boundary menu.
		Set SQ1-R1
		The plot is a view of a plate with a cavity after activating the Boundary Mode.
			`>>pderect([-1 1 -1 1])`
			`>>pderect([-.1 .1 -.6 .6])`
			After this enter SQ1-R1 in the Set formula field.

L-shaped plate with a rounded corner.
The sizes as per the second column.
	• Mark the Grid and Snap options and adjust the Axis limits and Grid spacing as follows:
	x-axis to
	`-1.5:0.1:1.5`
	and ***y***-axis to
	`-1:0.1:1.`
	• Click on the polygon button and draw the polygon with the following (*x*,*y*) corner coordinates: (-1,-1), (-1,1), (0,1), (0,0), (1,0), (1.-1), and (-1,-1) to close this shape.
		C1
		R1
		P1
			`>>pdepoly([-1 -1 0 0 1 1],[-1 1 1 0 0 -1])`
			`>>pdecirc(.4,.4,.4)`
			`>>pderect([0 .4 0 .4])`
			After this enter R1- P1+R1-C1
			In the Set formula field.

continued on following page

Table 3. Continued

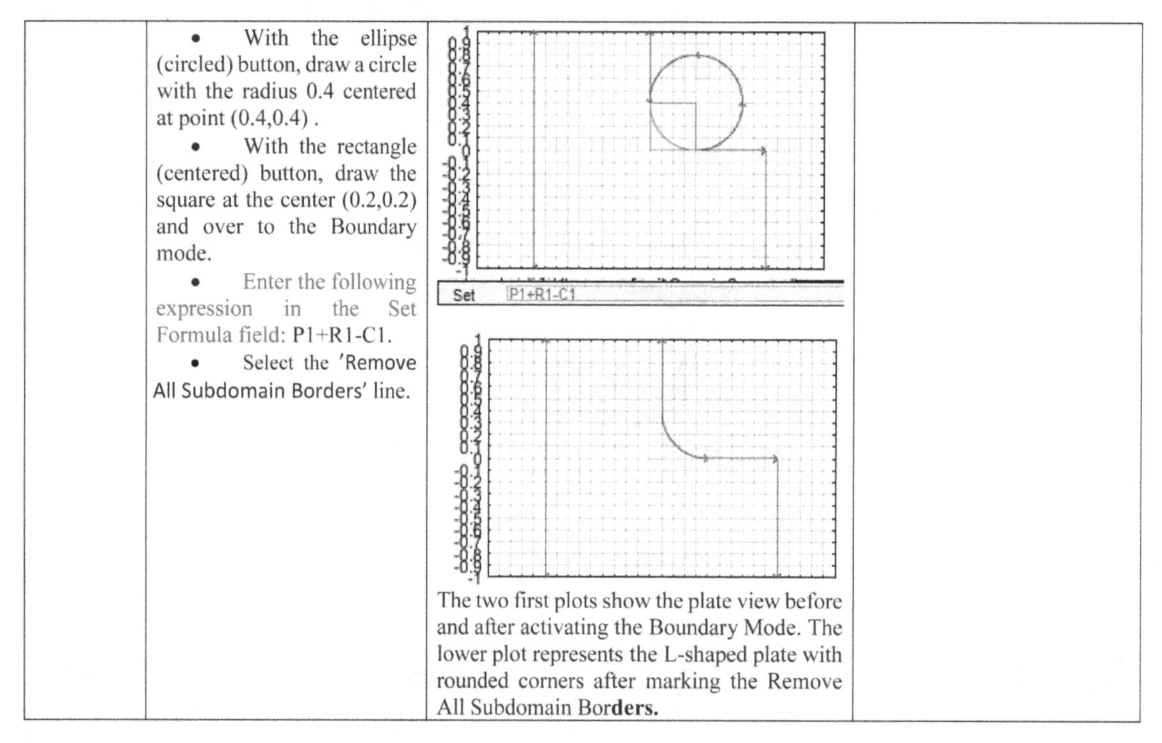

	• With the ellipse (circled) button, draw a circle with the radius 0.4 centered at point (0.4,0.4) . • With the rectangle (centered) button, draw the square at the center (0.2,0.2) and over to the Boundary mode. • Enter the following expression in the Set Formula field: P1+R1-C1. • Select the 'Remove All Subdomain Borders' line.	Set P1+R1-C1 The two first plots show the plate view before and after activating the Boundary Mode. The lower plot represents the L-shaped plate with rounded corners after marking the Remove All Subdomain Bor**ders.**	

Note:

• To easy draw a square or circle using the Rectangle (centered) or Ellipse (centered) buttons, respectively, move the mouse pointer to the center point and drag the mouse keeping the right button down to any desired square point or the desired circle radius point;

• To check/correct the drown square or circle, place the mouse pointer within the figure and click, in appeared window you may see and correct the required coordinates of the figure; in case of a polygon, the same action open window showing coordinates of the vertices.

8.4. APPLICATION EXAMPLES

8.4.1. 2D Pressure Distribution in a Lubricating Film Between Two Surfaces When One of Them Is Covered by Hemispherical Pores

The two-dimensional Reynolds equation (see Subsection 6.4.2.2) describes the hydrodynamic behavior of a thin lubricating film between two surfaces, when one of the surfaces is covered with hemispherical pores and is stationary, and the second is smoothed and moving. The equation has the form (Etsion & Burstein, 1995)

$$\frac{\partial}{\partial X^2}\left(H^3\frac{\partial P}{\partial X}\right)+\frac{\partial}{\partial Z^2}\left(H^3\frac{\partial P}{\partial Z}\right)=\frac{dH}{dX}$$

where P is hydrodynamic pressure arising in the film, X – coordinate, H – gap between the surfaces; all variables in this equation are dimensionless.

The gap $H(X)$ and its derivative $\frac{dH(X)}{dX}$ for a hemispherical pore and adjacent surface parts (called pore cell) are given with the following equations

$$H = \begin{cases} 1+\psi\sqrt{1-(X^2+Z^2)}, (X^2+Z^2)<1 \\ 1, otherwise \end{cases}$$

$$\frac{dH}{dX} = \begin{cases} -\frac{\psi X}{\sqrt{1-(X^2+Z^2)}}, (X^2+Z^2)<1 \\ 0, otherwise \end{cases}$$

where ψ is the pore radius to the gap ratio.

The hydrodynamic pressures at boundaries of a rectangular pore cell, X€[-ξ,ξ] and Z€[-ξ,ξ], are assumed to be zeros; ξ is the ratio of the cell dimension to the pore radius. In other words the boundary conditions are:

P=0 at X, Z=+/-ξ

Problem: Using the PDE Modeler, solve the Reynolds equation for a rectangular pore cell X€[-ξ,ξ] and Z€[-ξ,ξ] with Dirichlet (default of the PDE Modeler) boundaries: P=0 along four pore cell boundaries. Assume ξ =2 and ψ =8. Represent the pressure distribution $P(X,Z)$ in a 3D plot. Transfer defined P-values and mesh parameters to the workspace and display the P(X,Z) table for the orthogonal grid with seven equally spaced X and Z, each in the range $-\xi \ldots \xi$. Find the maximal pressure value. Save the automatically created program in a file named Ch_8_ApExample_8_1.

Determine firstly the type of the solving equation by matching this equation with the required standard form (Table 1); it is easy see that the Reynolds equation is an elliptic type of the PDE as in the standard form reads

$$\nabla\left(H^3\nabla P\right) = \frac{dH}{dX}$$

The equation is identical to the standard PDE form when u=P, Z=Y, m=0, d=0, c=H^3, a=0, and f= $\frac{dH}{dX}$. Further, we must remember that y and u of the PDE Modeler notations are coordinate Z and pressure P of the Reynolds equation, respectively.

Open now the PDE Modeler window with the command

```
>>pdeModeler
```

Mark the Grid and Snap lines in the popup menu of the Options button of the modeler menu, and type the x and y limits as [-2.5 2.5] and [-2.5 2.5], respectively (after selecting the Axis Limits line). Use the general Application option - Generic Scalar.

Then, go to the Draw Mode and draw a rectangle with the ▭ rectangle button. Now place the mouse arrow on point (-2,-2) and drag to point (2,2) of the plot. Check this rectangle geometry with the Object Dialog panel that appears after clicking within the rectangle. Draw now a circle clicking for this the ⊕ Ellipse/circle (centered) button. Place the mouse at the (0,0) point and, keeping the right mouth button, drag the mouse diagonally to the point (1,1), Check the circle geometry parameters with the Object Dialog panel that appears after clicking within the circle.

Click on the Boundary button and, in the appeared "Boundary Conditions" panel, select the Dirichlet option and enter 1 in the h field and 0 in the r field; do this for each side of the rectangle (these actions are optional, since the Dirichlet boundaries and the corresponding h and r values are set by default).

Then select the PDE Mode from the popup menu of the PDE Modeler main menu button. Studied body has two domains – a rectangle and a circle inside it, their numbers can be shown when the 'Show domain labels' options of the popup menu is marked. Place the mouse pointer on the circular domain and press the mouth button. In the appeared PDE Specification panel check the Elliptic type of PDE (default) and type:

(1+8*sqrt(1-x.^2-y.^2)).^3 in the *c* field;

0 in the *a* field;

and -8*x./sqrt(1-x.^2-y.^2) in the *f* field.

After that, move the mouse pointer outside the circular domain and open the PDE Specification panel for the rectangular domain; check/mark the Elliptic type of PDE and enter 1, 0, and 0 into the fields c, a, and f respectively.

Initialize the triangle mesh in Mesh Mode (line in the popup menu of the main menu Mesh button) and refine mesh two times.

Select the Solve PDE line in the popup menu of the Solve button of the main menu. The 2D solution with colored bar appears as below.

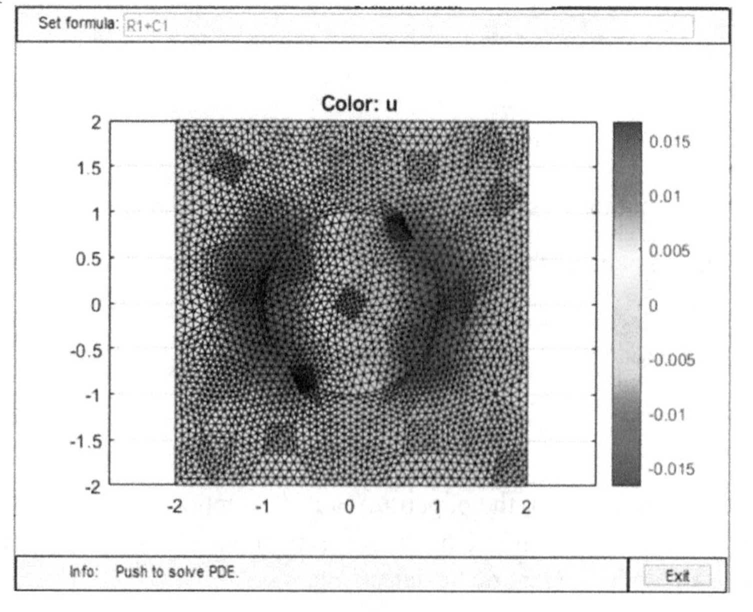

To generate this 2D graph, the color map selected as “jet” in the appropriate field of the “Parameter …” panel, which appears after selecting the appropriate line from the popup menu of the “Plot” button of the main menu. After marking the “Height(3D plot) “ and “Show mesh“ boxes on this panel, the following 3D plot will appear with the solution:

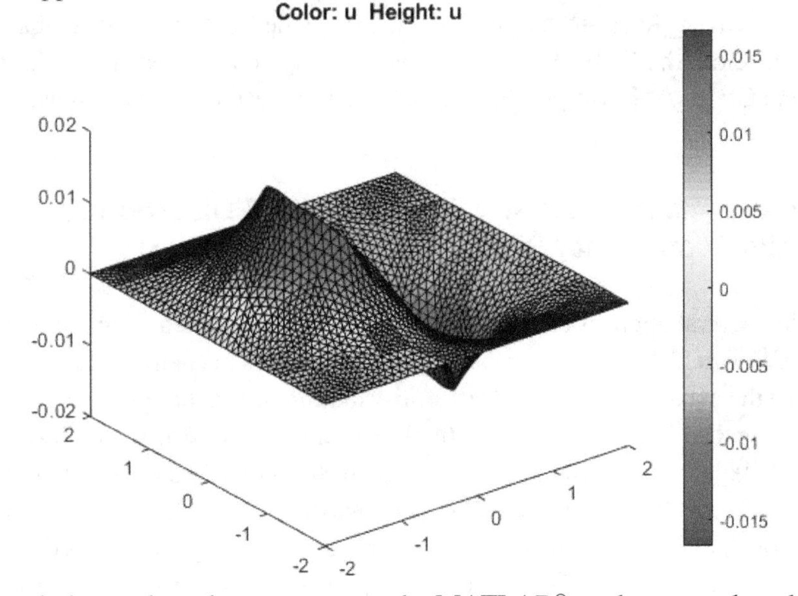

To transfer the solution and mesh parameters to the MATLAB® workspace, select the lines Export Mesh and Export Solution in the popup menus of the corresponding Solve and Mesh buttons of the main menu. If you have not changed the variable names, then the solution is in the *u* matrix, and the mesh parameters are in the *p*, *e* and *t* matrices. To obtain now the *u*-values at the required orthogonal grid points, the following commands should be entered in the Command Window to display the resulting table.

```
>> x=linspace(-2,2,7);y=x;
>> T=tri2grid(p,t,u(:,end),x,y)
T =
   0        0        0         0         0         0     0
   0   0.0876   0.1160    0.0013   -0.1141   -0.0871     0
   0   0.2180   0.3861    0.0012   -0.3827   -0.2179     0
   0   0.3086   0.4117   -0.0000   -0.4117   -0.3089     0
   0   0.2177   0.3827   -0.0012   -0.3861   -0.2176     0
   0   0.0871   0.1141   -0.0013   -0.1159   -0.0878     0
   0        0        0         0         0         0     0
```

To find for example the maximal pressure, type in Command Window

```
>> max(u)
ans =
    0.0167
```

This value can be compared with the maximum pressure value of 0.0198 obtained in the one-dimensional case (subsection 6.4.2.2). As it can be seen, the discrepancy is close to 19%.

The PDE Modeler generates the program with commands that perform all steps of the solution. To save the program in file, select the Save As line in the popup menu of the File button of the main menu and type the name Ch_8_ApExample_8_1 in the File name field of the panel that appears. To check the generated file, close the PDE Modeler window, then open the saved file in the MATLAB® Editor window and type Ch_8_ApExample_8_1 in the function definition line instead of the default name located there.

8.4.2. "Structural Mechanics" Application of the PDE Modeler: Stresses in the Rectangular Plate With Elliptic Hole

PDE Modeler has several applications that are adapted the PDE nomenclature and terminology used in the relevant scientific field. Here we describe how the "Structural mechanics, plain stress" application can be used to define the maximum stress and stress distribution in a rectangular plate with an elliptic hole.

Problem: Using the PDE Modeler "Structural mechanics, plain stress" application, define maximal stress value for a 14x8 (unit) rectangle with elliptical hole of width $2a$ and height $2b$, when a=2 and b=1 (unit). The vertical rectangle boundaries and boundaries of the hole are stress-free. Stresses σ act on the two horizontal boundaries and are equal 1 and -1 for positive and negative y-axis directions respectively. Assume Young's modulus E is 200Ʊ10^3, and once we use the "Refine mesh" option. The problem is static that means the elliptic PDE is solved. Represent the final y-stress values in the 2D plots, which should show the contour lines and color bar; use the "jet" color map. Save the automatically created program in the file named Ch_8_ApExample_8_2.

The "Structural mechanics, plain stress" application is designed for thin plates and in such a way that the user does not need to formulate the PDE and then match it with the standard form, it is enough to assign the type of the PDE: this problem is elliptical.

The application Structural Mechanics, Simple Stress is designed for thin plates and so that the user does not need to formulate a PDE and compare it with a standard form, it is enough to assign the type of PDE, this problem is an elliptical PDE.

There are two types of boundary conditions in this PDE Modeler application — displacements and surface stresses, which are classified as Dirichlet and Neumann, respectively. In the studied problem, all boundary conditions are Neumann conditions. In the general case, two stress components can act on boundary segment – in the x and in y directions, denoted by g_1 and g_2. Each of these components typically includes normal and shear stresses. In our problem, there are stresses acting only in y-directions, therefore g1=0 and g2=±1.

To solve the problem, open the PDE Modeler window with the command

```
>> pdeModeler
```

In the popup menu of the Options menu button, mark the Grid and Snap lines and type the x and y limits as [-0.5 14.5] and [-0.5 8.5] respectively (after selecting the Axis Limits line of the Options menu). Set the step to 1 after selecting the Grid spacing line of the current menu and then select the Application line and check the "Structural Mechanics, plain stress" option in the appearing popup menu.

Draw now the studied body, go for this to the Draw Mode and draw a rectangle with the rectangle button by pressing the mouse pointer at point (0,0) and dragging it to the (14,8) point. Check this rectangle geometry with the Object Dialog panel that appears after clicking within the rectangle. Draw now an ellipse clicking for this the Ellipse/circle (centered) button. Place the mouse and press its button at the (7,4) point and, keeping the right mouth button, drag the mouse diagonally to the point (6,8), Check the ellipse center and geometry parameters with the Object Dialog panel that appears after clicking within the circle, and if necessary correct the ellipse width and height typing 2 and 1 in the A- and B-semiaxes fields respectively. In the Set formula field change the expression R1+E1 to the R1-E1 to "cut" the elliptic hole.

Click on the Boundary button of the main menu and select the Boundary mode line. Place the mouse pointer on the left rectangle edge and click the left mouse button, in the appeared Boundary Conditions panel, select the Neumann option and check that the fields g1 and g2 contain zeros; do this for the opposite side of the rectangle and for the four boundaries of the elliptical hole. Click on the top rectangle edge of the rectangle and in the appeared Boundary Condition panel type 1 in the g2 field; the g1 field should contain 0 in our problem that coincides with the default value of this field. Perform the same actions for the bottom rectangle edge, but type now -1 in the g2 field.

Now, select the PDE Mode from the popup menu of the PDE main menu option, select the PDE Specification and mark the Elliptic box on the appeared PDE Specification panel. Enter the E value in the corresponding field and accept the default Poisson ratio (nu=0.3) and the default mass density (rho=1, not used in the static problem).

Initialize the triangle mesh in the Mesh Mode (by clicking the corresponding line in the popup menu of the main menu Mesh button) and then refine the mesh by selecting the Refine Mesh line.

Now select the Solve PDE line in the popup menu of the main menu Solve button. After this the PDE Modeler performs the solution.

To display the y-stress results in the best illustrated form, open the Plot Selection panel and select the "Parameters …" line of the Plot button of the main menu. Select y-stress line from the popup menu of the last box in the Property column; mark the Contour box and select the "jet" color from the popup menu of the Colormap box. Click then the Plot button on the current panel. Displayed result are shown in the following figure:

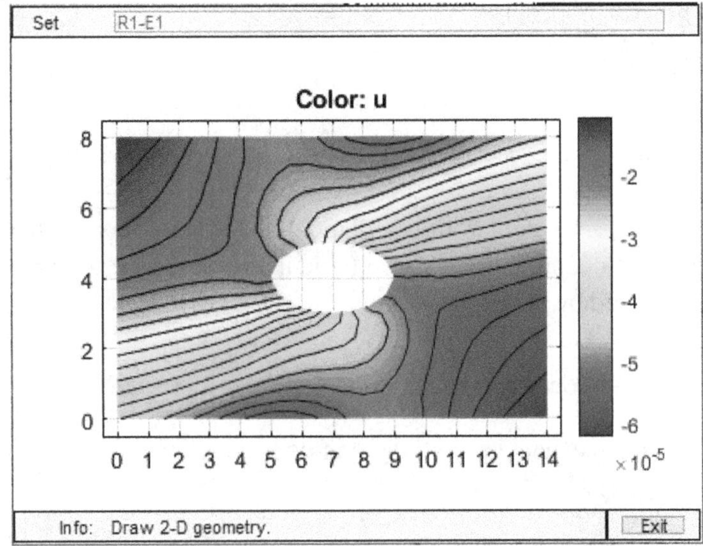

To obtain the maximal stress σ_{yy} we need firstly to export in the workspace the solution, mesh parameters, and PDE coefficients. It can be made by selecting the "Export Solution …", "Export Mesh …," and "Export PDE Coefficients …" lines using respectively the Solution, Mesh, and PDE main menu buttons. These panels allow you to specify variable names for the transmitted data. Default names are u (solution; the stresses in our case), p, t, e (mesh parameters), a, c, and d (PDE coefficients). These names were left unchanged. To obtain maximal stress, the special pdesmech command should be used. This command entered in the Command Window looks like this:

```
>> StressY=pdesmech(p,t,c,u,'tensor','syy');
```

Now the maximal stress value σ_{yy} can be defined:

```
>> syy= max(StressY)
syy =
    4.1732
```

Note, that the theoretical value obtained for an infinite plate with an elliptic hole is 5 (Boresi & Schmidt, 2003). It shows that solution for an infinite plate does not quite satisfactorily describe the real cases when the plate has finite dimensions.

To save the program, select the Save As line form the popup menu of the Save button of the main menu and type the name Ch_8_ApExample_8_2 in the File name field of the panel that appears. You can open the saved file in the Editor window and write Ch_8_ApExample_8_2 in the function definition line instead of the automatically created name (the default name is Model, it is generated automatically).

8.4.3. 2D Transient Heat Equation With Temperature-Dependent Material Properties

The dimensionless heat equation with temperature dependent properties of material reads

$$\frac{\partial T}{\partial t} = \frac{\partial}{\partial x}\left[\alpha\left(T\right)\frac{\partial T}{\partial x}\right] + \frac{\partial}{\partial y}\left[\alpha\left(T\right)\frac{\partial T}{\partial y}\right]$$

Where T is the temperature of an object; x and y are the Cartesian coordinates; t – time; α is the temperature-dependent diffusivity coefficient.

Problem: Solve the heat equation with the PDE Modeler for a squared 6x6 plate (with center at 0,0). Select the Generic Scalar application mode (default). Assume the thermal diffusivity is function of temperature and described by the expression

$\alpha=0.3T+0.4$

Consider the case when the plate has initially non-equal surface temperatures

$$T = \begin{cases} 1, at\left(x^2 + y^2\right) \le 1 \\ 0, otherwise \end{cases}$$

The boundary conditions are

$$\frac{\partial T}{\partial x} = \frac{\partial T}{\partial y} = 0$$

at all the plate sides (Neumann boundaries).

Set times in range 0 …1with step 0.1. Use the Refined Mesh option twice, in addition to the initial triangle mesh. Represent the final *u*- values in the 2D and 3D plots, which should show the four contour levels and the 'Parula' color map as well as contain the color bar. The resulting *u* and triangle *p*, *e*, *t*-values should be exported to the MATLAB® workspace; display *u* at *t*=1 at the rectangular mesh when the *x* and *y* coordinates are equal to -3, -2, …, 3 each. Save the automatically created program in the file named Ch_8_ApExample_8_3.

Matching our heat equation, rewritten in compact form $\frac{\partial u}{\partial t} - \nabla\left(\alpha\left(T\right)\nabla T\right) = 0$, with the standard equation (Table 1) we can detect that these equations are identical when *T*=*u*, *c*= α(*T*), *f*=0, and *d*=0; thus the parabolic PDE should be considered.

To solve the problem, open the PDE Modeler window with the command

```
>> pdeModeler
```

In the popup menu of the Options menu button, mark the Grid and Snap lines and type the *x* and *y* limits as [-3.5 3.5] each (after selecting the Axis Limits line). Set the Grid spacing with step 0.5 and mark the Axis Equal line. Select the general Application option - Generic Scalar.

To draw the square, click this ⊞ rectangle (centered) button, place the mouse arrow on the point (0,0) and drag to point (1,1) of the plot. Check this square geometry with the Object Dialog panel, which appears after clicking within the square.

Select the Boundary mode line of the popup menu of the Boundary main menu option and click on the appropriate boundary line. After the Boundary Conditions panel appears, mark the Neumann condition box and enter zeros in the *g* and *q* fields; do this for all four square sides. Note that c into the Neumann equation presented into the panel is the same as in the heat PDE and should not be specified at this stage.

Now select the PDE Mode of the popup menu of the PDE main menu option, select the PDE Specification and the Parabolic type of PDE found there; type 0.3*T+0.4 within the *c* field, 1 in the *d* field, and 0 within the *a* and *f* fields.

Initialize the triangle mesh in the Mesh Mode (a line in the popup menu of the Mesh button). Click the Refine Mesh line twice.

Select the "Parameters …" line in the popup menu of the Solve main menu option and type 0:0.05:1 in the Time field of the appeared Solve Parameters panel. In the same panel, the following string with initial conditions should be typed in the u(t0) field: '(x.^2+y.^2)<=1'. The result of this logical operation is u0 matrix of logical 1s and 0s with u0=1 when *x* and *y* are within the required range and u0=0 out of the range; the logical values are used as conventional numbers. After inputting the initial conditions, click the Solve PDE line of the Solve button and select the Parula color in the Plot Selection panel (opened with selecting the "Parameters ..." line of the popup menu of the Plot main menu option). The 2D solution with colored bar appears as below:

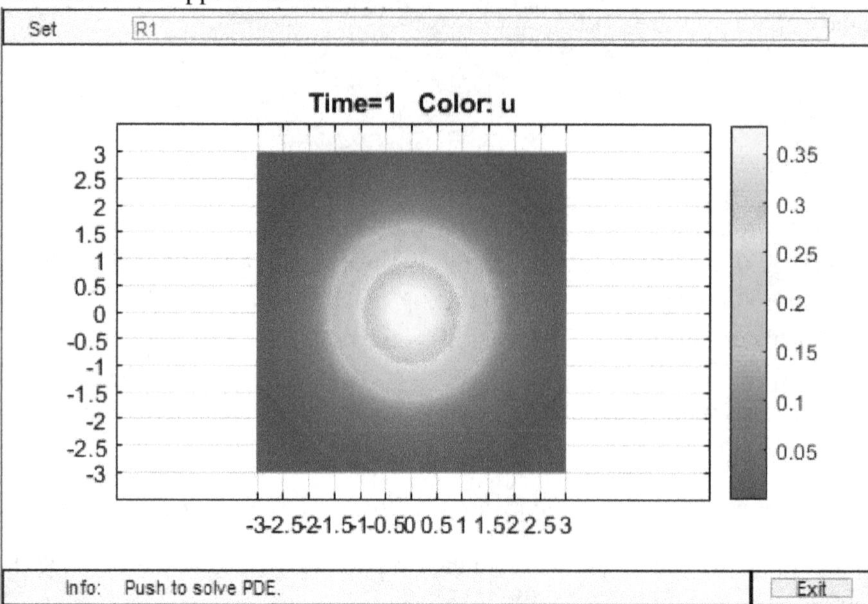

To generate the 3D plot, mark the Height(3D) and Show mesh boxes. The generated graph looks as below.

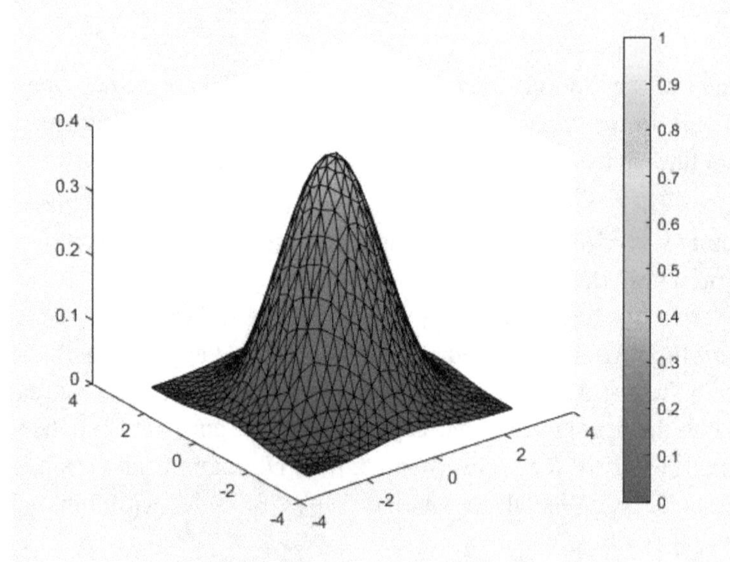

To transfer the solution and mesh parameters to the MATLAB® workspace, select the Export Mesh and Export Solution lines within the popup menus of the respective Mesh and Solve menu buttons. If the variable names were not changed, the solution is in the u matrix (each matrix column corresponds to the given times, i.e. $u(:,1)$ corresponds to t=0 and $u(:,11)$ to t=1) while the mesh parameters are within the p, e and t matrices. After this, to obtain the u-values at the required orthogonal grid points, the following commands must be entered, with the resulting table displayed in the Command Window.

```
>> x=-3:3;y=x;uxy=tri2grid(p,t,u(:,end),x,y)
uxy =
 -0.0000  0.0000  -0.0000   0.0000   0.0000  -0.0000   0.0000
 -0.0000 -0.0000   0.0000  -0.0007  -0.0001  -0.0000   0.0000
  0.0000 -0.0002   0.0899   0.4594   0.0960   0.0000   0.0000
  0.0000 -0.0004   0.3981   0.9776   0.4265  -0.0005   0.0000
  0.0000  0.0001   0.0753   0.4579   0.0562   0.0000   0.0000
  0.0000 -0.0000   0.0000  -0.0009   0.0000   0.0000   0.0000
  0.0000  0.0000   0.0000   0.0000  -0.0000   0.0000   0.0000
```

Any asymmetry in the temperature values is the result of the triangulation of the square plate and the circle with the initial temperatures, and the subsequent interpolation from the triangle to the orthogonal grid.

To save the automatically generated program, select the Save As line of the popup File menu and type the name Ch_8_ApExample_8_3 in the File name field of the panel that appears. After this, open the saved file in the MATLAB® Editor and write Ch_8_ApExample_8_3 in the function definition line instead of the default name produced automatically by the PDE Modeler.

8.4.4. Rectangular Membrane Vibrations

The transverse vibrations in the 2D plane describe by the wave equation

$$\frac{\partial^2 u}{dt^2} = \frac{d}{dx}\left(\frac{du}{dx}\right) + \frac{d}{dy}\left(\frac{du}{dy}\right)$$

where u is the wave perturbation; x, y and t are the horizontal, vertical coordinates and time, respectively. All dimensions here are assumed dimensionless. In case of membrane, u represents the deflections of each of the (x,y)-membrane points from its equilibrium position.

Problem: Consider this equation with the PDE Modeler for 2x3 rectangular membrane (the left bottom corner at (0,0)), which is fixed at all four rectangle sides. Select the Generic Scalar application mode. For all fixed sides, the Dirichlet boundaries, u=0, are valid. The initial condition that corresponds to the all boundary conditions and represents the initial membrane deformation is $u_0=0.5xy(2-x)(3-y)$. Set 41 equally spaced values of the time in range 0 … 5. Represent the final u- values in the 2D and 3D plots when 2D plot shows the ten contour levels, chose the 'jet' color map. Resulting u and triangle p, e, t- values should be exported to the MATLAB® workspace; display u at t=5 at the orthogonal mesh when the x and y coordinates are equal to 0, 0.5, …, 2 and 0,0.5, …, 3 respectively. Generate two 3D plots in the same window – u(x,y) at time t=2.5 and at time t=3. Create two three-dimensional graphs in the same window - u (x, y) at time t = 0 (initial deformation) and at time t = 5 (instantaneous position of the membrane in the last calculated time).

Matching the wave equation, rewritten in compact form $\frac{\partial^2 u}{\partial t^2} - \nabla\left(\nabla u\right) = 0$, with the standard equation (Table 1) we can detect that these equations are identical when c= 1, f=0, and d=1; thus the hyperbolic PDE should be solved.

To solve the problem, open the PDE Modeler window with the command

```
>> pdeModeler
```

In the popup menu of the Options button of the main menu, mark the Grid and Snap lines and type the limits [-0.5 3.5] for *x* axis and [-0.5 3.5] for the *y*-axis (after selecting the Axis Limits line). Set the Grid spacing with step 0.5 and mark the Axis Equal line. Select the general Application option - Generic Scalar.

To draw the rectangle, click this ☐ rectangle button, place the mouse arrow on the point (0,0) and drag to point (3,2) of the plot. Check this rectangle geometry with the Object Dialog panel, which appears after clicking within the rectangle.

Select the Boundary mode line of the popup menu of the Boundary main menu option and click on the appropriate boundary line. After the Boundary Conditions panel appears, check the Dirichlet condition box and enter 1 and 0 in the *h* and r fields respectively; do this for all four rectangle sides.

Now select the PDE Mode of the popup menu of the PDE main menu option, select the PDE Specification and the Hyperbolic type of PDE found there; type 1 in the *c* and *d* fields, and 0 in the *a* and *f* fields.

Initialize the triangle mesh in the Mesh Mode (a line in the popup menu of the Mesh button) and click then the Refine Mesh.

Select the "Parameters ..." line in the popup menu of the Solve main menu option and type linspace(0,5,31) in the Time field of the appeared Solve Parameters panel. In the same panel, the following string with initial conditions should be typed in the u(t0) field: '0.5*x.*y.*(x-2).*(x-3)'. After inputting the initial conditions, click the Solve PDE line of the Solve button and select the Jet color in the Plot Selection panel (opened with selecting the "Parameters ..." line of the popup menu of the Plot main menu option). Mark further the Contour box and type 10 in the 'Contour plot levels' field. The resulting 2D plot containing of the calculated *u* and the color bar is as follows:

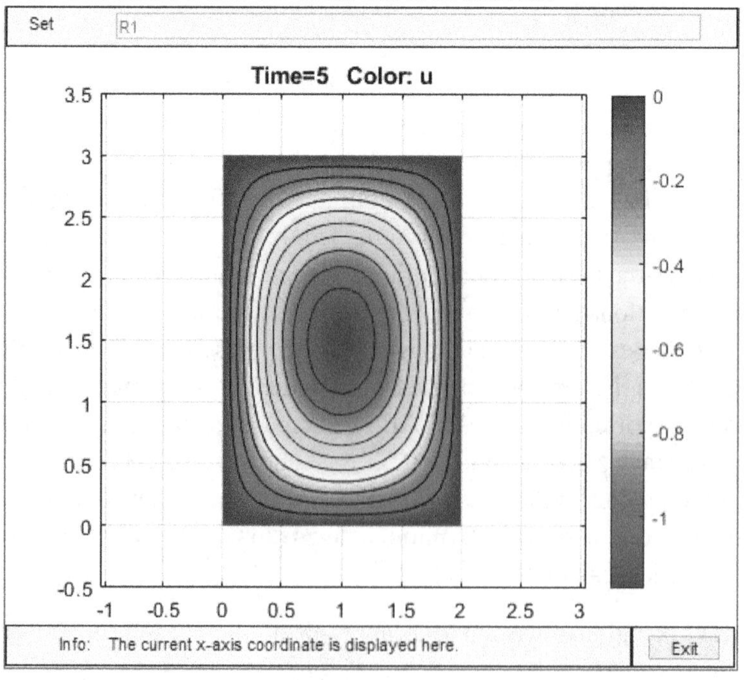

Mark the ‘Height(3D)’ and ‘Show mesh’ boxes and uncheck the ‘Contour’ box to generate the necessery 3D plot (see below).

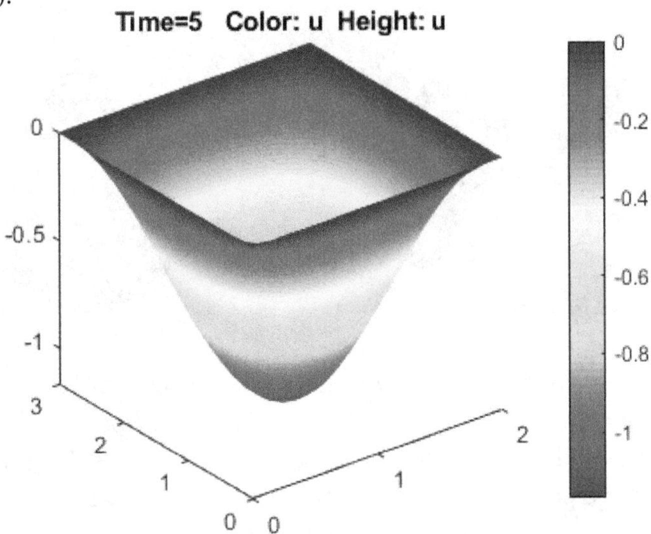

To transfer the solution and mesh parameters to the MATLAB® workspace, select the Export Mesh and Export Solution lines within the popup menus of the respective Mesh and Solve menu buttons. If the variable names were not changed, the solution is in the u matrix (each matrix column corresponds to the given times, i.e. $u(:,1)$ corresponds to t=0 and $u(:,31)$ to t=5) while the mesh parameters are within the p, e and t matrices. After this, to obtain the u-values at the required orthogonal grid points, the following commands must be entered, with the resulting table displayed in the Command Window.

```
>> x=linspace(0,2,7);y=0:.5:3;uxy=tri2grid(p,t,u(:,end),x,y)
uxy =
      0        0          0          0          0          0          0
      0  -0.3013    -0.5117    -0.5873    -0.5155    -0.3031          0
      0  -0.5256    -0.8979    -1.0237    -0.8963    -0.5261          0
      0  -0.6014    -1.0226    -1.1674    -1.0220    -0.5994          0
      0  -0.5265    -0.8986    -1.0220    -0.8962    -0.5266          0
      0  -0.3041    -0.5143    -0.5861    -0.5125    -0.3012          0
      0        0          0          0          0          0          0
```

To generate in one window the required two graphs, for the times t = 2.5 and 3 (this corresponds to the 16th and 19th columns of the matrix u), enter the following commands:

```
>>x=linspace(0,2,40);y=linspace(0,3,40);
>>uxy=tri2grid(p,t,u(:,16),x,y);subplot 121,surf(x,y,uxy)
>>title('Membrane deflections at t=2.5'),axis square >>xlabel('x'),ylabel('y')
,zlabel('u')
>>uxy=tri2grid(p,t,u(:,22),x,y);subplot 122,surf(x,y,uxy)
>>title('Membrane deflections at t=3'),axis square >>xlabel('x'),ylabel('y'),z
label('u')
```

These commands generate the following graphs

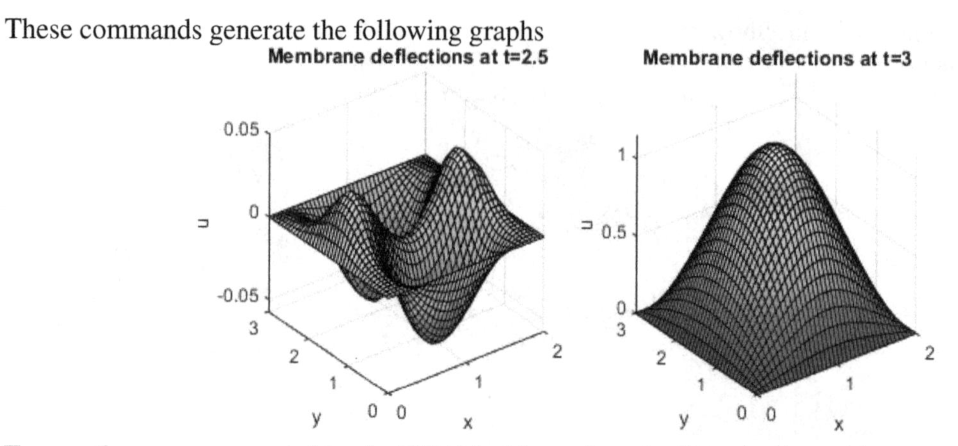

To save the program created by the PDE Modeler, select the Save As line of the popup File menu and type the name Ch_8_ApExample_8_4 in the File name field of the panel that appears. After this, open the saved file in the MATLAB® Editor and write Ch_8_ApExample_8_4 in the function definition line instead of the default name produced automatically by the PDE Modeler.

8.5. CONCLUSION

The PDE Modeler tool interface from the PDE toolbox is presented and verified on a spatial 2D PDE. Modes and possible application options are explored and used to solve the following M&T problems:

- Laminar flow in horizontal pipe,
- Instantaneous pressure distributions in lubricating film between two surfaces when one of them is covered by hemispherical pores,
- Stresses in the rectangular plate with elliptic hole,
- 2D transient heat equation with temperature dependent material properties,
- Rectangular membrane vibrations.

The tabulated equation reference forms (elliptical, parabolic, hyperbolic and eigenvalue PDEs) (Dirichlet and Neumann) presented in this chapter, together with boundary conditions, facilitate the interface understanding and enable users to solve M&T problems by reducing usual PDE and BC forms to their standard.

REFERENCES

Boresi, A. P., & Schmidt, R. J. (2003). *Advanced Mechanics of Materials* (6th ed.). John Wiley & Sons.

Burstein, L. (2020). *A MATLAB primer for technical programming in material science and engineering*. Cambridge: Elsevier-WP.

Etsion, I., & Burstein, L. (1996). A Model for Mechanical Seals with Regular Microsurface Structure. *Tribology Transactions*, *39*(3), 677–683. doi:10.1080/10402009608983582

ENDNOTES

[1] Based on chapter 5 Table 2 in Burstein, 2020.

[2] Based on chapter 5 Table 3 in Burstein, 2020.

APPENDIX

Table 4. List of examples, problems, and applications discussed in the chapter

No	Example, Problem, or Application	Subsection
1	Laminar flow in a horizontal pipe, two-dimensional solution.	8.2.1.
2	2D pressure distribution in a lubricating film between two surfaces when one of them is covered by hemispherical pores.	8.4.1.
3	"Structural mechanics" application of the PDE Modeler. Stresses in the rectangular plate with elliptic hole.	8.4.2.
4	2D transient heat equation with temperature-dependent material properties.	8.4.3.
5	Rectangular membrane vibrations.	8.4.4.

Note, some small examples, mostly related to non-M&T issues, are not included in the list.

Chapter 9
Curve Fitting for Mechanical and Tribological Problems

ABSTRACT

This chapter devoted to matching the data with mathematical expressions. Here the functions using fitting by polynomial and non-polynomial expressions is represented by examples from the mechanics and tribology (M&T) fields. The Basic Fitting tool and examples of its use are described. Single and multivariate fitting through optimization are discussed. Application examples are demonstrate the curve fitting for the following data: fuel efficiency-velocity, yield strength-grain diameter, friction coefficient-time, and machine diagnostic parameter.

INTRODUCTION

Professionals of various technological specialties and, of course, M&T specialists must often analyze practical data by describing them with an empirical or theoretical expression. If we have any mathematical relationship that accurately matches the actual data, then this allows you, for example, to predict the residual service line of the equipment, reduce the frequency and number of tests to ensure product quality, predict the readjustment time of the machine or production line, select the material with the best properties, evaluate the discrepancies between different measurements, determine the error of the most accurate data, etc.

In this chapter, we will present the basic commands and a specialized tool that a M&T scientist/engineer can use to describe data using a mathematical expression. The implementation of polynomials and other relations connecting the two variables will be described. It also explains how to use the optimization commands to find the relationship between one or more than two variables for non-polynomial equations.

The presented applications fit the data for fuel efficiency, yield strength, friction coefficient, and machine diagnostic parameter.

Unlike the program in previous chapters, the computing programs presented in this chapter incorporate more comments for its better understanding.

DOI: 10.4018/978-1-7998-7078-4.ch009

9.1. POLYNOMIAL AND NON-POLYNOMIAL FITTING

Process of finding the expression that best matches the data points is called curve fitting or regression analysis. An empirical or theory-based expression can be chosen to fit the data. Curve fitting with polynomial and some non-polynomial expressions is discussed below.

9.1.1. Polynomial Fitting

The process consists in finding the values of the coefficients of a mathematical expression, which should best fit the experimental data. The polynomial expression using for this reads

$$y = a_{n-1}x^{n-1} + a_{n-2}x^{n-2} + \ldots + a_1 x + a_0$$

where x is the independent variable that should be connected with another, dependent, variable - y, n and a_i are the number of polynomial terms and coefficients of polynomial respectively, i is the a-coefficients serial number that changes form 0 to n-1.

The largest value of the exponent, n-1, is termed the degree of the polynomial, e.g. the straight line $y=a_1x+a_0$ is a polynomial of the first-degree that has two coefficients, a_1 and a_0, which can be determined by given pairs of the x and y values (the x,y –pair denotes the data point for fit).

The n coefficients of a_i can be determined by the n of (x,y) points solving the set of $y(x)$ equations written for each of the of points. These equations are linear in respect to the coefficients a_i. Such a fit, when the number of coefficients is equal to the number of points, is not always efficacious. Although in this case a complete correspondence is observed with the used data points, this often leads to a waviness of the curve and to resulting inconsistencies with unused data points located between the fitted points.

The linear $y(x)$ dependence – first order polynomials - is another case of possible inefficiency of such procedure. Indeed, two points suffice to determine the two a-coefficients, but experimental data contain a larger number of points, therefore you can generate more than 2 linear equations to find a_i that forms the so-termed overdetermined system when number of equations is more than unknowns.

The most widely used fitting method is the least squares method. The coefficients a are determined in this method by minimizing the sum of the squared differences between the data and the fitting equation – $min\{\Sigma[y_i\text{-}f(x_i,\boldsymbol{a})]^2\}$; the $y_i\text{-}f(x_i,\boldsymbol{a})$ differences are called residuals and are denoted by R_i for each of the i-points. The minimization is carried out by taking the partial derivative of the R_i^2 with respect to each of the a-coefficients and equating the derivatives to zero, so for example for a straight line $\frac{\partial \sum R_i^2}{\partial a_1} = 0$ and $\frac{\partial \sum R_i^2}{\partial a_0} = 0$, and for obtaining the a_1 and a_2 values, the set of these two equations should be solved.

MATLAB has the polyfit command for fitting by a polynomial. Simplest form of this command is

```
a=polyfit(x_data,y_data,m)
```

where the input arguments x_data and y_data are the equal-size vectors with values of the x and y coordinates of the data points to be fitted, and m is the degree of the ftting polynomial; the output argument a is the (m+1)-element vector of the defined fitting coefficients. The first element a(1) is the a_{n-1} coefficient, the second a(2) is a_{n-2}, …, and the final a(m+1) is a_0.

For example, the first degree polynomial with the defined **a** coefficients looks as $y=a(1)*x+a(2)$, the second degree polynomial as $y=a(1)*x^2+a(2)*x+a(3)$, and so on.

After the polyfit command was used, the y-values can be calculated at any x within the fitted interval. It can be done using the polyval command, whose simplest form reads

```
y_poly=polyval(a,x)
```

where a is the vector of polynomial coefficients obtained with the polyfit command, x is that of the coordinates at which the y_poly values are calculated.

For example, the elongation-force test data, $\Delta l(F)$, obtained for a rubber are: F=14, 18, 24, 28, 32, 35, 38 N and Δl = 0, 16, 61, 92, 122, 153, 183 mm (stretch-free rubber length was 275 mm).

Problem: write the live script named ExElongFit that fits the data by the first-degree (m=1) and second degree (m=2) polynomials, and generate two fit curves together with the data in the same plot. Use the Output Inline option of the Live Editor.

The ExElongFit live script that solves this problem is shown in the Figure 1 together with the obtained fitting coefficients and the generated graph.

In the presented live script, two polyfit commands determine the fitting coefficients a_linear and a_quadratic for polynomials of the 1st- and 2nd-degree respectively; defined coefficients are displayed. The y_linear and y_quadratic vectors contain the elongation values obtained by the above polynomial coefficients at one hudreed points from the fitting range. The data (circled points) and two fitting curves (solid and dashed lines) generated with the plot command; the legend command explains the plotted data and series.

In accordance to the fitting coefficients located in the a_linear and a_quadratic vectors, the fit polynomials are:

$$\Delta l_1 = 7.6851F - 117.9264$$

$$\Delta l_2 = 0.1139F^2 + 1.7742F - 48.9935$$

where Δl_1 and Δl_2 are the stretched rubber elongations, in mm, fitted by the 1st – and 2nd degree polynomial respectively; F is the stretched force, in N.

Note that fittining expressions are not always true in the fitted range; in this example the 1st degree polynomial shows negative elongations at low tensile forces, which can not really be.

Figure 1. The ExElongFit *live script that performs the 1st and the 2nd degrees fit, shows fitting coefficients, and generats a plot*

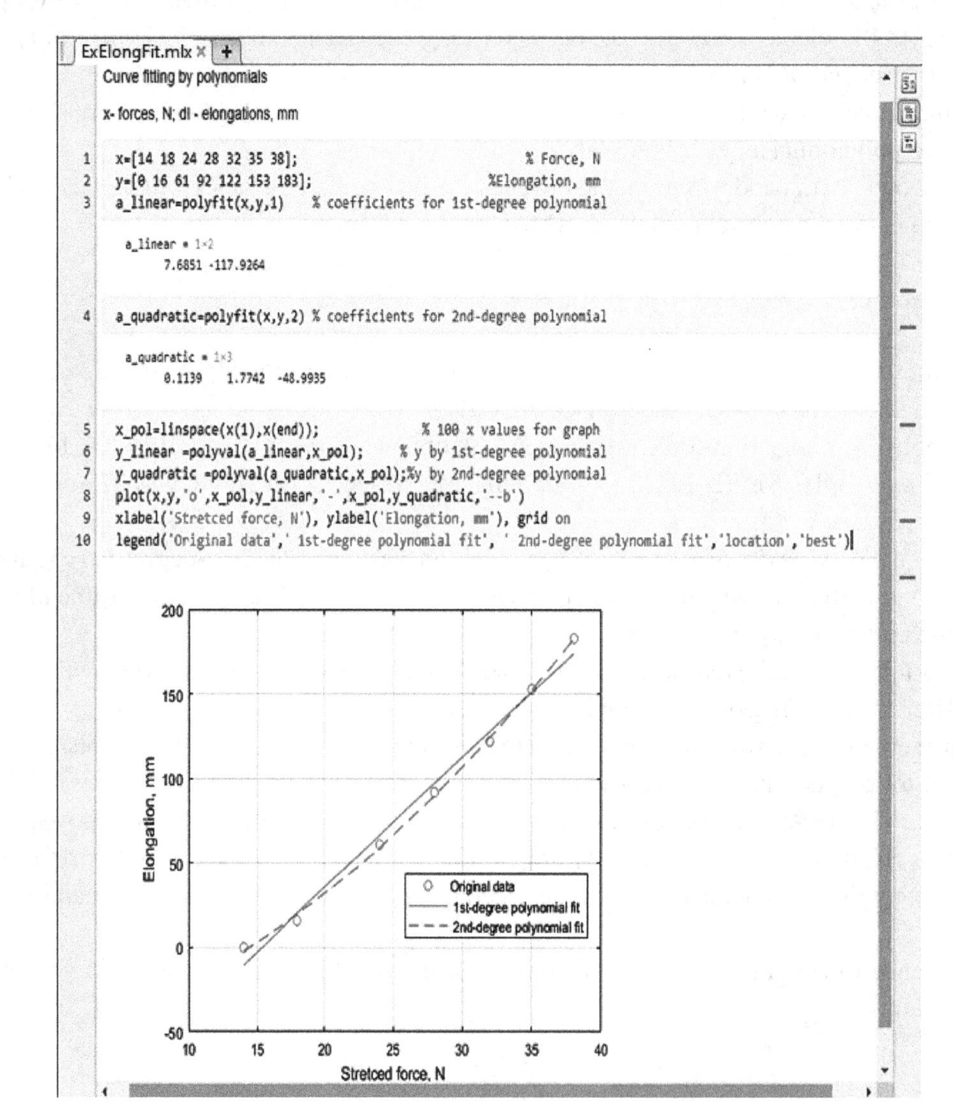

9.1.2. Using Polynomial Fitting With Some Non-Polynomial Functions

Many applications in the M&T fields require a fit that uses an exponential, power, or a logarithmic function:

$$y = a_0 e^{a_1 x}$$

$$y = a_0 x^{a_1}$$

$$y = \frac{1}{a_1 x + a_0}$$

$$y = a_0 + \frac{a_1}{\sqrt{x}}$$

$$y = a_1 \ln x + a_0 \text{ or } y = a_1 \log x + a_0$$

Each of these functions, after a simple conversion, can be fitted by polynomial with the polyfit command. It can be done if the functions can be modified to the first degree polynomial $y=a_1x+a_0$. The last three functions with square root and logarithms already have the reguired form, since $1/\sqrt{x}$, ln$x$, and log$x$ are obviously the modified x. The first two functions, exponential and power, can be performed in this form by taking logarithm of both parts of the equations, that gives:

$$\ln y = a_1 x + \ln a_0$$

$$\ln y = a_1 \ln x + \ln a_0$$

And the third function, reciprocal, can be presented as

$$\frac{1}{y} = a_1 x + a_0$$

That also is the first degee polynomial as $1/y$ is modified y.

All of the above together with representation these equations by the polyfit functions, and relationships between coefficients a_1, a_2 of the original equations and coefficients a defined by the polyfit are shown in Table 1.

To select the suitable function for fitting from one of the function of the Table 9.1, it is useful to estimate graphically how they match the data. For this, the data should be built in the certain axes - linear, reciprocal ($1/y$), logarithmic or semi logarithmic. The first two input arguments of the polyfit function represent the appropriate scaling of the axes, e.g. for the $y = a_0 x^{a_1}$ function (represented by the polyfit(log(x),log(y),1) command), both x and y axes should be logarithmic, and for the $y = a_0 e^{a_1 x}$ function (the polyfit(x,log(y),1) command) the x axis should remain linear and y should be logarithmic.

Consider, for example, the fit with the exponential $y = a_0 e^{a_1 x}$ function of the data volumetric wear rate– normal load, $w(N)$. The w-values for a casting alloy (4, 5, 6.8, 8, 10) 10^{-12} m^3/sec were tested on a pin-on-disk test bench at loads N=10, 15, 20, 25, and 30 N respectively.

Table 1. Some non-polynomial function, their conversion and representation by the polyfit function

Equation	Conversion to Polynomial	Representation in the Polyfit Function	The a_0 and a_1 via Vector a
$y = a_0 e^{a_1 x}$	$\ln y = \ln a_0 + a_1 x$ For polyfit: x=*x*; y= $\ln y$	a=polyfit(x,log(y),1)	a₁= a(1), a₀=exp(a(2))
$y = a_0 x^{a_1}$	$\ln y = \ln a_0 + a_1 \ln x$ For polyfit: x= $\ln x$; y= $\ln y$	a=polyfit(log(x),log(y),1)	a₁=a(1), a₀=exp(a(2))
$y = a_0 + \frac{a_1}{\sqrt{x}}$	$y = a_0 + a_1 X$ For polyfit: X= $\frac{1}{\sqrt{x}}$; y=*y*	a=polyfit(1./sqrt(x), y,1)	a₁=a(1), a₀=a(2)
$y = \frac{1}{a_1 x + a_0}$	$\frac{1}{y} = a_1 x + a_0$ For polyfit: x=*x*; y= $\frac{1}{y}$	a=polyfit(x,1./y,1)	a₁=a(1), a₀=a(2)
$y = a_1 \ln x + a_0$	For polyfit: x= $\ln x$; y=*y*	a=polyfit(log(x),y,1)	a₁=a(1), a₀=a(2)
$y = a_1 \log x + a_0$	x=*logx*; y=*y*	a=polyfit(log10(x),y,1)	a₁=a(1), a₀=a(2)

Problem: Fit these data by the polyfit commands, and generate two plots at the same Figure window: the *N* - log*w* curve (solid line with the circled points) and the *w*(*N*) curve (solid line) together with data (circles) in the same graph.

The following commands, saved as the ExExponentFit script file, solve the problem.

```
%            Curve fitting for the power function
 %            N - load, N; w - wear rate, m^3/sec
N=10:5:30;
w=[4 5 6.8 8 10]*1e-12;
a=polyfit(N,log(w),1);
a0=exp(a(2))
a1=a(1)
wf=a0*exp(N*a1);
subplot(1,2,1)
semilogy(N,wf,'-s'),grid on
```

```
xlabel('Normal load, N')
ylabel('Volumetric wear rate, m^3/sec')
title('Wear vs load')
subplot(1,2,2)
plot(N,w,'s',N,wf),grid on
xlabel('Normal load, N')
ylabel('Volumetric wear rate, m^3/sec')
title('Fit w=a_0e^{a_1N}')
legend('Original data','fitting with power function')
```

In the commands above, variable a represents the vector of the coefficients defined by the polyfit commands. The variables a0 and a1 are the coefficients of the exponential equation calculated by the a(1) and a(2) values of the a-vector (see Table 9.1); obtained values are displayed. The wf variable denotes the vector with volumetric wear rates obtained for the given loads N by the exponential function (with the defined coefficients a0 and a1); the N, w and wf vectors are used in the plot command for generating the data and fitting curve in the graph. The semilogy command plots data in coordinates with regular *x*-axis and logarithmic axis *y* to demonstrate how close is the is *N*-log*w* curve to a straight line.

After typing and entering the file name in the Command Window, the following coefficients are displayed and the graph is generated:

```
>> ExExponentFit
a0 =
  2.5546e-12
a1 =
  0.0461
```

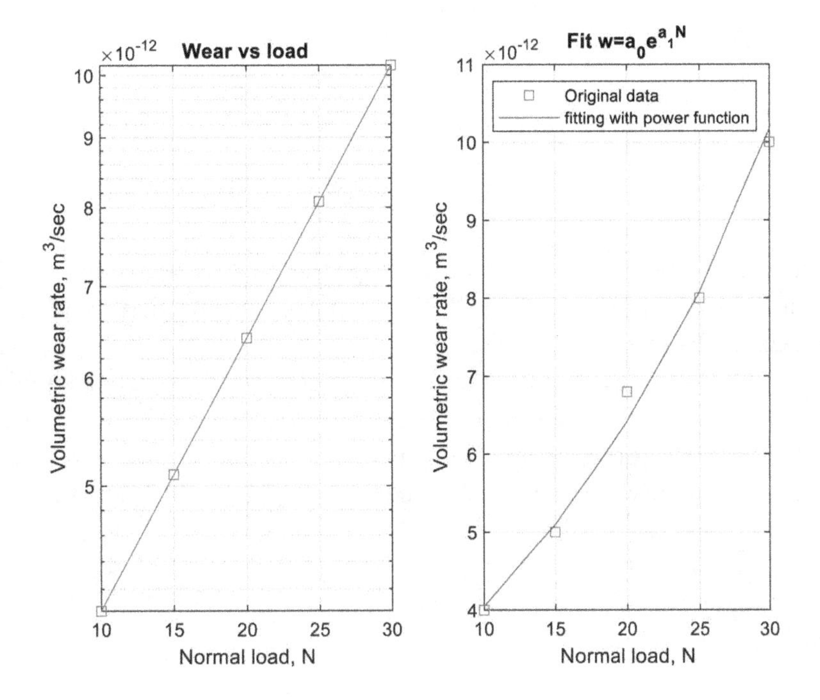

9.1.3. Goodness of Fit

The discrepancy between the obtained equation and the source data characterizes the goodness of fit. Unfortunately, there is not a single criterion for this purpose. One of the commonly used criterions is the so-called coefficient of determination or *R*-squared, R^2, that reads

$$R^2 = 1 - \frac{\sum_{i=1}^{n}\left(y_i - y_{fiti}\right)^2}{\sum_{i=1}^{n}\left(y_i - \overline{y}\right)^2}$$

where y and y_{fit} are the values respectively the observed and predicted with the modelled equation; $\overline{y}$ is the average of the observed values; n – amount of observed points to fitting; i is the current point.

R-squared ranges from zero to one. A fit is considered best if the criterion value is closest to 1.

For example, calculate the *R*-squared for the 1st and 2nd degrees polynomials that were defined with the ExElongFit live script program (example of the Subchapter 5.1.1). To do this, the following commands should be added to this program (Figure 2)

Figure 2. Aditional section of the ExElongFit *live script with commands for calculating the R-squared*

R-squared for the 1st- and 2nd-degree polynomials

```
12  y_lin=polyval(a_linear,x);        % predicted y, 1st-degree polynomial
13  y_qua=polyval(a_quadratic,x);     % predicted y, 2nd-degree polynomial
14  s1_lin=sum((y-y_lin).^2); % squared residuals, 1st-degree polynomial
15  s1_qua=sum((y-y_qua).^2); % squared residuals, 2nd-degree polynomial
16  s2=sum((y-mean(y)).^2);
17  R2_linear=1-s1_lin/s2                % R-squared, 1st-degree polynomial

    R2_linear = 0.9892

18  R2_quadratic=1-s1_qua/s2             % R-squared, 2nd-degree polynomial

    R2_quadratic = 0.9989
```

After running the modified ExElongFit live script program, the *R*-squared values for the 1st and 2nd-degree polynomials are displayed in addition to the defined fitting coefficients and graph (Figure 1).

As can be seen, the *R*-squared values for polynomials of the 1st and 2nd degree are very close to 1, but the *R*-squared for 2nd-degree polynomial is nevertheless closer, thus the quadratic polynomial better matches the original data for rubber elongation. The same conclusion can be drawn by observing a graph showing the data points and curves of both fitting polynomials (see the plot in Figure 1).

Another commonly used criterion of the fit goodness is the norm of residuals:

$$R_{norm} = \sqrt{\sum_{i=1}^{n}\left(y_i - y_{fiti}\right)^2}$$

where all notifications are the same as for the *R*-squared.

In contrary to the *R*-squared, the lower value of this R_{norm} reflects a better fit.

Calculate now the norm of residuals for the 1st and 2nd degree polynomials, continuing the example above. For this the following commands should be added to the **ExElongFit** live script - Figure 3.

Figure 3. Additional section of the **ExElongFit** *live script with commands for calculating the norm of residuals*

```
Norm of residuals for the 1st- and 2nd-degree polynomials
19   Rnorm_linear=sqrt(s1_lin)          % R_norm, 1st-degree polynomial
     Rnorm_linear = 17.4101
20   Rnorm_quadratic=sqrt(s1_qua)       % R_norm, 2nd-degree polynomial
     Rnorm_quadratic = 5.6079
```

After running modified the **ExElongFix** live script program, the values of the calculated norm of residuals (Figure 3) together with values and graph generated by the previous command of this program (Figures 1 and 2).

As it can be seen, the norm of residuals criterion is more sensitive to the descrepances between the predicted and observed data than the *R*-square: defined values R_{norm} for 1st and 2nd degree polynomials is differ above 3 times. The obtained **Rnorm_quadratic** value is significantly smaller then **Rnorm_linear**, therefore 2nd order fit is better than linear. Therefore the R_{norm} criterion brings the same conclusion as the *R*-squared.

Note: Described two criterions, as well as other available criteria, are imperfect, because they can show complete correspondance of the polynomial expression to the original points, but fluctuate between these points. Therefore, it is rational to plot the original points along with the fit curve to ensure the goodness of fit.

9.2. BASIC FITTING TOOL

In the subchapters above the polynomial fitting with commands was described. In addition to this, MATLAB has an interactive tool for fitting that called Basic Fitting. The tool allows:

- Fitting the data points using linear, quadratic, cubic and up to the 10th degree polynomials;
- Displaying the obtained equations and plotting resulting fit curve/s for better compare;
- Plotting the graph of residuals and displaying the values of the norm of residuals;
- Predicting the *y*-values by the *y*(*x*) fit polynomial at various *x*-points that can be inputted by the user.

To activate the tool, it is first necessary to input the data to be fitted and then plot them in graph as points without a line between them (the last option is not mandatory). After the Figure window appears with plot, select the Basic Fitting line from the popup menu of the Tools button of this window menu (left part of the Figure 4). Use to plot, for example, the elongation data from subsection 9.1.1 that were fitted by polynomials with commands to compare furher with the fit by the Basic Ftiing tool. Immediately after the action above, the first narrow panel of the Basic Fitting window appears (right panel in the Figure 4).

Figure 4. Activating the Basic Fitting tool; Figure window with the Tool button and popup menu containing the Basic Fitting line (to the left) and starting panel of the Basic Fitting window.

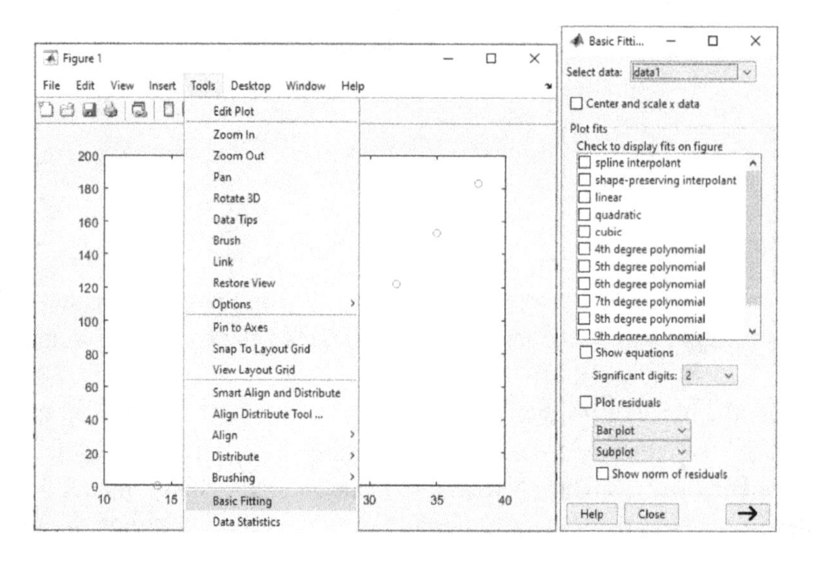

Click the 'Show next panel' button → to extend the 'Plot fits' panel (see left part of Figure 4). The 'Numerical Results' panel appears. The next click on the button → expands the last panel and the entire Basic Fitting window appears (Figure 5). To return back to the previous panel, the 'Hide last panel' button ← should be clicked.

Figure 5. The Basic Fitting window after two sequential extensions.

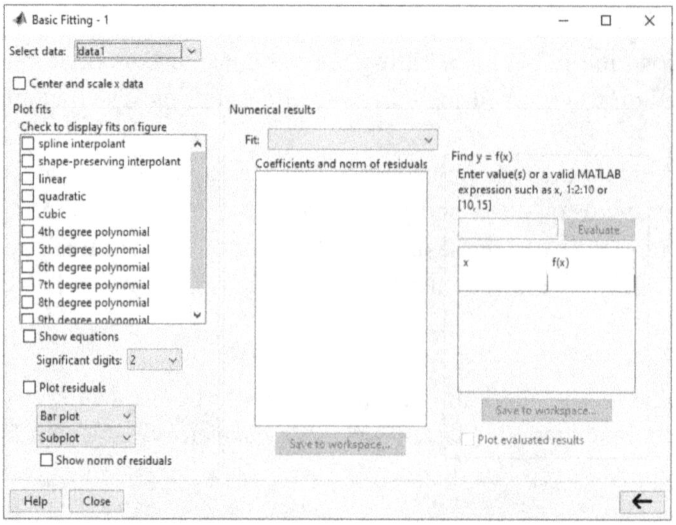

Items located in each panel of the Basic Fitting Window are explained below.

9.2.1. The 'Plot Fits' Panel

'Select Data' Field

The field permits to select the data set and give a name to it (e.g. 'data 1' –default); only one data set presented in the Figure window should be chosen for the fit by this item.

'Center and Scale x Data' Box

Simetimes the system recommends (by the appearing nameplate) to center and scale data. It should be made for a better fitting, mark for this the 'Center and scale x data' box. In this case, each the x_i value recalculates with the equation $z_i=(x_i-\mu)/\sigma$, in which μ and σ are the average and standard deviation of the x values respectively, z is the x value after centering and scaling, and i denotes the serial number of value. Frequently, when this option is not marked but the high-degree polynomial is selected to fit or/and x-values are very small or very large, the special nameplate appears with announce about poor conditionality of the polynomial and with advice to center and scale the data.

'Check to Display Fits on Figure' Area

This area contains a list of boxes allowing the user to select fit by polynomials from the first (linear) degree to the tenth degree, as well as interpolation by the spline or Hermite method. More than one fit box can be marked in this item. When the box of the desired polynomial degree has been marked, the fit line appears immediately on the plot, and at the same time, the polynomial equation, its coefficients, and the norm of residuals appear in the 'Coefficients and norm of residuals' field of the Numerical results' panel (see below). Note, you can not select the polynom with degree exeeiding the number of data points.

'Show Equation' Box and 'Significant Digits' Field

When the first of these items was marked, the fit equation is displaied on the plot. The second item represents possibility to choose the number of significant digits of the fitting coefficients that will be displayed within the equation in the plot.

'Plot Residuals' Box With Two Adjustent Fields

To plot residuals mark the 'Plot residuals' box. Residuals, R_i, are calculated as $y_i - y_{fiti}$, where the original y_i and the pedicted y_{fiti} are values associated with the i-point. The first of the adjustent fields, if its down arrow is selected, opens a list with possible resulting plot types: bar, scatter or line. The second allows to select the place where the residuals should be plotted - in separate Figure window or as a subplot in the same window where the dataset and the fitted curve were generated.

'Show Norm of Residuals' Box

After the box was marked, the goodness of the fit is displayed as the norm of residuals. The norm is calculatied as $\sqrt{\sum_{i=1}^{n} R_i^2}$ in accordance with the description in Subsection 9.1.3.

9.2.2. The 'Numerical Results' Panel Items

'Fit' Dropdown List

This item is the dropdown list that helps to explore, independently from the previous choises, a degree of fitting polynomials and as result to see the fit equation, coefficients and residuel norm for one of the selected fits; the data are displayed in the 'Coeeficients and norm of residuals' box.

'Coeeficients and Norm of Residuals' Field

As stated above, this item displays the selected polynomial, fitting coefficient values and norm of residuals.

'Save to Workspace' Button

After clicking this button, a dialog box appears with boxes and fields. The boxes to be marked allow to save the fit, residuals, and norm of residuals in the MATLAB workspace. The default names of these parameters displayed in the fields next to the boxes can be changed by user.

'Find y=f(x)' Items

At the beginning, it contains the the explanatory text 'Enter value(s) or a valid MATLAB expression…'[1]. Then there are: the Evaluate field and button, the resulting data table, the 'Save to workspace' button, the 'Plot evaluated results' box.

'Evaluate' Field and Its Button

The desirable values of *x* can be typed in this field under to calculate corresponding *y* value/s by the defined expression. The entered and calculated *y*-values are displaied in the table below after clicking the 'Evaluate' button located to the right of the Evaluate field.

'Save to Workspace' Button

After clicking this button, the dialog box will appear in which the variables and its names can be selected/typed for saving in the MATLAB workspace.

'Plot Evaluated Results' Box

When this item is marked, the evaluated *y*-values are displayed in the previously generated together with data and a fit curve.

9.2.3. Basic Fitting Tool by Example

Here we use the data on rubber elongation, mm, - force, N, that was fitted with the polyfit command in subsection 1.1.3 to demonstrate how the same and some additional results can be achieved without commands with the Basic Fitting tool.

Problem: Define linear and square equations (1st and 2nd degree polynomials, in other word) that match the $\Delta l(F)$ dataset; show fitted coefficients and norm of residuals in the 'Numerical results' panel and on the plot, evaluate rubber elongation at test force F=40 N by the fit equation with the lesser norm of residuals; save defined equations, norm of residuals and evaluated Δl value in the workspace.

To solve the problem, enter first the dataset in the Command Window and plot the data with the following commands.

```
>>x=[14 18 24 28 32 35 38];                              % Force, N
>>y=[0 16 61 92 122 153 183];                          %Elongation, mm
>>plot(x,y,'o')                              % plots data points as circles
```

After inputting these commands, the Figure window appears with the plot represented data as circled points. Select now the Basic Fitting line from the popup menu of the Tools menu option of the Figure window. This oppens the starting panel of the Basic Fitting window. This panel that should be extended in full winow with the 'Show next panel' button as described in subsection 9.2.

Now, mark the linear box in the 'Check to display fits on figure' fiels of the list of boxes (the 'Plot fits' panel), the defined linear equation y=p1*x+p2, coefficients p1=7.6851 and p2=-117.63, and norm of residuals 17.41 appear into the 'Coefficients and norm of residuals' box of the 'Numerical results' panel appears.

Mark then the following boxes: 'Show equations', 'Plot Residuals' and 'Show norm of residuals'. Two subplots appear in the Figure Window - the fit curve with the typed linear equation and toriginal data, and the residual bar graph with residual norm value. Now mark the 'quadratic' box in the 'Check to display fits on figure' area that contains a list of boxes (in the 'Plot fits' panel). The defined quadratic equation y=p1*x^2+p2*x+p3, coefficients p1 = 0.11388, p2 = 1.7742, p3 = -48.994, and norm of residuals 5.6079 appear in the 'Coefficients and norm of residuals' box of the 'Numerical results' panel. These results are completely identical to those obtained in Subchapters 9.1.1 and 9.1.3. To each of the previously created plots, the results achieved for a quadratic fit are added. So, in the upper subplot of the Figure window, the linear and quadratic polynomial curves appear together with the fit equations (by default the fitting coefficients have two significant digits only). In the lower plot of the Figure window, the two groups of the residual bars appear with the values of their norm of residuals.

As can be seen, the norm of residuals of the quadratic polynomial is more than three time lower than the norm of residuals for the linear polynomial. Thus, the linear polynomial better describes the data and therefore can be used for evaluation of the expected dielectric constant at the 40 N stretched force. Enter the number 40 in the 'Enter value(s) …' field and press the 'Evaluate' button.

To mark the evaluated point in the upper plot of the Figure window, check the 'Plot evaluate results' box in the bottom of the 'Find y=f(x)' panel.

The two figures below show the Basic Fitting window with the desribed quadratic fitting and an evaluated elongation value, and the Figure window that represents both linear and quadratic fits and obtained residuals

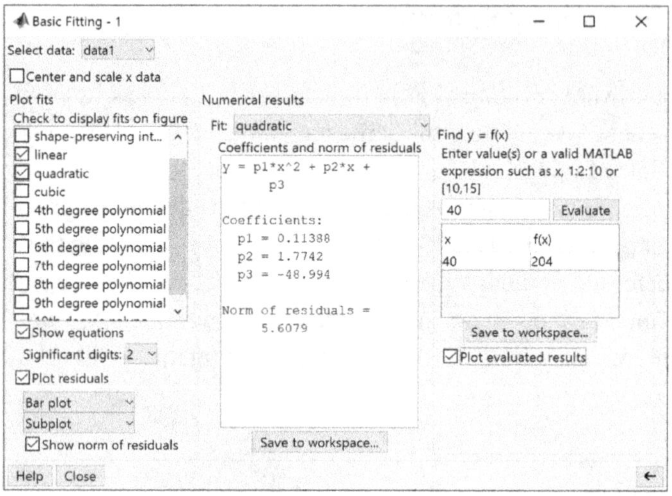

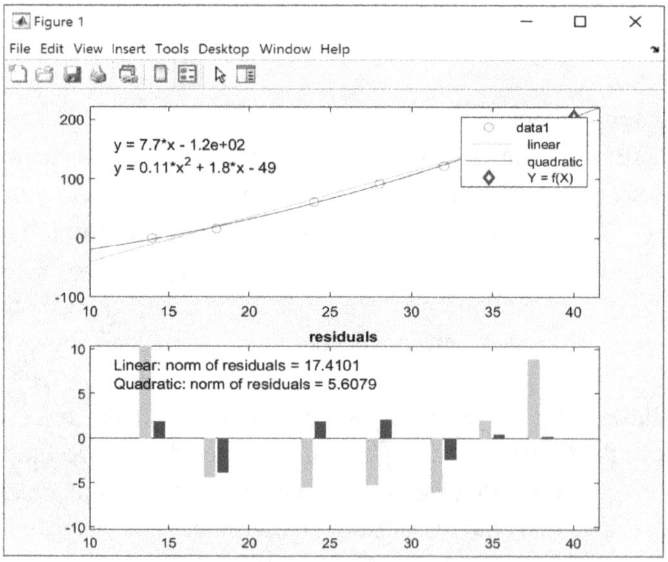

Basic Fitting - 1
Select data: data1
Center and scale x data
Plot fits
Check to display fits on figure
shape-preserving int...
linear
quadratic
cubic
4th degree polynomial
5th degree polynomial
6th degree polynomial
7th degree polynomial
8th degree polynomial
9th degree polynomial
Show equations
Significant digits: 2
Plot residuals
Bar plot
Subplot
Show norm of residuals
Numerical results
Fit: quadratic
Coefficients and norm of residuals
```
y = p1*x^2 + p2*x +
     p3

Coefficients:
  p1 = 0.11388
  p2 = 1.7742
  p3 = -48.994

Norm of residuals =
     5.6079
```
Save to workspace...
Find y = f(x)
Enter value(s) or a valid MATLAB expression such as x, 1:2:10 or [10,15]
40 Evaluate

x	f(x)
40	204

Save to workspace...
Plot evaluated results
Help Close

Figure 1
File Edit View Insert Tools Desktop Window Help
$y = 7.7*x - 1.2e+02$
$y = 0.11*x^2 + 1.8*x - 49$
data1
linear
quadratic
Y = f(X)
200
100
0
-100
10 15 20 25 30 35 40
residuals
Linear: norm of residuals = 17.4101
Quadratic: norm of residuals = 5.6079
10
5
0
-5
-10
10 15 20 25 30 35 40

To export the resuts of the fittting to the Workspace, press the 'Save to Workspace ...' button of the 'Numerical results' panel. The following dialog box appears.

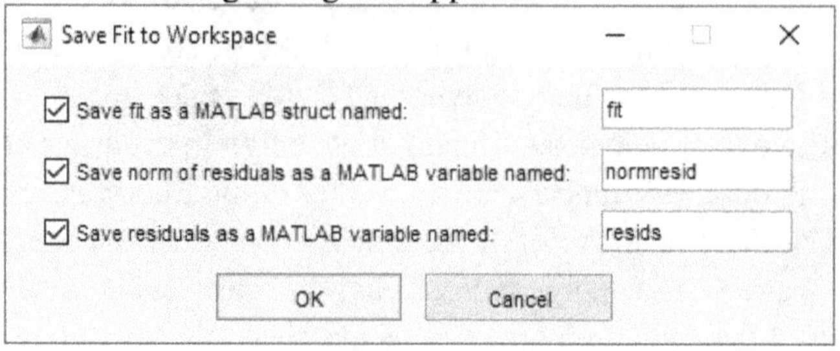

Save Fit to Workspace
Save fit as a MATLAB struct named: fit
Save norm of residuals as a MATLAB variable named: normresid
Save residuals as a MATLAB variable named: resids
OK Cancel

After clicking the 'OK' button, the fit, normresid, and resid variables are exported to the workspace and their icons appear in the Workspace window. The fit variable has a datatype called structure (see Section 5); this concept lies out of the scope of this book. To receive and use in further calculations the defined coefficients from the fit structure, the fit.coeff variable name should be entered in the Command Window.

```
>> fit.coeff
ans =
  0.1139   1.7742 -48.9935
```

The same actions should be done to save the evaluated elongation value with the given force: press another the 'Save to Workspace ...' button located under the table with results of evaluation; then in the appeared dialog box click OK. This exports the x and fx variables with the given force and evaluated elongation values to the Workspace window.

9.3. USING OPTIMIZATION FOR FITTING

Tthe fitting discussed so far has used polynomials or modified to polynomial functions of one variable. This subchapter describes nonlinear multidimensional fitting that also includes single-variate fit. Such sort of fit may be done with the optimization command that searches minimum of multivariate function by minimizing the sum of the square errors:

$$\underbrace{minimize}_{a} \sum_{i=1}^{n} \left[ydata_i - y_{fiti}\left(a, x1data_i, x2data_i, \ldots\right) \right]^2$$

where y_i and y_{fiti} are respectively the given (observed in the experiment or taken from the table) and fitting functions; of the $x1data_i, x2data_i, \ldots$ independent parameter values at which the *ydata* were observed; n –number of points, i –current point, a is the vector of fitting coefficients that should be defined.

There are many of optimization functions belongs to the Optimization toolbox, neverthless here we represent the basic MATLAB command that can be used for fitting via optimization. This is the **fminsearch** command described eaarly (Table 4.1), one of it forms are:

```
a=fminsearch(@fun,a0,[],x1data,x2data,…,ydata)
```

ydata is the vector of the observed values;

x1data, x2data, ... are the one, two or more vectors of the observing parameters; these vectors should have the same length as the ydata vector;

@fun is the call to a **fun** function that should include the equation to fit, **fun** is the name, e.g. **myfun, myfit**, or any other; for purposes of fitting the fun should be organized as sum of squared differences between the *y*-data and *y*-fit equation, e.g. for the fitting equation $y_{fit}=a+be^{cx}$ the minimized function in fun should be written as sum((y-(a(1)+a(2)*exp(a(3)*x)).^2), where vector a denotes the *a*, *b*, and *c* coefficients of the equation; whole fun in this case should be written as follows

```
function sse=myfun(a,x,y)
sss= sum((y-(a(1)+a(2)*exp(a(3)*x)).^2);
```

[] denotes the place where various additional options (e.g. absolute or/and relative tolerance and others) should be placed; if the square brackets are empty – default options are set.
a0 is vector of the initial values of the a coefficients;
a is the output vector of coefficients presented in the fitting equation which are searched with the optimization procedure by the derivative-free method[2]; defined *a*-values are transmitted to a.

As an example, we fit the dynamic viscosity μ data that was measured at various temperatures *T* and pressures *P*. The equation that is expected to fit this data is:

$$\mu = aP^{c}\exp\left(\frac{b}{T}\right)$$

where *a*, *b*, and *c* are the coefficients that should be defined.

Problem: Write the live script program with name ExViscosityP_T that determines values of the *a*, *b* and *c* coefficients by the fminsearch command and the following data: *T*=323, 323, 333, 333, 343, 343, 353, and 353 K, *P*=0.1, 20, 40, 60, 80, 100, 0.1, and 100 MPa, μ=0.5 0.65 0.7 0.85 0.91 1.06 0.37 and 0.96 mPa·s. The initial values of *a*, *b* and *c* coefficients are 1, -50, and 1 respectively. The program should display defined coefficients, *R*-squared, and a table containing the temperature, pressure, and dynamic viscosity data, together with defined viscosities and residuals.

The following commands written in the Live Editor window solve this problem – Figure 6.

Represented commands execute the following operations:

- Assign the initiating values of coefficients *a*, *b*, and *c* to the a0 vector; assign the temperature, pressure and viscositiy values to the T, P and mu vectors;
- Fit the $\mu(T,P)$ data with the fminsearch optimization command and assign defined coefficients to the a_b_c vector; the optimization is performed by searching minimum of the sse - squared sums - $\Sigma[\mu - aP^{c}\exp(b/T)]^2$; sse is calculated in the myfun sub-function that has the a, T, P, and mu variables as its input and the sse as output;
- Assign the defined coefficients to the *a*, *b*, and *c* parameters of the fitting expression;
- Calculate the mu_fit values by the fitting equation using the defined *a*, *b* and *c* coefficients;
- Calculate and assign the residuals to the res variable;
- Display with the fprintf command, the table caption and five columns with the T, P, mu, mufit and res values;
- Calculate the *R*-squared and assign this value to the output variable R_squared.

The resulting table, the fitting coefficients, and *R*-squared can be seen in the Figure 9.6.

Figure 6. Live script for the viscosity μ(T,p) data fitting via optimization

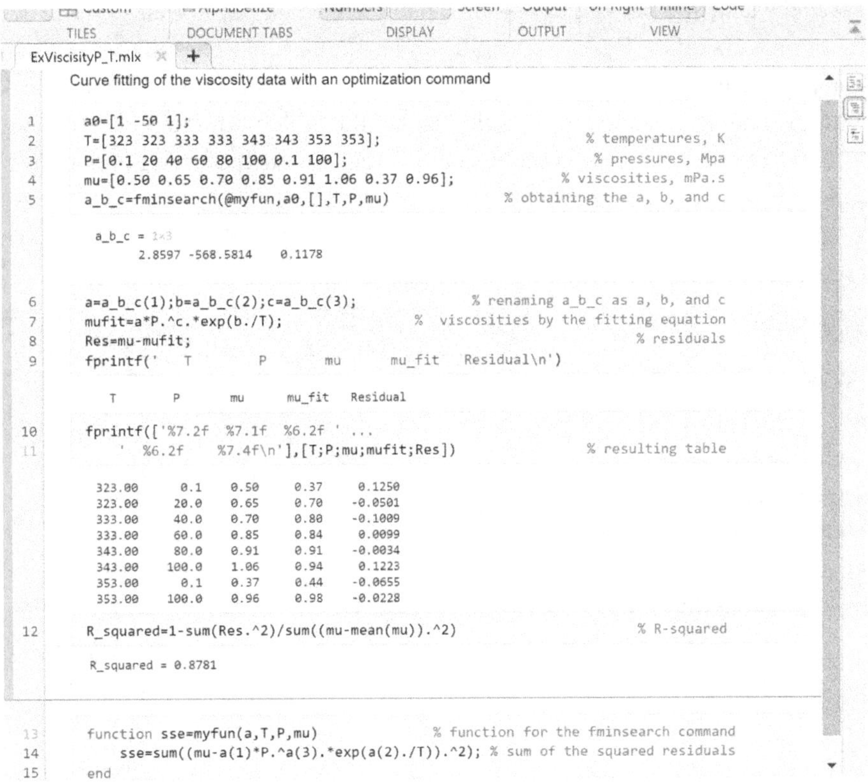

```
TILES          DOCUMENT TABS          DISPLAY          OUTPUT          VIEW
ExViscisityP_T.mlx
Curve fitting of the viscosity data with an optimization command

1   a0=[1 -50 1];
2   T=[323 323 333 333 343 343 353 353];                        % temperatures, K
3   P=[0.1 20 40 60 80 100 0.1 100];                              % pressures, Mpa
4   mu=[0.50 0.65 0.70 0.85 0.91 1.06 0.37 0.96];            % viscosities, mPa.s
5   a_b_c=fminsearch(@myfun,a0,[],T,P,mu)                % obtaining the a, b, and c

    a_b_c = 1x3
        2.8597  -568.5814    0.1178

6   a=a_b_c(1);b=a_b_c(2);c=a_b_c(3);                  % renaming a_b_c as a, b, and c
7   mufit=a*P.^c.*exp(b./T);                          %  viscosities by the fitting equation
8   Res=mu-mufit;                                                        % residuals
9   fprintf('    T          P          mu        mu_fit    Residual\n')

      T         P        mu       mu_fit    Residual

10  fprintf(['%7.2f  %7.1f  %6.2f ' ...
11      '  %6.2f     %7.4f\n'],[T;P;mu;mufit;Res])                % resulting table

    323.00      0.1    0.50     0.37      0.1250
    323.00     20.0    0.65     0.70     -0.0501
    333.00     40.0    0.70     0.80     -0.1009
    333.00     60.0    0.85     0.84      0.0099
    343.00     80.0    0.91     0.91     -0.0034
    343.00    100.0    1.06     0.94      0.1223
    353.00      0.1    0.37     0.44     -0.0655
    353.00    100.0    0.96     0.98     -0.0228

12  R_squared=1-sum(Res.^2)/sum((mu-mean(mu)).^2)                      % R-squared

    R_squared = 0.8781

13  function sse=myfun(a,T,P,mu)                  % function for the fminsearch command
14      sse=sum((mu-a(1)*P.^a(3).*exp(a(2)./T)).^2); % sum of the squared residuals
15  end
```

The *R*-squared value is lesser than 0.9 that shows a not very good approximation of the viscosities with the approximation equation; it would be better apparently to use some other viscosity equation to fit the experimental data.

Note,

- This solution is sensitive to the a0 - initial values of coefficients; as a whole, the solution is more accurate when the values of the *R*-squared or R_{norm} criterions are better (Subsection 9.1.3).
- This or another fit can be also improved with the one of additional forms of the fminsearch command that consider accuracy, repeatability, amount of iterations, etc. To study more about these options, type `>>doc fminsearch` in the Command Window.

9.4. APPLICATION EXAMPLES

9.4.1. Fuel Efficiency - Velocity Data Fitting With the 2rd and 3th Degree Polynomials

The fuel efficiencies F_e measured at various car velocities v are: F_e=4.7, 9.4, 11.9, 12.5 12.8, 12.8, 11.5, 9.8 kml and v=8, 24, 40, 56, 72, 88, 104, 120 km/h.

Problem: fit the F_e (v) data with a 2rd and 3rd - degree polynomials and calculates discrepancies (i.e. residuals) between the data and the values obtained using the fitted polinomials. Write the program as a user-defined function named Ch_9_ApExample_9_1 without input parameters and with fitting coefficients as output parameters. The program should also generate graph F_e (v) showing the data points and two fitted curves; in addition the program should display the table with the fuel efficiency and velocity data, and with calculated efficiences and the residual values for each of polynomials.

The commands solving this problem are

```
function [coeff_3,coeff_4]=Ch_9_ApExample_9_1
                    %       fits fuel efficiency-velocity data by polynomials
               %             To run: >>[coeff_3,coeff_4]=Ch_9_ApExample_9_1
v=8:16:120;                                                              % km/h
Fe=[4.7 9.4 11.9 12.5 12.8 12.8 11.5 9.8];                               % kml
a_quadr=polyfit(v,Fe,2);
a_third=polyfit(v,Fe,3);
Fe_fit=[polyval(a_quadr,v);polyval(a_third,v)];
Residuals=Fe-Fe_fit;
vg=linspace(v(1),v(end));                          %vector velocities for graph
Feg_2=polyval(a_quadr,vg);                              % fuel effic. for graph
Feg_3=polyval(a_third,vg);                              % fuel effic. for graph
plot(v,Fe,'o',Tg,Feg_2,'-b',vg,Feg_3,'--r')                   % resulting plot
grid on
xlabel('Velocity, km/h'),ylabel('Fuel efficiency, kml')
legend('Data','2nd degree poly. fit',...
'3rd degree poly. fit','location','best')                        % best located
axis tight
figure                                         % opens separate Figure window
bar(v,Residuals)
xlabel('Velocity, km/h'),ylabel('Residual, kml')
legend('2nd degree poly.','3d degree poly.')
fprintf(' Velocity   Fe   Fe_fit1 Resid1 Fe_fit2 Resid2\n')
                                                %head of the resulting table
fprintf(' %9.4f %7.4f %7.4f %7.4f %7.4f %7.4f\n',...
[v;Fe;Fe_fit(1,:);Residuals(1);...
Fe_fit(2,:);Residuals(2)])                                   % resulting table
```

These commands perform the following actions:

- Define the Ch_9_ApExample_9_1 function with the output variables coeff_2 and coeff_3 that should contain the resulting fitting coefficients for polynomials of the 2nd and 3rd degrees respectively; this function has no input arguments;
- Explain the function purposes and represents the running command in the two help lines;

- Assign the velocity and fuel efficiency values to the v and Fe variables respectively;
- Fit the $F_e(v)$ data by the 2nd degree and 3rd degree polynomials with two polyfit commands and assign defined coefficients to the coeff_2 and coeff_3 output variables;
- Calculate two-row matrix with the fuel efficiencies Fe_fit computed using defined coefficients; in this matrix, the first row is the fuel efficiences for the 2rd degree polynomial and the second row for the 3rd degree polynomial. Note that the matrix is created by the two polyvalue commands placed within the square brackets;
- Calculate the residuals as a two-row matrix, the first and second rows of which are residuals for polynomials of the secod and the third degrees respectively;
- Assign, for further plotting, the hundreed-element velocity vector vg by the linspace command with minimal (as the first element of the v vector) and maximal (as the last element of the v vector) velocities; calculate two hundreed-element vectors of the fuel efficiencies Feg_2 and Feg_3 for the 2nd and the 3rd degrees polynomials respectively, using two polyvalue commands with the assigned vg;
- Plot the $F_e(v)$ data (circles) and two fitting curves for polynomials of the 2nd and 3rd degree (solid blue and dashed red lines, respectively), and also label the axes and show the legend in the 'better' place;
- plot in the separate Figure window the residuals for polynomial fittings of the second and the third orders, and also label the axes and show the legend;
- Display the table caption and six columns with the v, Fe, Fe_1 and Resid1 (the 2nd degree fit and residuals respectively), and Fe_2 and Resid2 (the 3rd degree fit and residuals respectively).

The running command, the resulting table, fitting coefficients for polynomials of the 2nd and 3rd -degrees and the generated graphs are:

```
>> [a_quadr,a_third]=Ch_9_ApExample_9_1
 Velocity    Fe     Fe_1      Resid1   Fe_2       Resid2
        8     4.7    5.4     -0.6667    4.9      -0.1894
       24     9.4    8.7      0.6548    9.1       0.3139
       40    11.9   11.2      0.7071   11.7       0.2299
       56    12.5   12.7     -0.2095   12.9      -0.4141
       72    12.8   13.3     -0.4952   13.1      -0.2907
       88    12.8   13.0     -0.1500   12.5       0.3273
      104    11.5   11.7     -0.1738   11.3       0.1671
      120     9.8    9.5      0.3333    9.9      -0.1439
a_quadr =
  -0.0018   0.2693   3.3283
a_third =
  0.0000  -0.0039   0.3794   2.1010
```

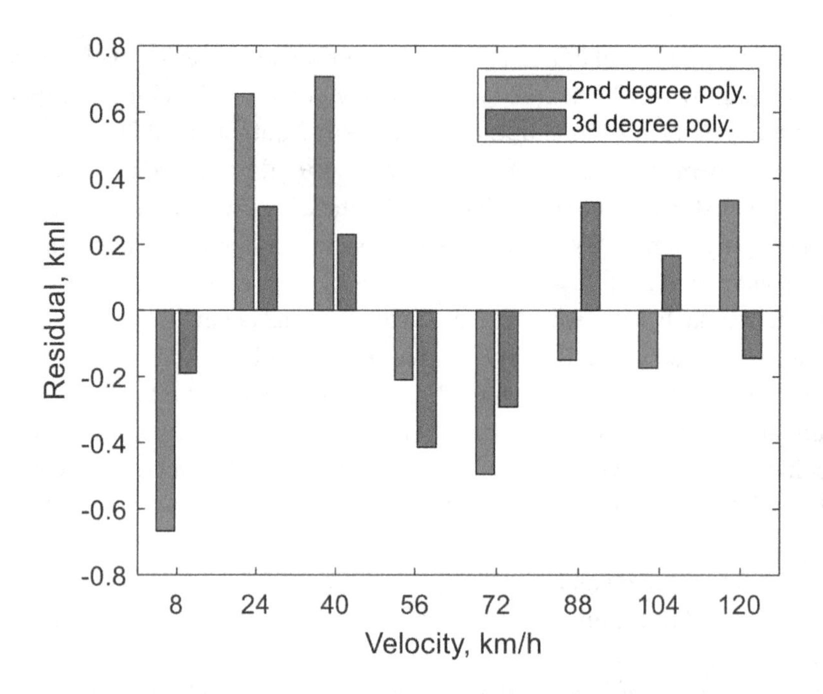

9.4.2. Fitting the Yield Strength-grain Diameter Data

The yield strength-grain diameter data, $\sigma_y(d)$, for a certain industrial metal are: σ_y= 0.205 0.150, 0.135, 0.097, 0.089, 0.080, 0.070, and 0.067 kPa and d=5, 9, 16, 25, 40, 62, 85, and 110 μm. This data can be fitted with the the non-polynomial Hall-Petch-type equation

$$\sigma_y = a_0 + a_1 . d^{-1/2}$$

Problem: fit the $\sigma_y(d)$ data with a first degree polynomial and calculates discrepancies (i.e. residuals) between the data and the values calculated by the fitting equation for the yield strength. Write the program as user-defined function named Ch_9_ApExample_9_2 without input parameters and with the defined coefficients and the norm of residuals as output parameters. The program should also generate graph $\sigma_y(d)$ with the experimental points and fitted curve, in addition the program should display the table with the d, σ_y data, and with σ_y calculated by the fitted equation; additionly the table should include the percentage of the discrepancies $|(\sigma_y - \sigma_{y\_fit})|/\sigma_y*100$ where σ_{y_fit} is the yield strength values calculated by the fitted equation.

The commands solving this problem are

```
function [a0,a1,Norm_res]= Ch_9_ApExample_9_2
                        %        fits the yield strength - grain diameter data
                        %        to run: >>[a0,a1,Norm_res]=Ch_9_ApExample_9_2
sigmay=[.205 .150 .135 .097 .089 .080 .070 .067];                       % kPa
d=[5 9 16 25 40 62 85 110];                                             % micron
```

```
a_1_2=polyfit(d.^(-1/2),sigmay,1);                    % polynomial fit
a0=a_1_2(2);a1=a_1_2(1);                               % fitting coefficients
sigmay_fit=a0+a1*d.^(-1/2);                            %sigma by fit equation
Res=sigmay-sigmay_fit;
R_percent=abs(Res)./sigmay*100;                        % discrepancies, percent
Norm_res=sqrt(sum(Res.^2));                            %norm of residuals
dg=linspace(d(1),d(end));                              % d for graph
sigmayg=a0+a1*dg.^(-1/2);                              %sigma for graph
plot(d,sigmay,'o',dg,sigmayg)
grid on
xlabel('Grain diameter, micron')
ylabel('Yield strength, kPa')
legend('Data','Fit','Location','best')                 % best location
fprintf(' d   sig_y   fit   Error,%%\n')
fprintf('%4.0f %7.3f %7.3f %7.2f\n',...
  [d;sigmay;sigmay_fit;R_percent])                     % resulting table
```

The commands of this user-defined function act as follows:

- Define the Ch_9_ApExample_9_2 function with the output variables a0, a1, and Norm_res; the function has no input arguments;
- Explain in the help lines the purposes and represents the command to run the function;
- Assign σ_y and d data to the variables sigmay and d, respectively
- Fit the $\sigma_y(d)$ data by the 1st degree polynomial using the polyfit command;
- Assign defined coefficients of fitting (contining in the a_1_2 vector) to the output variables a0 and a1;
- Calculate sigmay_fit, the yield strengths σ_{y_fit}, using the fitting expression with the defined coefficients;
- Calculate residuals, Res, discrepancies in percentages, R_percent, and norm of residuals, Norm_res.
- Assign the hundred-element vector of diameters dg for further plotting using the linspace command with minimal and maximal diameters (the first and the last elements of the d vector); calculate vector of the yield strengths sigmayg by the assigned dg using the fitting equation;
- Plot the $\sigma_y(d)$ data (circles) and fitting curve (solid line), label the axes and show the legend, that is located in the 'best' place of the graph;
- Display the table caption and four columns with the grain diameters, with the given and the fitted yield strengths, and with percentage of the residual values; note that two percent simbols are used in the fprintf command to print the % sign.

The running command, the resulting table, the defined fitting coefficients and the resulting plot are:

```
>> [a0,a1,Norm_res]=Ch_9_ApExample_9_2
     d      sig_y    fit      Error,%
       5    0.205    0.202    1.62
       9    0.150    0.158    5.02
      16    0.135    0.125    7.24
      25    0.097    0.106    9.11
      40    0.089    0.090    0.67
      62    0.080    0.078    3.08
      85    0.070    0.070    0.49
     110    0.067    0.065    2.60
a0 =
   0.0283
a1 =
   0.3877
Norm_res =
   0.0158
```

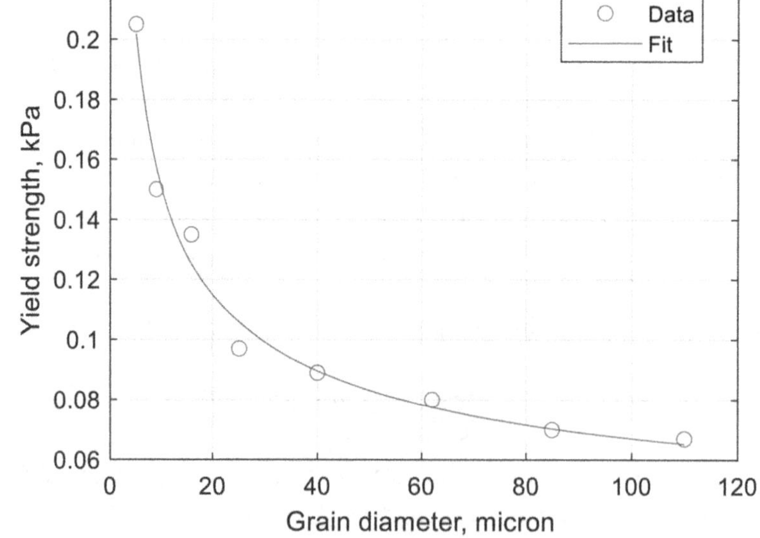

9.4.3. Using the Basic Fitting Tool to Fit Friction Coefficient Data

The friction coefficient μ of the smooth surface friction pair was measured at different times *t*. The data is: μ=0.045, 0.03, 0.029, 0.03, 0.031, 0.045, 0.059, 0.06, 0.059 0.057, and 0.057, at *t*=0, 200, 750, 1300, 1900, 2400, 2900, 3500, 3900, 4600, and 5300 sec.

Problem: Fit this data with the 3rd and 4th degree polynomials using the Basic Fitting interface and choose the polynomial with the lower "Norm of residual" criterion value. Represent the residual plot in the separate Figure window. Evaluate the friction coeficient value at *t*=4000 s with the best polynomial obtained.

To open and further use the Basic Fitting interface, first, the $\mu(t)$ graph should be plotted. For this, enter in the Command Window sequentially the t and μ values, the plot and formatting commands as follows:

```
>>t=[0 330 750 1300 1900 2400 2900 3500 3900 5000];
>>mu=[.045 .033 .029 .03 .035 .045 .055 .059 .059 .059];
>>plot(t,mu,'o')
>>xlabel('Time, s'),ylabel('Friction coefficient')
```

Thereafter, select the Basic Fitting line in the popup menu of the Tools button located in the main menu of the opened Figure window. The first panel of the Basic Fitting window appears, click now the "Show the next panel" button → and in the next panel click the → button again. The Basic Fitting window opens fully. Since the values of the friction coefficient are less than 1, for higher degrees of the polynomial terms this can yield to very small values, as a result of which the polynomial will be badly conditioned. To avoid this, we mark the "Center and scale x data" box. Now, perform the following actions in the 'Plot fits" panel:

- Mark in the 'Check to display fit on figure' field the 'Cubic' and '4th degree polynomial' boxes;
- Mark the 'Show equations', 'Plot residuals' and 'Show norm of residuals' boxes;
- Select the "Separate figure" line of the dropdown menu of the second of two adjacent fields below the " Plot residuals" box.

Now, the two Figure windows show the following two graphs:

As it can be seen, the 4rd degree polynomial has lower norm of residuals than the 3rd degree and therefore is more accurately fits the original friction coefficient data.

To evaluate the friction coefficient value at *t*=4000 s, with the 3rd degree polynomial, check/select the '4th degree polynomial' line in the drop-down list of the Fit field (in the "Numerical results" panel) and then type the value 4000 in the 'Enter value(s) or …' (shortened) field of the 'Find y=f(x)' panel and click the 'Evaluate button, the searching value 0.0604 appears in the 'f(x)' field.

The 'Basic Fitting' window with all signed fields/boxes and achieved results are represented below.

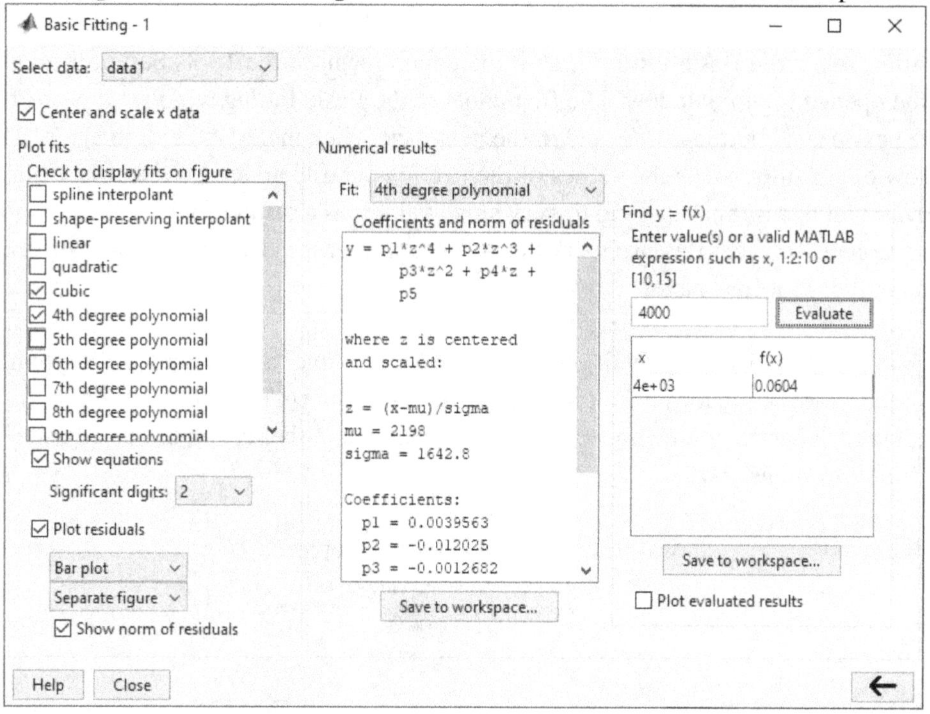

Basic Fitting - 1

Select data: data1

☑ Center and scale x data

Plot fits

Check to display fits on figure

☐ spline interpolant
☐ shape-preserving interpolant
☐ linear
☐ quadratic
☑ cubic
☑ 4th degree polynomial
☐ 5th degree polynomial
☐ 6th degree polynomial
☐ 7th degree polynomial
☐ 8th degree polynomial
☐ 9th degree polynomial

☑ Show equations

Significant digits: 2

☑ Plot residuals

Bar plot

Separate figure

☑ Show norm of residuals

Numerical results

Fit: 4th degree polynomial

Coefficients and norm of residuals

```
y = p1*z^4 + p2*z^3 +
     p3*z^2 + p4*z +
     p5

where z is centered
and scaled:

z = (x-mu)/sigma
mu = 2198
sigma = 1642.8

Coefficients:
  p1 = 0.0039563
  p2 = -0.012025
  p3 = -0.0012682
```

Save to workspace...

Find y = f(x)

Enter value(s) or a valid MATLAB expression such as x, 1:2:10 or [10,15]

4000 Evaluate

x	f(x)
4e+03	0.0604

Save to workspace...

☐ Plot evaluated results

Help Close

The obtained polynomial, the modified *x* (denoted as the centered and scaled *z* variable), the defined coefficients (not all coefficients are fully visible in presented figure) and the norm of residuals are represented in the "Coefficient and norm of residuals" field of the "Numerical Results" panel.

9.4.4. Diagnostic Parameter Coefficients of a Machine With Friction Using the Fitting

The warming temperatures of a reduction gear *T*=29.5, 29.7, 30, 30.3, 30.6, 31, 31.3, and 31.7 °C were measured at times t=1, 3, 5, 7, 9, 11, 13, and 15 sec. Uncertainty of the used thermometer was ΔT=±0.1 °C. Equation that is expected to fit this data (Burstein, 2010) is

$$\theta = \frac{e^{\gamma \hat{t}} - 1}{e^{\gamma \hat{t}} - C}$$

where $\hat{t} = t/t_0$, and $\theta = (T-T_0)/(Tc-T_0)$ are the non-dimensional time and temperature respectively; t_0 and T_0 are the initial (first value in their series) time and temperature respectively; *Tc* – is the final (last in the series) measured value of temperature; γ and *C* are the coefficients of a some diagnostic parameter.

The coefficients γ and *C* called the diagnostic parameter coefficients are used to calculate a diagnostic parameter and the residual service life of a machine with friction (Burstein, 2009), nevertheless now we demonstrate only how to determine them using the fitting of the $\theta(\hat{t})$ data.

Problem: Fit the $\theta(\hat{t})$ data with the above equation using the fminsearch optimization command. Write user-defined function with name Ch_9_ApExample_9_4 with the γ, C, and R-squared as output parameters and without input parameters. Take the initial γ and C values as 0, and -1 respectively. The function should generate a graph that shows the data with their uncertainties and the fitted curve. Numerical results should be displayed as a table containing the time and temperature data, together with the defined temperatures and residuals; all values must be dimensional.

The commands that solve this problem are:

```
function [gam,C,R_Squared]=Ch_9_ApExample_9_4
                                   %        fits the temperature versus time data
                              %     To run: >>[gam,C,R_Squared]=Ch_9_ApExample_9_4
t=1:2:15;                                                                    % sec
T=[29.5 29.7 30 30.3 30.6 31.1 31.5 31.8];                              % degree C
T_error=0.1*ones(1,length(T));                             % errors for each T-point
Tc=T(end);dT=(Tc-T(1));
Theta=(T-T(1))./dT;T_errorg=T_error./dT;                          %dimensionless T
a0=[-1 0];                                                  % initial coefficients
gam_C=fminsearch(@myfun,a0,[],t,Theta);                                  % fitting
gam=gam_C(1);C=gam_C(2);                                        % output variables
Thetafit=(exp(gam*t)-1)./(exp(gam*t)-C);                             % teta by eq.
Res=Theta-Thetafit;                                                    % residuals
R_Squared=1-sum(Res.^2)/sum((Theta-mean(Theta)).^2);                          %R^2
Tfit=Thetafit*(Tc-T(1))+T(1);                                      % dimensional T
errorbar(t,T,T_error,'o '),hold on                           %plots data with error
plot(t,Tfit),grid on                                        %plots the fitted curve
xlabel('Time, s'),ylabel('Temperature, ^oC')
legend('Data','Fit', 'Location','best'),hold off
fprintf(' t     T     T_fit  Residual\n')
fprintf('%5.0f %7.1f  %7.1f  %7.4f\n',[t;T;Tfit;Res])
function sse=myfun(a,t,Theta)                                  % minimized function
sse=sum((Theta-(exp(a(1)*t)-1)./(exp(a(1)*t)-a(2))).^2);
```

Represented commands execute the following:

- Define the Ch_9_ApExample_9_4 function with the output variables gam, C, and R_squared that should contain the γ, C, and R^2 values, respectively;
- Explain the purposes and represents the running command - in the two first help lines;
- Assign the time and temperature data to the variables t and T respectively;
- Calculates nondimensional temperatures θ; note, the first t value is equal to 1, therefore, nondimensional temperature values are equal to dimensional values and they do not need to be transformed;

- Fit the $\theta(t)$ data with the fminsearch optimization command and assign defined coefficients to the gam_C vector; the optimization is performed by searching minimum of the sse - squared sums - $\Sigma[\theta-(e^{\gamma t}-1)/(e^{\gamma t}-C)]^2$; sse wrote in the myfun sub-function that has the a, t, and Teta variables as its input and the sse as output;
- Assign the defined coefficients (located in the two-element vector gam_C) to the gam, and C output variables;
- Calculate the Thetafit values by the fitting equation using the defined γ and C coefficients;
- Calculate and assign the residuals to the res variable;
- Calculate the R-squared and assign this value to the output variable R_squared;
- Transform the defined dimensionless temperatures Thetafit to the dimensional Tfit with the equation $T_{fit}=T_0+\theta_{fit}(T_c-T_0)$;
- Generate resulting graph using two commands: errorbar and plot, the first command produces T-data points with the rerror bars and the second adds the fitting curve $T(t)$ to the same graph, which allows the hold on command to be performed;
- Display with the two fprintf commands, the table caption and four columns with the t, T, Tfit and Res values.

The running command, the resulting table and graph, the fitting coefficients, and R-squared are:

```
>> [gam,C,R_Squared]=Ch_9_ApExample_9_4
  t     T       T_fit  Residual
   1   29.5    29.5   -0.0161
   3   29.7    29.7    0.0157
   5   30.0    29.9    0.0443
   7   30.3    30.3    0.0143
   9   30.6    30.7   -0.0544
  11   31.1    31.2   -0.0223
  13   31.5    31.5    0.0204
  15   31.8    31.6    0.0746
gam =
  0.3935
C =
 -28.3988
R_Squared =
  0.9871
```

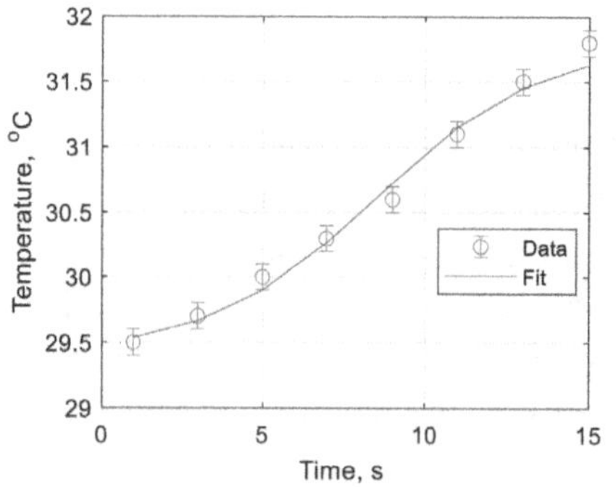

The *R*-squared value is more than 0.98 that shows a very good approximation of the warming temperatures with diagnostic parameter coefficients that can be used for defining the remaining service life of the tested gear box.

The solution is sensitive to the initial coefficients values; see comment in Subsection 9.3 on this subject.

9.5. CONCLUSION

Commands for single-variable polynomial, non-polynomial and multi-variable fitting as well as the Basic Fitting tool were studied and used to fit the following experimental M&T data:

- Volumetric wear rate– normal load,
- Rubber elongation - force,
- Dynamic viscosity – pressure and temperature,
- Fuel efficiency – velocity,
- Yield strength - grain diameter,
- Friction coefficient - time,
- Diagnostic parameter – time.

The developed programs, in addition to fitting, evaluate the goodness of fit by the *R*-squared and *R*-norm criteria. As a whole, the studied means are convenient for mathematical description and data analysis for M&T problems.

REFERENCES

Burstein, L. (2009). Machine friction as diagnostic tool: empirical and theoretical considerations. In J. P. Davim (Ed.), *Tribology research advances* (pp. 79–108). Nova.

Burstein, L. (2010). Machine thermal diagnostics latest advances. In E. N. Mallory (Ed.), *Nondestructive testing. Methods, analyses and applications* (pp. 147–183). Nova.

ENDNOTES

1 The text is shortened.

2 The method that do not use derivatives or finite differences; studied in the numerical analysis courses.

APPENDIX

Table 2. List of examples, problems, and applications discussed in the chapter

No	Example, Problem, or Application	Subsection
1	Fit the 'wear rate– normal load' data.	9.1.2, 9.1.3.
2	Fit the elongation- force data for a rubber using the Basic Fitting tool.	9.2.3.
3	Fit for data on the dynamic viscosity as function of temperatures and pressures using an optimization command.	9.3.
4	Fuel efficiency - velocity data fitting with the 2rd and 3th degree polynomials.	9.4.1.
5	Fitting the yield strength-grain diameter data.	9.4.2.
6	Using the Basic Fitting tool to fit friction coefficient data.	9.4.3.
7	Diagnostic parameter coefficients of a machine with friction using the fitting.	9.4.4.
8	Fit the 'wear rate– normal load' data.	9.1.2, 9.1.3.
9	Fit the elongation- force data for a rubber using the Basic Fitting tool.	9.2.3.

Note, some small examples, mostly related to non-M&T issues, are not included in the list.

Chapter 10
Symbolic Calculations

ABSTRACT

The chapter is devoted to symbolic calculations in which the variables and commands operate on mathematical expressions containing symbolic variables. The representation of a symbolic expression, its simplification, the solution of algebraic expressions, symbolic differentiation and integration, and conversion of the symbolic numbers to their decimal form are described. ODEs solutions are also presented. The final sections of the chapter give examples of the symbolic calculation implementation for some mechanical and tribological problems that were solved numerically in previous chapters, namely lengthening a two-spring scale, shear stress in a lubrication film, a centroid of a certain plate, and two-way solutions of the ODE describing the second order dynamical system – traditional and using the Laplace transform.

INTRODUCTION

The commands, calculations, and application examples discussed in all previous chapters were numerical, which means that all variables, matrix elements, operations, and calculation results were numbers (single number, vector or array of numbers). In science in general and also in engineering practice, operations are often performed with non-numeric variables, and the results are presented in the form of mathematical expressions containing the variable names (called in this case 'symbolic variables'). For example, to calculate $2*a*a$ in numerical calculations, you must first assign to the variable a some numerical value, e.g. $a = 2$, and the result of $2*a*a$ is the number 8. In contrast to this, in symbolic calculations, the result $2*a*a$ is the expression $2a^2$, and in this case is no need to pre-assign a number to the variable a. Another example, a numerical solution of the equation $ax+b=0$ when $a=2$ and $b=-3.12$ is $x=3.12/2=1.56$ when a symbolic solution is $x=-b/a$.

In MATLAB®, symbolic calculations are performed with the Symbolic Math Toolbox (SMTbx) that should be previously installed. This toolbox contains many commands and functions defining symbolic variables, performing symbolic math operations, achieving symbolic expressions and plotting results. When the toolbox operates with symbolic numbers, they are not converted to approximate values, how this is done in numerical computing, but carried out exactly as they are. For example, $y/2+y/3$ is $5/6y$

DOI: 10.4018/978-1-7998-7078-4.ch010

and not 0.8333*y*. SMTbx today uses MuPad® software to provide symbolic computing; until 2007, the Maple® software was used for this purpose.

In this chapter, the following commands and its usage for symbolic calculations are presented:

- Definition the variables and objects;
- Manipulations with math expressions, algebraic and matrix operations, extremum and limits of math functions;
- Solutions of the algebraic equations, differentiation, and integration;
- Solutions of the ordinary differential equations;
- Applications with the examples of the symbolic solutions to some mechanical and tribological problems that were previously solved numerically.

Since programming for symbolic calculations is different from previous calculations, the programs in this chapter have more comments than in preceding chapters.

10.1. DEFINING SYMBOLIC OBJECTS

In symbolic calculations the variables, group of variables and numbers are defined as symbolic objects and belong to the class of the symbolic variables. Due to this, all symbolic operations can be performed with such variables. Therefore, the variable should be defined first as symbolic that can be made with one of the two commands – syms and sym.

10.1.1 The syms Command

One or several symbolic variables or matrices can be created with two the syms commands that have the following forms:

```
syms variable_1 variable_2 ...
```

or

```
syms matrix_1 matrix_2 … [n m] …
```

After typing this command, the semicolon is not required to suppress the result displaying.

Examples of creating symbolic objects:

```
>> syms E b x                                  % creates 3 symbolic variables
>> E                                          % to display symbolic variable E
E =
E
>> syms v1 v2 [1 3]             % creates 2 symbolic vectors, 3 elements each
>> v1,v2                              % to display symbolic vectors v1 and v2
v1 =
```

```
[ v11, v12, v13]
v2 =
[ v21, v22, v23]
>> syms m [2 3]                              % creates symbolic 2x3 matrix
>> m                                         % to display symbolic matrix m
m =
[ m1_1, m1_2, m1_3]
[ m2_1, m2_2, m2_3]
```

In addition, the **syms** command can be used to define symbolic function with one or two variables, for example,

```
>>syms y(t)              %creates symbolic function y of the variable t
>>syms f(x,y)            %creates symbolic function f of the var-s x and y
```

The independent variables of such functions (t or x and y in these examples) are also symbolic and should not be specified separately.

Once the variables were created with the **syms** command, their type is symbolic that can be checked in the Workspace window when the Class option is selected.

To address some element of symbolic matrix the same rules should be used as for numerical matrices – the row and column numbers separated by coma should be written after matrix name in the parentheses. For example, an element in the second row and third column of the m-matrix can be obtained by entering the command m(2,3); or all elements in the second column can be obtained by entering m(:,2).

10.1.2. The sym Command

Another command that can be used to create a symbolic variable, as well as to create a symbolic number, or a group of variables (object) is the sim command, the simplest two forms of which are as follows:

```
Sym_object=sym('String')
```

or

```
Sym_object=sym('String',[n m])
```

Examples of the **sym** command usage are:

```
>> x=sym('x')                % creates object x containing symbolic variable x
x =
x
>> E=sym('Young')               % creates object E containing symbolic Young
E =
Young
>> y=sym('12.3546')             %creates object y containing symbolic number
```

```
y =
12.3546
>>V=sym('a',[1 4])                    %creates 1x4 vector V with symbolic a-terms
V =
[ a1, a2, a3, a4]
>>M=sym('m1',[2 3])                   %creates 2x3 matrix M with symbolic m1-terms
M =
[ m1_1, m1_2, m1_3]
[ m2_1, m2_2, m2_3]
```

Note that the sym command actually converts the string to a symbolic object, since the string in its right-hand side is converted to a symbolic object on the left-hand side.

10.1.3. Some Additional Information about the syms and sym Commands

- In case creating a matrix with the sym command, the matrix elements are **not separate symbolic variables** and can not be used separately from the matrix, e.g. m1_1, m1_2, … in the M matrix of the sym example are not separate variables and they not appear in the Workspace window;
- In case creating a matrix with the syms command, the matrix elements are separate symbolic variables appearing in the Workspace;
- While creating variables with the sym command, some assumption can be added to this command after the variable name, e.g. sym a positive defines a as symbolic variable that is greater than zero, or sym a b real defines a and b as symbolic and real variable; possible assumptions can be obtained by entering doc assume.
- When the number is defined with the sim command it can be written without single quotas, e.g. y=sym(12.3546).
- For definition of the quadratic matrix only one dimensional number can be written in the syms or in the sym command, e.g. >>syms m 2 produces 2x2 symbolic matrix m and M=sym('M',3) produces 3x3 symbolic matrix object M.

10.2. SYMBOLIC EXPRESSIONS, MATRICES, AND MANIPULATIONS WITH THEM

10.2.1. Expressions, Matrices and Algebraic Operations

If symbolic variables are defined, then the mathematical expressions can be written in symbolic notations. For this the same MATLAB® characters are used as in regular, numerical calculations. For example, quadratic algebraic equation can be entered so:

```
>>syms a b c x
>>y=a*x^2+b*x+c
 y =
 a*x^2 + b*x + c
```

If at least one of the variables is symbolic on the right side of the equation, then the variable on the left side of the equation is also symbolic, and it does not need to be defined using the syms or sym command. Therefore, in the above equation y is symbolic, although it is not defined in the previous sym command.

If we have more than one equation, they can be added, subtracted, or multiplied, divided one into another. For example, in addition to the quadratic equation above we have equation z=*a*e-x:

```
>>z=a*exp(-x)
```

Notice that here we did not define the a and x variables as symbolic because they were defined as those in the previous example.

Execute now the addition, multiplications, and division operations with the y and z expressions.

```
>>y+z,y*z,y/z
ans =
c + b*x + a*x^2 + a*exp(-x)
ans =
 a*exp(-x)*(a*x^2 + b*x + c)
ans =
(exp(x)*(a*x^2 + b*x + c))/a
```

The variables in resulting symbolic expressions are always ordered in decreasing order of power.

Numbers in symbolic expressions are transformed to the ratio of two integers. For example, the friction factor of the pipe surface are (subsection 2.1.7.2):

$$f = \left\{ \left(\frac{64}{Re} \right)^{8} + 9.5 \left[ln \left(\frac{\varepsilon}{3.7D} + \frac{5.74}{Re^{0.9}} \right) - \left(\frac{250}{Re} \right)^{6} \right]^{-16} \right\}^{\frac{1}{8}}$$

This expression presented symbolically reads

```
>> syms epsilon D Re
>>f=((64/Re)^8+9.5*(log(epsilon/(3.7*D)+5.74/Re^.9)-(250/Re)^6)^-16)^(1/8)
```

After entering these commands, the friction coefficient f looks like this

```
f =
 (19/(2*(log((10*epsilon)/(37*D) + 287/(50*Re^(9/10))) - 244140625000000/
Re^6)^16) + 281474976710656/Re^8)^(1/8)
```

Here 19/2=9.5, 9/10=0.9, 37/10=3.7. The computer also performed some additional transformations: the first term moved to the last, the number 250^6 is represented as 244140625000000, and the (..)^-16 part of the expression is rewritten as 1/(...)^16.

The computer also performed some additional transformations: the first term moved to last, the number 250 ^ 6 is represented as 244140625000000, and the expression (..) ^ - 16 is rewritten as 1 / (...) ^ 16.

Note, when an equation has the right part, it can be assigned to the variable as it with the double equal sign between the equation parts, for example, equation $ax^2+b\sin(x)=23x$ can be assigned to variable eq in such way

```
>> syms x a b
>> eq=a*x^2+b*sin(x)==23*x
eq =
a*x^2 + b*sin(x) == 23*x
```

The basic arithmetic/algebraic operations can be used also for matrices. Create with the sym command, for example, two 2x2 matrices and execute addition, multiplication, and division operations

```
>>M1=sym('M1',2)                                    % creates 2x2 matrix object M1
M1 =
[ M11_1, M11_2]
[ M12_1, M12_2]
>>M2=sym('M2',2)                                    % creates 2x2 matrix object M2
M2 =
[ M21_1, M21_2]
[ M22_1, M22_2]
>>M1+M2
ans =
[ M11_1 + M21_1, M11_2 + M21_2]
[ M12_1 + M22_1, M12_2 + M22_2]
>>M1*M2
ans =
[ M11_1*M21_1 + M11_2*M22_1, M11_1*M21_2 + M11_2*M22_2]
[ M12_1*M21_1 + M12_2*M22_1, M12_1*M21_2 + M12_2*M22_2]
>>M2*M1
ans =
[ M11_1*M21_1 + M12_1*M21_2, M11_2*M21_1 + M12_2*M21_2]
[ M11_1*M22_1 + M12_1*M22_2, M11_2*M22_1 + M12_2*M22_2]
>>M1/M2
ans =
[ (M11_1*M22_2 - M11_2*M22_1)/(M21_1*M22_2 - M21_2*M22_1), -(M11_1*M21_2 -
M11_2*M21_1)/(M21_1*M22_2 - M21_2*M22_1)]
[ (M12_1*M22_2 - M12_2*M22_1)/(M21_1*M22_2 - M21_2*M22_1), -(M12_1*M21_2 -
M12_2*M21_1)/(M21_1*M22_2 - M21_2*M22_1)]
```

This operations actually illustrate the rules of work with matrices, since it can be seen that M1*M2 is not the same as M2*M1, and M1/M2 is matrix division, but not elementwise division.

10.2.2. Simplification of Symbolic Expressions

A symbolic expression can be changed by combining identical terms, by expanding products, and by many other mathematical and trigonometrical transformations. Further, we present some commands that can be used to change form of a symbolic expression.

10.2.2.1. The collect Command

This command assembles the symbolic expression terms that represented by the variables with the same power. Two simplest forms of this command are

```
collect(s_expression)
```

Examples for the collect command usage are:

```
>>syms x                                  % define symbolic variables x and y
>>s1=(x^2+x-log(x))*(x-1);              % assign the symbolic expression to s1
>>collect(s1)                                    % collect s1 by power of x
ans =
x^3 + (- log(x) - 1)*x + log(x)
>>s2=x*cos(x)+x*sin(x);                % assign a new symbolic expression to s2
>>collect(s2)                                    % collect s2 by power of x
ans =
(cos(x) + sin(x))*x
>>s=s1+s2                                              % addition s1 and s2
s =
(x - 1)*(x - log(x) + x^2) + x*cos(x) + x*sin(x)
>>collect(s)                                      % collect s by power of x
ans =
x^3 + (cos(x) - log(x) + sin(x) - 1)*x + log(x)
>> s3=(x + y)*(y - 2*x + 3)                %expression with two symb. variables
s3 =
(x + y)*(y - 2*x + 3)
>> collect(s3)
ans =
- 2*x^2 + (3 - y)*x + y*(y + 3)
```

10.2.2.2. The expand Command

With this command products of terms in parentheses expends to the non-parentheses sum of terms. The simplest form of this command is

```
expand(s_expression)
```

The examples are:

```
>> syms x y                                % define symbolic variables x and y
>> s=(x+y)^4                                 % assign (x+y)^4 expression to s
s =
(x + y)^4
>>expand(s)                    % expands s and orders in decreasing power of x
ans =
x^4 + 4*x^3*y + 6*x^2*y^2 + 4*x*y^3 + y^4
```

This command can expand trigonometric expressions by using trigonometric identities, e.g.

```
>> expand(cos(x+y))            % expands cos according to trigonometr.identity
ans =
cos(x)*cos(y) - sin(x)*sin(y)
```

10.2.2.3. The simplify Command

The command is intended to transform a symbolic expression to its simplest form that has least number of terms. However, the command does not guarantee that the final expression is the most simplified form of expression, although actually this happens frequently.

The simplify command can operates with regular algebraic, trigonometric, logarithmic and exponential expressions. The form of this command is

```
simplify(s_expression)
```

Some examples:

```
>> syms x y a
>> s=x^3 + 3*x^2*y + 3*x*y^2 + y^3;
>> simplify(s)
ans =
(x + y)^3
>>s1=5+ sin(x)^2 + cos(x)^2
s1 =
cos(x)^2 + sin(x)^2 + 5
```

```
>> simplify(s1)
ans =
6
>>s2= exp(a*log(sqrt(x+y)))
>> simplify(s2)
ans =
(x + y)^(a/2)
```

10.2.2.4. Representing Symbolic Expression in Regular Mathematical Form

Symbolic expressions are presenting in MATLAB notations used signs and forms that do not the same as in mathematics, e.g. a*b instead of ab, sqrt(x) instead $\sqrt{x}$, and so on. The pretty command permits to display a symbolic expression like regular mathematical expression. The form of this command is

```
pretty(s_expression)
```

For example, the symbolic equation for the friction factor of the pipe surface (see subsection 10.2.1), in MATLAB notations, has complicate form

```
>> syms epsilon D Re
>> f=((64/Re)^8+9.5*(log(epsilon/(3.7*D)+5.74/Re^.9)-(250/Re)^6)^-16)^(1/8)
f =
 (19/(2*(log((10*epsilon)/(37*D) + 287/(50*Re^(9/10))) - 244140625000000/
Re^6)^16) + 281474976710656/Re^8)^(1/8)
```

After entering the **pretty(f)** command, the f expression view resembled to presented in Figure 1

Figure 1. Typical view of a symbolic expression after entering the pretty command

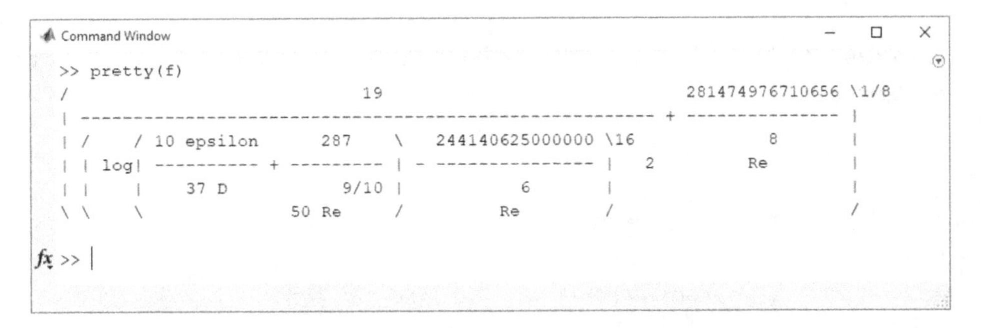

Although this form is much more readable and close to the form of a regular mathematical equation (than written in MATLAB notations above), it still does not look quite "pretty". More regular math form of this and any other symbolic equation can be obtained with the Live Editor; and this is not required usage of the pretty command – Figure 2.

Figure 2. Typical representation of symbolic expression using the Live Editor

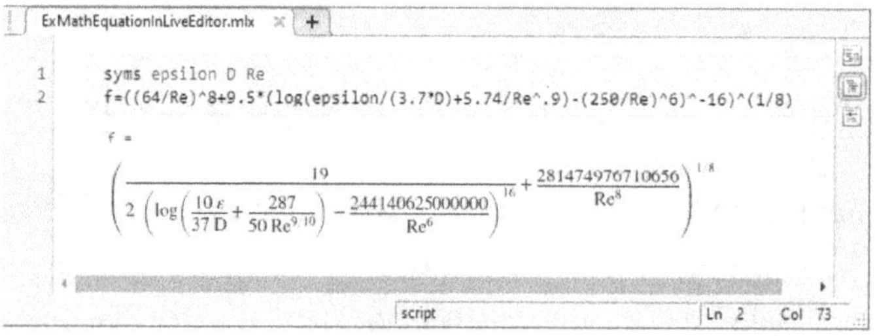

ExMathEquationInLiveEditor.mlx

```
1   syms epsilon D Re
2   f=((64/Re)^8+9.5*(log(epsilon/(3.7*D)+5.74/Re^.9)-(250/Re)^6)^-16)^(1/8)
```

f =

$$\left(\frac{19}{2\left(\log\left(\frac{10\,\varepsilon}{37\,D}+\frac{287}{50\,\mathrm{Re}^{9/10}}\right)-\frac{244140625000000}{\mathrm{Re}^6}\right)^{16}}+\frac{281474976710656}{\mathrm{Re}^8}\right)^{1/8}$$

script | Ln 2 Col 73

As it can be seen, the representation of the symbolic expression and other results of symbolic calculations in Live Editor is more understandable and will be used further in this chapter.

10.3. DECIMAL REPRESENTATION OF THE SYMBOLIC NUMBERS

As stated above, numbers in symbolic calculations are converted to the ratio of two integers. To save the decimal number notations of the original expression or convert symbolic numbers of the achieved expression to decimal, the vpa command can be used. Its simplest form reads

```
vpa(s_equation,n)
```

For example, the f expression above with entering the vpa(f,4) command converted the two integers ratio and long integer numbers in decimal numbers with four digits – Figure 3.

Figure 3. Converting symbolic numbers in the symbolic expression to the four-digit decimal numbers with the vpa command

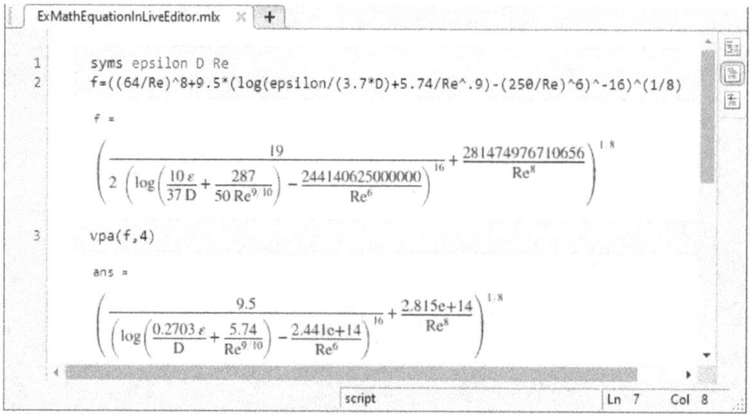

ExMathEquationInLiveEditor.mlx

```
1   syms epsilon D Re
2   f=((64/Re)^8+9.5*(log(epsilon/(3.7*D)+5.74/Re^.9)-(250/Re)^6)^-16)^(1/8)
```

f =

$$\left(\frac{19}{2\left(\log\left(\frac{10\,\varepsilon}{37\,D}+\frac{287}{50\,\mathrm{Re}^{9/10}}\right)-\frac{244140625000000}{\mathrm{Re}^6}\right)^{16}}+\frac{281474976710656}{\mathrm{Re}^8}\right)^{1/8}$$

```
3   vpa(f,4)
```

ans =

$$\left(\frac{9.5}{\left(\log\left(\frac{0.2703\,\varepsilon}{D}+\frac{5.74}{\mathrm{Re}^{9/10}}\right)-\frac{2.441\mathrm{e}{+}14}{\mathrm{Re}^6}\right)^{16}}+\frac{2.815\mathrm{e}{+}14}{\mathrm{Re}^8}\right)^{1.8}$$

script | Ln 7 Col 8

The **vpa** command can be used also to calculate quantity with more than 15 digits (as it the case in numerical calculations). For example 25! in numerical long format is 15th decimal digits number 1.551121004333099e+25 and entering the **vpa(factorial(25),25)** command we obtain the 25th digits of precision - 15511210043330986055303170.0.

10.4. SOLUTION OF ALGEBRAIC EQUATIONS

Symbolic solution of a single algebraic equation is dependence of any of variables from the others. Solution of a set of algebraic equations is dependence of desired variables form others. The solution can be executed with the solve command, which has the following simplest forms

```
s=solve(s_equation1,s_equation2,…)
```

```
s=solve(s_equation1,s_equation2,…, s_var1, s_var2,…)
```

Consider for example the inverse problem of the pipe's coefficient of friction λ, case of fully developed turbulent flow (see subsection 5.2). The equation reads (Colebrook formula, based on Hwang & Hita, 1987)

$$\lambda - \frac{1}{\left[2\log\left(3.7\frac{d}{k}\right)\right]^2} = 0$$

Here the ratio of pipe diameter d to surface roughness parameter k should be defined; all variables in this equation are real and positive.

The solution of this equation implemented in the Live Editor is as follows (hereinafter we give only commands and inline output, and not the full Live Editor window):

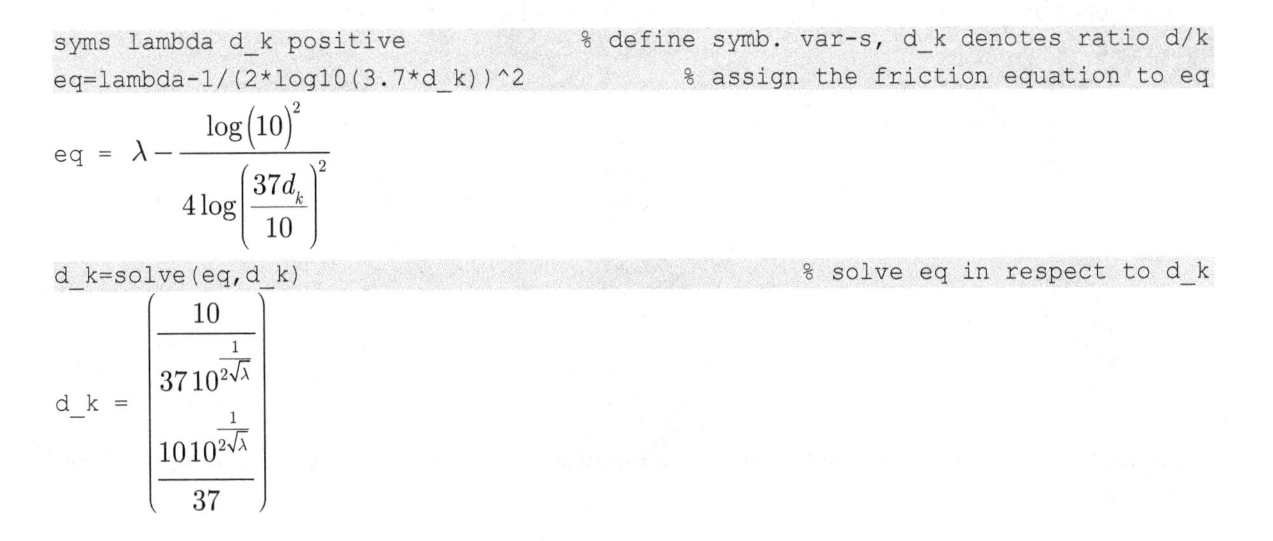

```
syms lambda d_k positive                 % define symb. var-s, d_k denotes ratio d/k
eq=lambda-1/(2*log10(3.7*d_k))^2              % assign the friction equation to eq
```

eq = $\lambda - \frac{\log(10)^2}{4\log\left(\frac{37 d_k}{10}\right)^2}$

```
d_k=solve(eq,d_k)                                    % solve eq in respect to d_k
```

d_k = $\begin{pmatrix} \frac{10}{37\,10^{\frac{1}{2\sqrt{\lambda}}}} \\ \frac{10\,10^{\frac{1}{2\sqrt{\lambda}}}}{37} \end{pmatrix}$

As it can be seen, there are two possible symbolic solutions; in the case of numerical calculation of *d*/*k* by defined two formulas, the most real *d*/*k* ratio should be selected, i.e. *d*/k should be greater than 1, because the pipe diameter *d* a priory is much larger than parameter of the roughness profile *k* .

To demonstrate how to use the **solve** command for a set of equations, consider now set of two equations that should fit some experimental data (Enachescu et. al., 1999) on the friction force *F* as a function of the contact area *s* (see Subsection 2.4.1.):

$$a_1 n + a_2 \sum_{i=1}^{n} s_i = \sum_{i=1}^{n} F_i$$

$$a_1 \sum_{i=1}^{n} s_i + a_2 \sum_{i=1}^{n} s_i^2 = \sum_{i=1}^{n} s_i F_i$$

here *n* is the number of data points, symbol Σ denotes summation of quantity over all data points; a_1 and a_2 are the fitting coefficient that should be determined symbolically.

Solution of this equations in Live Editor is as follows:

```
syms n a1 a2 sigma_s sigma_f sigma_s2 sigma_sf              % define symbolic var-s
eq1=a1*n+a2*sigma_s==sigma_f;                        % assign first equation to eq1
eq2=a1*sigma_s+a2*sigma_s2==sigma_sf;               % assign second equation to eq2
s=solve(eq1,eq2,a1,a2);                % solve eq1 and eq2 with respect to a1 and a2
a1=s.a1                                    % assign the first element of the s to a1
```

$$a1 = \frac{\sigma_f \sigma_{s2} - \sigma_s \sigma_{sf}}{n\sigma_{s2} - \sigma_s^2}$$

```
a2=s.a2                                   % assign the second element of the s to a2
```

$$a2 = \frac{n\sigma_{sf} - \sigma_f \sigma_s}{n\sigma_{s2} - \sigma_s^2}$$

This set of linear equations can be represented and solved in matrix form (see Subsection 2.4.1) as follows:

```
syms n a1 a2 sigma_s sigma_f sigma_s2 sigma_sf              % define symbolic var-s
A=[n sigma_s;sigma_s sigma_s2];                % assign left-part coefficients to A
B=[sigma_f;sigma_sf];                  % assign right-part coefficients to vector B
a=A\B                                          % matrix solution of the equation set
```

$$a = \begin{pmatrix} \frac{\sigma_f \sigma_{s2} - \sigma_s \sigma_{sf}}{n\sigma_{s2} - \sigma_s^2} \\ \frac{n\sigma_{sf} - \sigma_f \sigma_s}{n\sigma_{s2} - \sigma_s^2} \end{pmatrix}$$

As expected, both demonstrated solutions of the two-equation system reveal the same resulting expressions.

10.5. SUBSTITUTING NUMERICAL VALUES IN SYMBOLIC EXPRESSIONS

When symbolic solution achieved, it can be used to obtain actual numerical results, for which numerical values should be assigned to the symbolic variables and symbolic result should be converted to its numerical representation. While the second operation can be perform with the previously described vpa command, the first action should be executed with the subs command that has the following forms:

```
N=subs(s_expression, s_var, num_val)
```

```
N=subs(s_expression,{s_var1,s_var2,…},{num_val1,num_val2,…})
```

In case assigning numbers to the variables before the **subs** command, this command can be written in form

```
N=subs(s_expression)
```

In this case all previously assigned variables are automatically substituted in the symbolic expression.

We shall demonstrate the **subs** and the **vpa** commands use with the previous subsection examples where the symbolic expressions were obtained for ratio *d/k* and for fitting coefficients a_1 and a_2. Now we obtain numerical results using this examples.

The live program for defining the value of the ratio *d/k* for λ=0.024 is as follows

```
syms lambda d_k positive                                  % d_k denotes ratio d/k
eq=lambda-1/(2*log10(3.7*d_k))^2;                % assign the friction equation to eq
d_k=solve(eq,d_k);                                       % solve eq in respect to d_k
d_k=subs(d_k,lambda,0.024)              % in d_k expres. substitute 0.024 in lambda
```

$$d\_k = \begin{pmatrix} \frac{10}{37\,10^{\frac{\sqrt{3}\sqrt{125}}{6}}} \\ \frac{10\,10^{\frac{\sqrt{3}\sqrt{125}}{6}}}{37} \end{pmatrix}$$

```
d_k_numeric=vpa(d_k,7)             % convert symbolic d_k to numeric with 7 digits
```

$$d\_k\_numeric = \begin{pmatrix} 0.0001600708 \\ 456.3356 \end{pmatrix}$$

As the d/k ratio can not be less than 1 thus only second result is valid (compare with numerical calculations in Subsection 5.2).

The numerical values of the coefficients in the fitting equations for friction force and contact area values are F =3, 4, 5, 6, and 7 and s =0.77, 1.00 1.29 1.51, and 1.71

```
syms n a1 a2 sigma_s sigma_f sigma_s2 sigma_sf             % define symbolic var-s
eq1=a1*n+a2*sigma_s==sigma_f;                          % assign first equation to eq1
eq2=a1*sigma_s+a2*sigma_s2==sigma_sf;                 % assign second equation to eq2
s=solve(eq1,eq2,a1,a2);                % solve eq1 and eq2 with respect to a1 and a2
a1=s.a1;                                    % assign the first element of the s to a1
a2=s.a2;                                    % assign the second element of the s to a
s=3:7;                                                           % Contact area values
F=[0.77 1.00 1.29 1.51 1.71];                                   % friction force values
n=length(s);
sigma_s=sum(s);
sigma_f=sum(F);
sigma_s2=sum(s.^2);
sigma_sf=sum(s.*F);
a1=subs(a1)                                     % substitute numerical values in a1
```

a1 = $\frac{61}{1000}$

```
a2=subs(a2)                                     % substitute numerical values in a2
```

a2 = $\frac{239}{1000}$

```
a1_numeric=vpa(a1,5)                  %convert symbolic number a1 to the numeric form
a1_numeric =
0.061
a2_numeric=vpa(a2,5)                  %convert symbolic number a2 to the numeric form
a2_numeric =
0.239
```

Obtained coefficients a_1 and a_2 are identical to those calculated numerically in the subsection 2.4.1.

10.6. DIFFERENTIATION AND INTEGRATION

Symbolic differentiation can be performed with the **diff** command that has the following forms

```
diff(s_expression)
diff(s_ expression,var)
diff(s_ expression,n)
diff(s_ expression,var,n)
```

These commands can be written in an assigning expression, e.g. dsdt=diff(s_expression,var,n).

The first form of the diff commands performs differentiations with respect to default variable – *x*, *y*, or *t* (in given precedence)

Some live script examples of differentiation of the $y=a\cdot\sin(bt)$:

```
syms a b t                                     % define symbolic variables a, b, and t
y=a*sin(b*t)                               % expression for further differentiations
```
 y = $a\sin(bt)$
```
dydt1=diff(y,t)                                   % first derivative with respect to t
```
 dydt1 = $ab\cos(bt)$
```
dydt2=diff(y)                          % first derivative with respect to t by default
```
 dydt2 = $ab\cos(bt)$
```
dydt3=diff(y,a)                                   % first derivative with respect to a
```
 dydt3 = $\sin(bt)$
```
dydy4=diff(y,b)                                   % first derivative with respect to b
```
 dydy4 = $at\cos(bt)$
```
dydt5=diff(y,3)                             % third order derivative with respect to y
```
 dydt5 = $-ab^3\cos(bt)$
```
dydt6=diff(y,b,2)                          % second order derivative with respect to b
```
 dydt6 = $-at^2\sin(bt)$

Note that the **diff** command is also used to obtain differences in numerical calculations (Subsection 5.6); it performs symbolic calculations only when **s_expression** is symbolic.

Symbolic integration can be performed with the int command, that can be used as for indefinite so for definite integrations. The available forms of this command are

```
int(s_expression)
int(s_expression, var)
int(s_expression,a,b)
int(s_expression,var,a,b)
```

Some examples of integration with the same $y{=}a{\cdot}\sin(bt)$ function that was previously used for differentiation:

```
syms a b y t                                             % define symbolic variables
y=a*sin(b*t)                               % expression for further differentiations
```
 y = $a\sin(bt)$
```
I1=int(y,t)                                     % indefinite integral with respect to t
```
I1 = $-\frac{a\cos(bt)}{b}$
```
I2=int(y)                            % indefinite integral with respect to t by default
```
I2 = $-\frac{a\cos(bt)}{b}$
```
I3=int(y,a)                                     % indefinite integral with respect to a
```
I3 = $\frac{a^2\sin(bt)}{2}$
```
I4=int(y,b)                                     % indefinite integral with respect to b
```
I4 = $-\frac{a\cos(bt)}{t}$

```
I5=int(y,0,pi)                     % integral from 0 to pi with respect to t by default
```

$$I5 = \frac{2a\sin\left(\frac{\pi b}{2}\right)^2}{b}$$

```
I6=int(y,b,0,1)                    % integral with respect to b in limits 0 and 1
```

$$I6 = -\frac{a\left(\cos(t)-1\right)}{t}$$

If some integral does not have a solution or MATLAB may not find a solution, the following message appears - Explicit integral could not be found.

10.7. 2D AND 3D PLOTTING OF SYMBOLIC EXPRESSIONS

To generate two-dimensional graph for symbolic expression with one independent variable the ezplot command can be used. The two simplest forms of this command are

```
ezplot(s_expression)
```

and

```
ezplot(s_expression,[x_min x_max])
```

s_expression means the same as above but in addition it can be written as string, i.e. 's_expression';
x_min and x_max are the range (domain) of the x-axis determined by the minimal and maximal values of the independent variable x, default range is $-2\pi\ldots2\pi$.

- For expression with one symbolic variable, s_expression(var), the var values are accepted as the x-axis and the corresponding s_expression values are the y-axis.
- In case the expression has more than one symbolic variables, s_expression(var1,var2,var3,...) and you want to plot s_expression(var1) graph, thus the second variable, var2, the third, var3, and others must be replaced with numeric values to generate a 2D plot.
- The ezplot command generates a graph with a title (containing the plotted expression) and with the labeled x-axis; you can change these captions and also perform other formatting of the graph using the same formatting commands that are used for numeric expressions (see subsections 4.1.4, 4.2.4).

Generate, for example, graph of the piston position x in the piston-crank mechanism

$$x = -r\cos\theta + \sqrt{l^2 - r^2\sin^2\theta}$$

where r and l are the lengths of the crank arm and the connecting rod respectfully, θ is the crank angle. Assume r=0.15 m and l=0.4 m.

The live script with the generated $x(\theta)$ graph is presented below.

```
syms theta r l                                         % define symbolic variables
x=-r*cos(theta)+sqrt(l^2-r^2*sin(theta)^2);            % simbolic x(teta,r,l) expr
x=subs(x,{r,l},{0.15,0.4})             % substitutes r=0.15 and l=0.4 into the x
```

$$x = \sqrt{\frac{4}{25} - \frac{9\sin(\theta)^2}{400}} - \frac{3\cos(\theta)}{20}$$

```
ezplot(x)                                                 % generates x(theta) plot
```

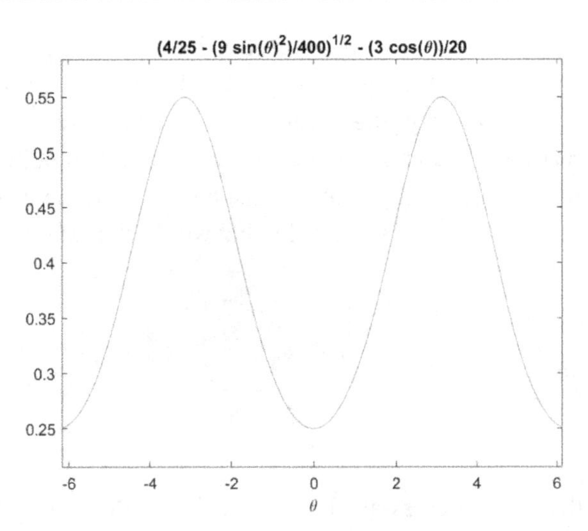

To format this plot with the grid and some added and changed captions the following commands are included in the live script

```
grid on
xlabel('Theta, rad')
ylabel('Position, m')
title('Piston pin position, meters')
```

With the formatting commands, the plot has view

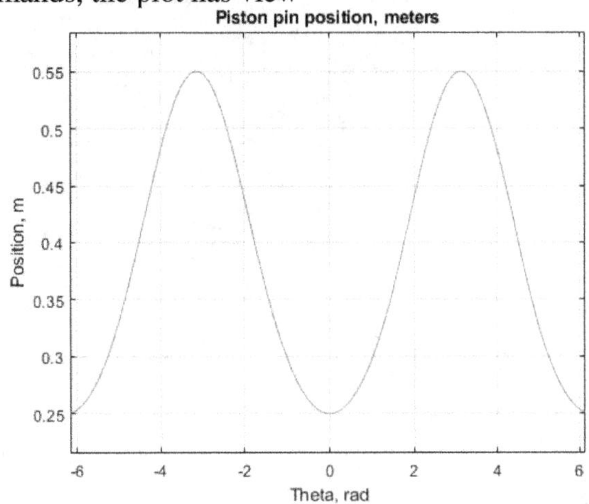

Symbolic expressions with two independent symbolic variables can be graphically represented with the **ezsurf** command that has the following simplest forms

```
ezsurf(s_expression)
```

and

```
ezsurf(s_expression,[x_min x_max y_min y_max])
```

s_expression means the same as above but in addition it can be written as string, i.e. '**s_expression**';

the four-element vector **[x_min x_max y_min y_max]** denotes the *x,y* domain with minimal, **x_min** and **y_min**, and maximal, **x_max** and **y_max**, values of the independent variables *x* and *y*; by default, 60 points are taken within ranges for *x* and *y*, and the **ezsurf** command builds surface with 60x60 mesh.

As an example, plot the 3D mesh graph for the distance *e* from the centroid to neutral axis of the circular beam section (Budynas and Nissbett, 2011), that was built previously in numerical calculations (subsection 4.2.2):

$$e = r_c - \frac{R^2}{2\left(r_c - \sqrt{r_c^2 - R^2}\right)}$$

where R is the radius of the circular cross-section; r_c – radius of the centroid axis. To create the $e(R, r_c)$ graph, assign R,r_c domain as R= 2.5 … 7.5 cm and r_c =7.5 … 12.5 cm. To add/change the graph captions to those in subsection 4.2.2 we use the same format commands that were used previously.

The live script with commands generating the necessary graph is:

```
syms r_c R                                                % define symbolic varianbles
e=r_c -R^2/(2*(r_c-sqrt(r_c^2-R^2)));                   % simbolic e(rc,R) expression
r_c_min=7.5;r_c_max=12.5;                          % assign range values to the symbolic
r_c R_min=2.5;R_max=7.5;                         % assign range values to the symbolic R
v=[R_min R_max r_c_min r_c_max];                  %vector of ranges for the x and y axes
ezsurf(e,v)                                                    % generates e(R,rc) plot
xlabel({'Beam cross-section';'radius, cm'})
ylabel({'Centroidal axis';'    radius, cm'})
zlabel('Eccentricity,cm')
```

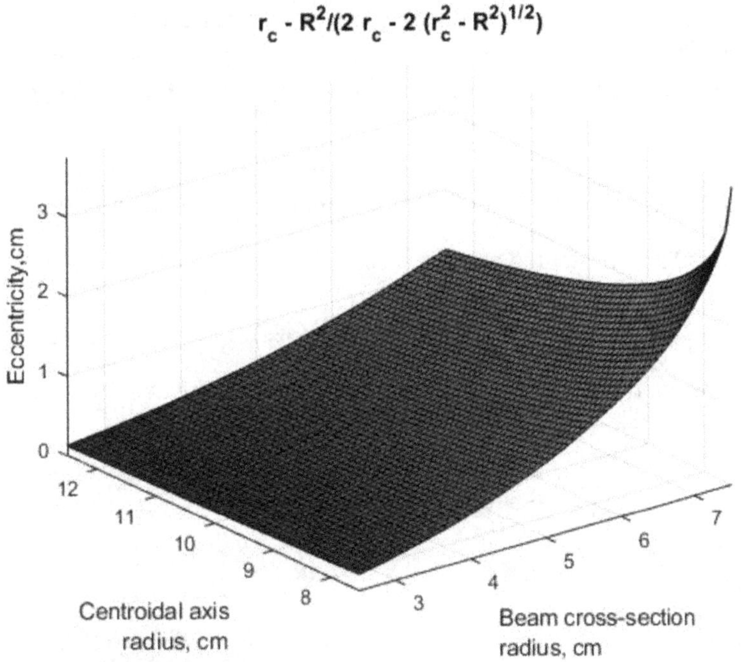

By default, the x and y axis are swapped in this plot comparing to the graph in Figure 14 in chapter 3.

Please note that in the recent MATLAB versions, the fplot command is recommended instead of ezplot, and the fsurf command is recommended instead of ezsurf; the use of recommended commands is the same as for the commands described above, for example, in the latter live program the fsurf(e,v) command can be used instead of the ezsurf(e,v).

10.8. SYMBOLIC SOLUTIONS OF THE ORDINAL DIFFERENTIAL EQUATIONS

In Chapter 7, numerical solutions of the ordinal differential equations, ODEs, were discussed and we assume familiarity with the ODE basic terms and notations. Here symbolical solutions of the single ODE with the dsolve command are described. The command has the following forms

```
dsolve(ODE,var)
```

or

```
dsolve(ODE, cond)
```

- ODE is a symbolical form of the ordinal differential equation in which the double assign symbol (==) should be written between the left and write equation parts, and the differential should be represented using the diff command and double assign symbol, e.g. diff(y,x)==y represents the equation $\frac{dy}{dx} = y$, and diff(y,x,2)=10*y^2 represents equation $\frac{d^2y}{dx^2} = 10y^2$.

- cond is the row vector of initial or boundary conditions, for example, for the first order ODE with the following initial conditions $y=0$ at $t=0$, this vector has an one element and can be written as y(0)=0, and for a second order ODE with the boundary conditions $y=0$ at $t=0$ and $t=1$, this vector should be written so [y(0)=0 y(1)=0].

We demonstrate the dsolve command usage for symbolic solutions and its graphical representations of the two first order ODEs presented as examples in Chapter 6 Table 1: $\frac{dy}{dt}=10t^2$ with initial condition y=1 at t =0 and $\frac{dy}{dt}=2t^2\sin t$ with initial condition $y=0$ at $t=0$. The live script containing solutions of these equations together with the graphical representation of the defined solutions is:

```
syms y(t)                                          %define symbolic function y(t)
ode=diff(y,t)==2*t^2*sin(t);              % assign OD equation to the variable ode
s=dsolve(ode,y(0)==0)               % solve ode with initial condition y=0 at t=0
```

s = $4t\sin(t)-2\cos(t)\left(t^2-2\right)-4$

```
subplot(1,2,1)                            % for the first plot in the same Figure
ezplot(s,[0 4])                                    % plot obtained y(t) solution
grid on
subplot(1,2,2)                           % for the second plot in the same Figure
ode=diff(y,t)==10*t^2;                    % assign OD equation to the variable ode
s=dsolve(ode,y(0)==1)               % solve ode with initial condition y=0 at t=0
```

s = $\frac{10t^3}{3}+1$

```
ezplot(s,[0 2])                                    % plot obtained y(t) solution
grid on
```

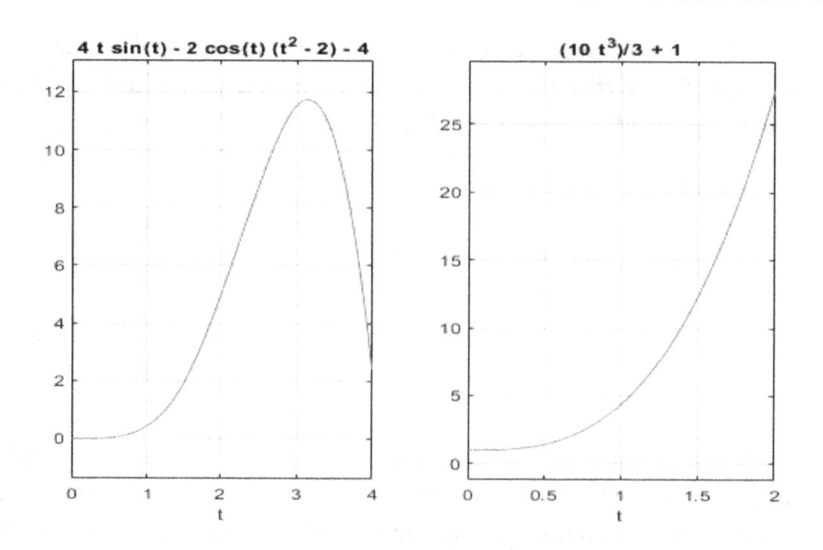

10.9. SUPPLEMENTARY COMMANDS FOR SYMBOLIC CALCULATIONS

Some additional commands that can be used for symbolic calculations are presented in the Table 1.

Table 1. Some additional commands that can be used in symbolic calculations

Command	Parameter/s	Description	Example (Live Script)
symvar(s)	s - symbolic expression	Finds symbolic variables in a symbolic expression and represents its in alphabetical order.	syms a b c x y=a*x^2+b*x+c; symvar(y) ans = $\begin{pmatrix} a & b & c & x \end{pmatrix}$
double(s)	s - symbolic expression	Converts symbolic expression to numerical form.	s=sym(0.2*sqrt(15.1)) s = $\frac{\sqrt{10}\sqrt{151}}{50}$ double(s) ans = 0.7772
factor(p)	p – symbolic polynomial expression	Transforms a symbolic polynomial expression to a product of polynomials of a lower degree; result is the vector containing defining factors.	syms x p=x^3+6*x^2+11*x+6; factor(p) ans = $\begin{pmatrix} x+3 & x+2 & x+1 \end{pmatrix}$
taylor(f,x,a)	f - a mathematical function in a symbolic representation; x – an independent variable with respect to which Taylor expansion is performed; a - a single point above which the expansion is performed.	Fifth order Taylor series expansion of f with respect to x about the point a .	syms x T_5=sin(2*x); taylor(T_5,x,pi) ans = $2x-2\pi-\frac{4(x-\pi)^3}{3}+\frac{4(x-\pi)^5}{15}$
P=piecewise(cond1,ex1,cond2,ex2,...)	cond1, cond2, ... – condition 1, 2, ...; v1, v2, .. – expression 1, 2, ... ; P – piecewise expression, created by the piecewise function	sets a math function defined by several sub-functions, each of which is in a certain range.	syms x psi H=piecewise(abs(x)<1,1+psi*sqrt(1-x^2),x>=1,1) H = $\begin{cases} \psi\sqrt{1-x^2}+1 & if\ \lvert x\rvert<1 \\ 1 & if\ 1\le x \end{cases}$

continued on the following page

Table 1. Continued

Command	Parameter/s	Description	Example (Live Script)
vpasolve(eq,x)	eq – symbolic equation; x - an independent variable with respect to which Taylor expansion is performed.	Solves the symbolic equation numerically.	syms x s_eq=cos(2*x)-0.5; vpasolve(s_eq,x) ans = -226.71826983406341204238743082667
limit(ex,x,a)	ex - a symbolic expression; x – a symbolic variable; a – a limiting value.	Takes the limit of a symbolic expression – $\lim_{x \to a} s(x)$	syms a b x ex=(x+a)/(x-b); limit(ex,x,inf) ans = 1
L=laplace(f)	f - a mathematical function of the independent real variable t (default) in a symbolic representation; L – transformation of f(t) into the L(s) form; when f is a function of s (not t), then in L, instead of s, the variable z is used.	Returns the Laplace transform L(s) of the symbolic function f(t)	syms a t f=exp(a*t); L=laplace(f) L = $-\frac{1}{a-s}$
f=Ilaplace(L)	L – function of the independent variable s (default); f – transformation of L(s) into the f(t) form.	Inverse Laplace transform of a symbolic function of the s variable to the f(t) function	syms a s L=-1/(a-s); f=ilaplace(L) f = e^{at}
rewrite(ex,'str')	ex - a symbolic expression; 'str' – a string that defining the target change in the expression.	Rewrites the ex in terms of another function written as string.	syms x eq=cos(2*x); rewrite(eq,'sin') ans = $1-2\sin(x)^2$

10.10. APPLICATION EXAMPLES

Below are examples of the use of symbolic calculations for M&T problems performed earlier in numerical calculations. The symbolic solution allows you to obtain both a symbolic expression and a numerical result, and the latter can be compared with that obtained in numerical calculations.

10.10.1. Lengthening of a Two Springs Scale Determined Using Symbolic Calculations

The problem was discussed in Subsection 5.7.2 (Figure 5.7) to illustrate usage of the fzero command for M&T calculations and here we demonstrate solution using commands of the Symbolic Math Toolbox™.

The weight W applied to the system of two springs causes its elongation x. These parameters are related by the following equation

$$W - 2F_s \frac{X}{L} = 0$$

where

$$F_s = K_1\left(L - L_0\right) + K_2\left(L - L_0\right)^3$$

$$L_0 = \sqrt{a^2 + b^2}$$

$$L = \sqrt{a^2 + X^2}$$

$$X = b + x$$

In these equations K_1, K_2, a, and b are the spring constants, and L_0 and L are respectively the initial and current length of the joint point of the springs.

Problem: Write live script named Ch_10_ApExample_1 that present symbolic solution and then calculates spring elongation x when the weight applied to the springs are W=350N, and spring constants K_1, K_2, a, and b are 1500, N/m, 90000 N/m^3, 0.23 m, and 0.07 m respectively; take x_0=1 (initial x-value). The live script should generate a plot illustrating the solution in the same form on Figure 7 in chapter 4.

The live script solving the problem with resulting x and generated plot are:

```
syms X x real
a=0.23;b=0.07;
K_1=1500;K_2=90000;
W=350;
L_0=sqrt(a^2+b^2);
L=sqrt(a^2+(X)^2);
F_s=K_1*(L-L_0)+K_2*(L-L_0)^3;
eq=W-2*F_s*(X)/L;
X_sol=solve(eq,X)
X_sol =
```

$$\frac{\sqrt{4\,\mathrm{root}\left(z^8 - \frac{51\sqrt{2}z^7}{50} + \frac{25423z^6}{30000} - \frac{247843\sqrt{2}z^5}{1500000} + \frac{1289047z^4}{90000000} + \frac{38152913\sqrt{2}z^3}{5625000000} - \frac{599488777757z^2}{20250000000000} + \frac{28638811031\sqrt{2}z}{112500000000000} - \frac{190747942009}{1125000000000000}, z, 8\right)^2 - \frac{529}{2500}}}{2}$$

```
x_sol=vpa(X_sol-b,4)
x_sol =
0.1802
```

```
eq=subs(eq,X,x+b);
ezplot(eq,[0 0.25])
grid on
xlabel('Elongation, m')
ylabel('W_0-2F_s(x+a)/L, N')
hold on
plot(x_sol,0,'o')
legend('W_0-2F_s(x+a)/L','Solution point')
legend('W_0-2F_s(x+a)/L','Solution point')
hold off
```

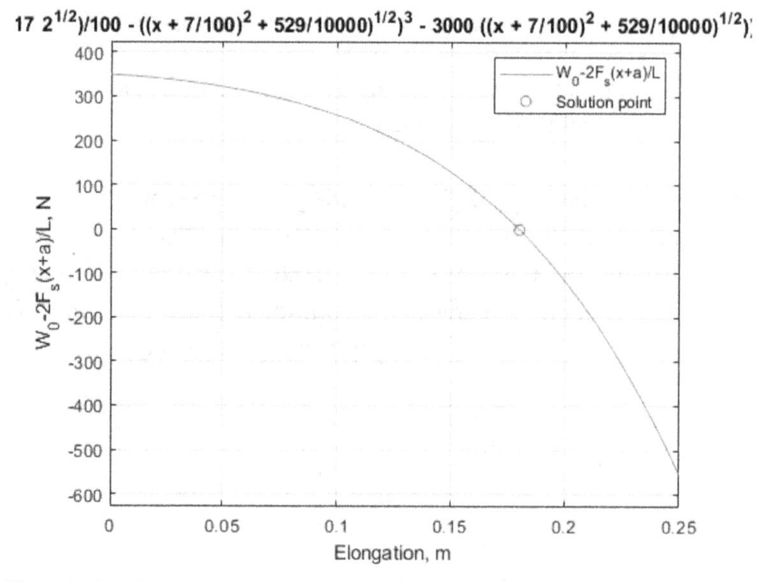

The Ch_10_ApExample_1 live script is arranged as follows:

- The first four lines contain commands for defining symbolic variables and for assigning numerical values to numerical variables, the latter is done to simplify the further equation solution;
- The next four lines represent the components of the equation and the equation itself for a symbolic solution;
- The **solve** command solves the **eq** equation and assign the solution to the symbolic variable X_sol; the solution contains the variable z that designates here the root of the seventh order polynomial used to obtain the symbolic solution;
- To convert symbolical numbers in the received solution to single digital number with 4 significant digits, the vpa command is used; the result is assigned to the x_sol variable to be used further in the graphical representation of the solution point;
- To generate a graph with an abscissa representing the x (and not X) variable, the subs command are used; the command changes the X with the $x+b$ in the symbolic equation eq;
- The ten commands at the end of this program generate and format the resulting graph so that it matches the plot in Figure 7 in chapter 4.

As it can be seen, the elongation values obtained in both numerical (Subsection 5.7.2) and symbolic calculations are identical.

10.10.2. Shear Stress of the Lubricating Film Determined Using Symbolic Calculations

Shear stress τ arises in lubricating film between two parallel surfaces, unmoved covered with semicircular pores and moved smooth surface. For one pore cell (pore and adjacent surface), the shear stress on the top surface can be described with the following expression (see Subsection 5.7.5, Burstein, 2016):

$$\tau = 3\frac{dP}{dX}H + \frac{1}{H}$$

where P is the hydrodynamic pressure, X- coordinate, and H is the gap between surfaces; all variables in this equation are dimensionless.

In the simplest case, when the pore radius to gap ratio ψ is 1 (the pore diameter is close to the cell size), the hydrodynamic pressure in the lubricating film between the single pore and the opposite surface, abs(X)<1, is determined by the equation

$$P = \frac{\sin\theta\left(2\cos\theta + 1\right)}{3\left(1+\cos\theta\right)^2} - \frac{5}{3}\left(\frac{2-3\xi}{4-5\xi}\right)\frac{\sin\theta\left(6\cos\theta + \cos 2\theta + 3\right)}{10\left(1+\cos\theta\right)^3}$$

where θ=asin(X), and ξ is the ratio of pore cell dimension to pore radius.

The gap can be calculated with the following equation

$$H = 1 + \psi\sqrt{1 - X^2}$$

Problem: write live script named Ch_10_ApExample_2 that represents symbolic expressions for the derivative dP/dX and for the shear stress in lubricating film when ξ=1.1 and ψ=1Generate and format the τ (X) plot. Use the Output Inline option to display symbolic, graphical results, and the τ value for the point X=0 with 5 significant digits. Use the fplot commands to generate a graph.

The Ch_10_ApExample_2 program that solves the problem are

```
syms X real
theta=asin(X);
xi=1.1;
psi=1;
d=1+cos(theta);
C=-5/3*(2-3*xi)/(4-5*xi);
P=sin(theta)*(2*cos(theta)+1)/(3*d^2)+C*sin(theta)*(6*cos(theta)+ ...
 cos(2*theta)+3)/(10*d^3);
dPdX=diff(P,X);
dPdX=simplify(dPdX)
```

$$\text{dPdX} = -\frac{50X^2 + 40\sqrt{1-X^2} - 90\left(1-X^2\right)^{3/2} - 50}{90\left(4X^4 - 12X^2 + \sqrt{1-X^2} + 6\left(1-X^2\right)^{3/2} + \left(1-X^2\right)^{5/2} + 8\right)}$$

```
H=1+psi*sqrt(1-X^2);
tau=3*dPdX*H+1/H;
tau=simplify(tau)
```

$$\text{tau} = \frac{1}{\sigma_1 + 1} - \frac{\left(\sigma_1 + 1\right)\left(50X^2 + 40\sigma_1 - 90\left(1-X^2\right)^{3/2} - 50\right)}{30\left(4X^4 - 12X^2 + \sigma_1 + 6\left(1-X^2\right)^{3/2} + \left(1-X^2\right)^{5/2} + 8\right)} \quad \text{where } \sigma_1 = \sqrt{1-X^2}$$

```
fplot(tau,[-0.99 0.99])
grid on
xlabel('Coordinate, nondim')
ylabel('Shear stress, nondim')
title('Shear stress')
title('Shear stress')
```

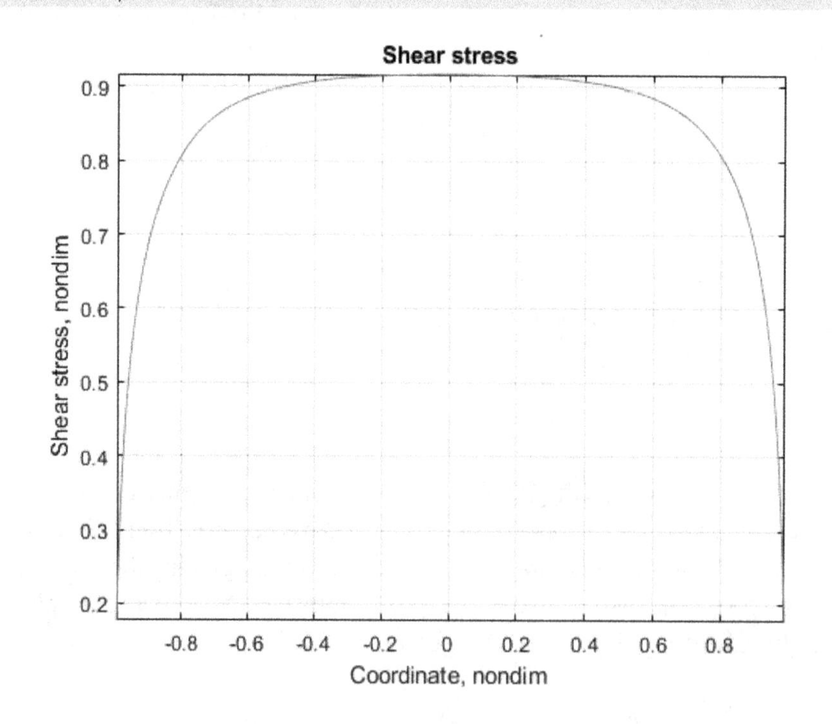


```
tau_0=subs(tau,X,0);
Shear_stress_at_X_equals_0=vpa(tau_0,4)
Shear_stress_at_X_equals_0 =
0.9167
```

The Ch_10_ApExample_2 live script is arranged as follows:

- First four lines contain commands for definition of the symbolic variable X, designation theta variable via X, and assigning numerical values to the xi and psi variables;
- The next three commands enter the symbolic P equations that should be differentiated;
- The diff command differentiate P with respect to X, the result dPdX is simplified next with the simplify command and displayed;
- The three next commands enter H and compute and display the symbolic expressions of the shear stress τ;
- The fplot command displays the τ(X) graph in the range -0.99 … 0.99 as the problem is solved at abs(X)<1 (not equal 0); next four lines contain commands for plot formatting;
- Final three commands convert symbolic number to regular number with the four significant digits (commands subs and vpa), and display the obtained value.

10.10.3. Centroid of the Cross-Section of a Mechanical Part, Determined Using Symbolic Calculations

The problem was discussed in Subsection 5.7.4 to illustrate usage of the quad command for M&T calculations and here we demonstrate solution using commands of the Symbolic Math Toolbox.

The location of the geometric center, centroid, for the cross section of a some mechanical plate can be calculated with the following expressions:

$$x_c = \frac{\int_{x_1}^{x_2} x\left(y_2 - y_1\right)dx}{\int_{x_1}^{x_2}\left(y_2 - y_1\right)dx}$$

$$y_c = \frac{\int_{x_1}^{x_2}\left(y_2^2 - y_1^2\right)dx}{2\int_{x_1}^{x_2}\left(y_2 - y_1\right)dx}$$

Assume the plate is described with the line y_1=2x and cubic parabola y_2=2x, where x is the coordinate.

Problem: Write a live script with name Ch_10_ApExample_3 that derives symbolic expressions for x_c and y_c and calculates the centroid coordinates when x_1 and x_2 are 0 and $\sqrt{2}$ respectively. The program should generate plot of the cross section and show the defined centroid place, in the same way as in the Subsection 5.7.4. Use for plot generation the fplot commands.

The Ch_10_ApExample_3 live script program that solves the problem:

```
syms x x_1 x_2
y_1=2*x;
y_2=x^3;
I_1=int(x*(y_2-y_1),x_1,x_2);
I_2=int(y_2-y_1,x_1,x_2);
I_3=int(y_2^2-y_1^2,x_1,x_2);
x_c=simplify(I_1/(I_2))
```

$$x\_c = \frac{4\left(-\frac{x_1^5}{5}+\frac{2x_1^3}{3}+\frac{x_2^5}{5}-\frac{2x_2^3}{3}\right)}{-x_1^4+4x_1^2+x_2^4-4x_2^2}$$

```
y_c=simplify(I_3/(2*I_2))
```

$$y\_c = \frac{2\left(-\frac{x_1^7}{7}+\frac{4x_1^3}{3}+\frac{x_2^7}{7}-\frac{4x_2^3}{3}\right)}{-x_1^4+4x_1^2+x_2^4-4x_2^2}$$

```
x_1=0;
x_2=sqrt(2);
x_c=subs(x_c);y_c=subs(y_c);
range=[x_1 x_2];
fplot(y_1,range)                                        % plots the 2x line
hold on                                                 % hold current graph
fplot(y_2,range)                                        % plots the x^3 parabola
plot(x_c,y_c,'*')               % draws the centroid point with an asterisk
xlabel('x')
ylabel('y')
legend('y_1=2x','y_2=x^3','Centroid point',...
   'location','northwest')                              %left top angle location
hold off
```

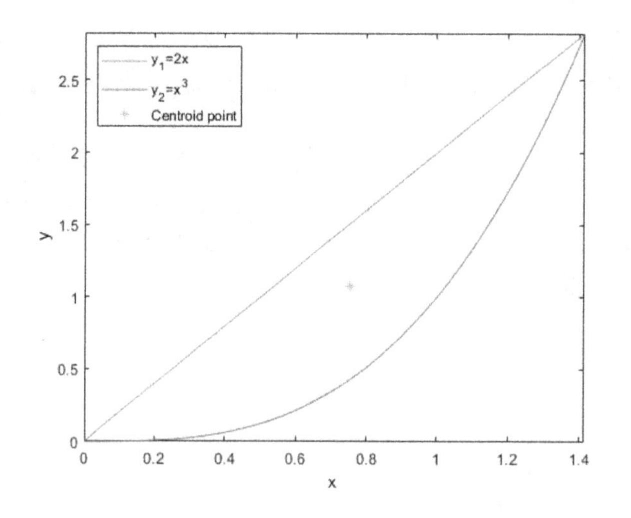

```
x_c=vpa(x_c,4),y_c=vpa(y_c,5)
x_c = 0.7542
y_c = 1.0775
```

The Ch_10_ApExample_3 live script is arranged as follows:

- The first three lines contain commands for defining the symbolic variable x and inputting symbolic expressions describing the shape of the plate;
- The following three commands symbolically calculate the integrals I_1, I_2, and I_3 that should be determined in accordance with the x_c and y_c expressions; for this the int commands are used;
- Two subsequent commands compute and display symbolic expressions for the centroid coordinates x_c and y_c; the resulting equations are simplified using the simplify commands;
- The further four commands enter numbers for the integration limits x_1 and x_2, substitute them into the determined above coordinates x_c and $y_{c,}$ and define the vector range with the *x*-limits values;
- The following eight commands (fplot, plot, and others) generate and format the resilting graph representing the shape of the studied plate and its centroid point;
- The two final vpa commands convert the symbolic x_c and y_c values to regular numbers with four and five significant digits, respectively.

10.10.4. Second Order Dynamical System: A Symbolic Solution

The second order dynamical system is described with the following ODE (see Subsection 6.1.2.1):

$$a\frac{d^2y}{dt^2}+b\frac{dy}{dt}+cy=f\left(t\right)$$

The *y* in this equation is the dependent variable denoting the seeking function at time *t* (independent variable) and *a*, *b* and *c* are the constants, and *f*(*t*) is a case-oriented function; all variables, constants and functions in this equation are dimensionless.

Problem: Write a live script with name Ch_10_ApExample_4 that symbolically solve the above ODE. Consider this differential equation for *a*=0.3, *b*=0.09, *c*= 3, and *f*(*t*)=sin(π*t*/90). The time range at which we want to display solution are 0 … 24; the initial *y* values are *y*(0)=15 and *y*'(0)=0. The program should generate the *y*(*t*) plot in the same form as in Subsection 6.1.2.1. Use the fplot commands to generate a graph.

The Ch_10_ApExample_4 live script program that solves the problem:

```
syms y(t) a b c y0 Dy0
f=sin(pi*t/90);
Dy2=diff(y,t,2);
Dy=diff(y,t);
ode=Dy2+b/a*Dy+c/a*y==f/a;
cond=[y(0)==y0,Dy(0)==Dy0];
y_sym=dsolve(ode,cond);
y_sym=simplify(y_sym);
y_sym=subs(y_sym,{a,b,c,y0,Dy0},{.3,.09,3,15,0});
y_sym=vpa(y_sym,4)
```

y_sym = $0.3334\sin\left(0.03491t\right) - 0.0003492\cos\left(0.03491t\right) + e^{t(-0.15+3.159i)}\left(7.5 - 0.3543i\right)$

```
fplot(y_sym,[0 24])
title('Symbolic solution for the second order dynamical system')
grid on
ylabel('y(t)'),xlabel('time')
```

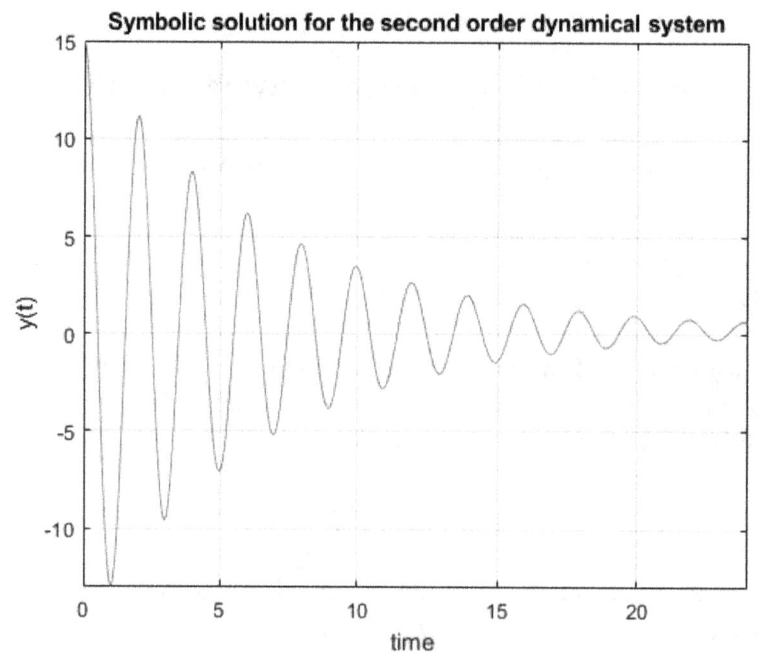

The Ch_10_ApExample_4 live script is designed as follows:

- The first six lines contain commands for defining of the symbolic function y(x), variables a, b, c, y0, Dy(0) and assigning the ODE to the ode variable;
- The next command assigns to the vector cond the initial values for y and for its derivative dy/dt;
- The subsequent two commands solve ODE (the dsolve command) and simplify (the simplify command) the achieved results y_sym. Note, symbolic result, as in this example, can be presented in the complex form despite that in traditional mathematics it is often written in the real form; in the next section we give solution of this ODE that gives a real solution;

- The **subs** command substitutes into the resulting expression **y_s** the decimal numbers instead of the variables **a**, **b**, **c**, **y0**, and **Dy(0)**; this then allows you to draw a two-dimensional resulting plot;
- The final five commands (**fplot**, and others) generate and format the resulting plot in the same form as in Subsection 6.1.2.1; this plot represents a second order ODE graphical solution describing the behavior of a dynamical system.

10.10.5. Second Order Dynamical System: A Symbolic Solution With Laplace Transform

The symbolic solution performed in the previous subsection represents the correct graphic results, but give complex expression as symbolic solution of the studied ODE. Now we solve this ODE using the Laplace transform, with the help of which the dynamic systems are usually analyzed. The Laplace transform is an integral transform that converts the real function $f(t)$ in its complex image $F(s)$. Such a transformation allows us to solve ODE as an ordinary algebraic equation. The solution should be then converted to the real variable using the inverse Laplace transform. The Laplace transform is studied in special mathematical courses, and its explanation is beyond the scope of this book. Here we demonstrate how to use the **laplace** and **ilaplace** functions (Table 10.1) to solve ODE.

Problem: Write a live script with name **Ch_10_ApExample_5** that solve symbolically the ODE described behavior of the second order dynamical system with the same parameters and conditions that were studied in the previous subsection. Use the Laplace transform. Represent graphical results analogously to the graph above.

The **Ch_10_ApExample_5** live script program that solves the problem is:

```
Syms a b c t s Y y(t)
f=sin(pi*t/90);
Dy2=diff(y,t,2);
Dy=diff(y,t);
ode=Dy2+b/a*Dy+c/a*y==f/a;                                    % ODE to solution
ode_L=laplace(ode,t,s);                            % Laplace transform of the ODE
                 % Denote the laplace(y,t,s) variable as Y and substitute the IC
Y_L=subs(ode_L,{laplace(y,t,s),y(0),Dy(0)},{Y,15,0});
y_L=solve(Y_L,Y);                                         % Solve for the unknown Y
y_L=subs(y_L,{a,b,c},{.3,.09,3,});                       % Substitute all constants
y=ilaplace(y_L,s,t);                                     % Inverse Laplace Transform
y=vpa(y,4)                       % for decimal representation of symbolic numbers
y =
```

$$0.3334\sin\left(0.03491t\right) - 0.0003492\cos\left(0.03491t\right) + 15.0e^{-0.15t}\left(\cos\left(3.159t\right) + 0.04724\sin\left(3.159t\right)\right)$$

```
fplot(y,[0 24])
grid on
ylabel('y(t)')
xlabel('time')
title('Second order dynamical system, solution with the Laplace transform')
```

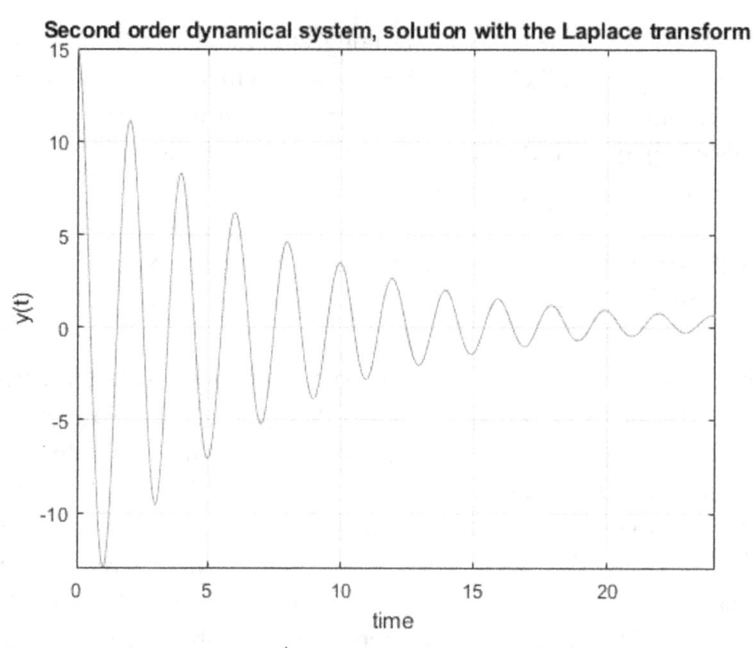

The Ch_10_ApExample_5 live script is constructed as follows:

- The first four lines contain commands for defining of the symbolic variables a b c t s Y, symbolic function y(x), and entering the solving ODE;
- The laplace command transforms the ODE into the algebraic equation ode_L in accordance with the Laplace transform and in terms of the laplace(y,t,s) function;
- The subs command redefines the laplace(y,t,s) function (that appears in the ode_L equation) as a new variable Y_L, and also substitute the initial conditions in ode_L;
- The solve command solves the algebraic equation ode_L with respect to the unknown Y_L, and then the subs command assigns numerical values to the constants a, b, and c;
- The ilaplace command converts the Laplace solution y_L to the *y(t)* form, and then the vpa command converts the symbolic numbers contained in the solution (each of which is the ratio of two integers) into their decimal forms;
- The final five commands (fplot, and others) generate and format the graph so that it looks the same as in Subsection 10.10.4.

10.11. CONCLUSION

Symbolic commands of the Symbolic Math Toolbox and manipulations with have been explored, and their action was demonstrated with the many new examples and examples used in previous chapters containing numerical calculations. The symbolic calculations were used to obtain analytical expressions and numerical results for the following M&T problems:

- Friction coefficient of the pipe,
- Instantaneous piston position,
- Lengthening of a two-spring scale,

- Shear stress of the lubricating film,
- Centroid of the cross-section of a mechanical part,
- Solving second order dynamical system solving using regular and Laplace transform methods.

In addition, symbolic graphic commands are performed to demonstrate the obtained analytical solutions. In general, symbolic calculations reveal their effectiveness and should be used for the M&T problem solutions.

REFERENCES

Budynas, R. G., & Nisbett, J. K. (2011). Shigley's mechanical engineering design (9th ed.). McGraw-Hill.

Burstein, L. (2016). Friction Force of the Sliding Surface with Pores Having a Semicircular Cross Section Form. *International Journal of Surface Engineering and Interdisciplinary Materials Science*, *4*(2), 1–15. doi:10.4018/IJSEIMS.2016070101

Enachescu, M., van den Oetelaar, R. J. A., Carpick, R. W., Ogletree, D. F., Flipse, C. F. J., & Salmeron, M. (1999). Observation proportionality between friction force and contact area at the nanometer scale. *Tribology Letters*, *7*(2-3), 73–78. doi:10.1023/A:1019173404538

APPENDIX

Table 2. List of examples, problems, and applications discussed in the chapter

No	Example, Problem, or Application	Subsection
1	Friction factor of the pipe surface in symbolic representation.	10.2.1.
2	Friction factor in a resembled math format in Editor and in Live Editor.	10.2.2.4, 10.3.
3	Inverse problem for coefficient of friction of the fully developed turbulent flow.	10.4.,10.5.
4	Defining fitting coefficients for the friction force – surface relation.	10.4., 10.5.
5	2D graph of the piston pin position by symbolic equation for the piston-crank mechanism.	10.7.
6	3D mesh graph by symbolic equation for the distance from the centroid to neutral axis of the circular beam section.	10.7.
7	Two examples for symbolic solution of ordinal differential equations.	10.8.
8	Lengthening of a two springs scale, determined with symbolic calculations.	10.10.1.
9	Shear stress of the lubricating film, determined using symbolic calculations.	10.10.2.
10	Centroid of the cross-section of a mechanical part, determined using symbolic calculations.	10.10.3.
11	Second order dynamical system: a symbolic solution.	10.10.4.
12	Second order dynamical system: a symbolic solution with Laplace transform.	10.10.5.

Note, some small examples, mostly related to non-M&T issues, are not included in the list.

Chapter 11
Statistical Calculations

ABSTRACT

This chapter presents statistical commands and its applications to various problems of the mechanics and tribology (M&T). Descriptive statistics, data statistics tool, specialized statistical graphs, probability distributions, and hypothesis tests are discussed. The solutions of various applied problems are given. In particular, surface roughness indices are calculated by the measured data using the descriptive statistics command; the histogram generated by a runout data are matched with the theoretical distribution; capability plot generation is shown by the data for the piston ring gaps; friction torques for two different oil additives are compared using a hypothesis test.

INTRODUCTION

Statistical calculations are widely used in M&T sciences and engineering to analyze and present data obtained in scientific or manufacturing processes, in particular to describe the mechanical or tribological characteristics of materials, surface roughnesses, mechanical part tolerances, as well as to quality assurance and fabrication process control, and in many other areas.

In this chapter, the basic MATLAB® toolbox commands for descriptive statistics and the Statistics and Machine Learning Toolbox™ commands for statistical graphs, probability distributions and hypotheses tests are highlighted. All discussed commands are provided with problem-oriented examples. In the final subsection the following application are presented:

- Surface roughness indices by the descriptive statistics commands;
- Histogram and theoretical pdf curve for a runout data;
- Process capability calculations for piston ring gaps;
- Comparing the friction torque data for the two oil additives;
- Sample size evaluation for desired precision of the diameter of a machine part.

DOI: 10.4018/978-1-7998-7078-4.ch011

11.1. DESCRIPTIVE STATISTICS

A summary description of the obtained data, called descriptive statistics, occupies an important place in science and technology in general and, in particular, in the areas of M&T. MATLAB® has significant set of commands and specialized means for this purpose. Below we describe the commands and the tool "Data Statistics", designed for statistical processing of samples or the entire population of data.

11.1.1. About Commands for Quantifying Data

To calculate descriptive statistics the basic commands max, mean, median, min, mode, std and var are used, most of these commands were presented in the Chapter 2 (Table 4). The simplest forms of new two commands are

```
mode(a)
```

and

```
var(a)
```

These commands compute respectively the most frequently value and variance (dispersion $\frac{\sum_{i=1}^{N}\left|a_i - mea(a)\right|^2}{N-1}$ of the N values); a – is a vector or matrix, in the latter case these command return vector of values calculated for each of the column in a.

Problem: In a laboratory of engine parts, the bore (inner diameter) of a worn cylinder was tested. The data is 60.1, 60.19, 60.25, 60.62, 60.47, 60.35, and 60.83 mm. Write program named ExDescrStat that calculate and outputs all of the above descriptive statistics indices and determine outliers (if any) - values that are more than three standard deviations than the average. In case of absence of the outliers, the program should type 'Data does not have outliers'. Use the fprintf command to output names and resulting values of the descriptive statistics.

The ExDescrStat program that solves this problem are.

```
d=[60.1 60.19 60.25 60.62 60.47 60.35 60.83];
Aver=mean(d);
Minim=min(d);
Maxim=max(d);
Mod=mode(d);
StDev=std(d);
Rng=range(d);
varia=var(d);
outliers=d(abs(d-Aver)>3*StDev);
if outliers>0
  outliers
```

```
else
  fprintf('\n  Data does not have outliers\n')
end
DescrStat=[Aver Minim Maxim Mod StDev Rng varia]';
fprintf('\n  Mean Minimum Maximum Mode StDev Range Variance\n')
fprintf('%7.2f %7.2f %7.2f %7.2f %5.3f %5.3f %7.4f\n',DescrStat)
```

After starting, the program outputs the following results:

```
>> ExDescrStat
  Data does not have outliers
  Mean Minimum Maximum Mode StDev Range Variance
 60.40  60.10  60.83  60.10 0.257 0.730 0.0662
```

In case the outliers were found among the data, this program outputs the found outlier values.

11.1.2. Descriptive Statistics With the Data Statistics Tool

Another means to obtain the descriptive statistics indices is the Data Statistics tool. This tool represents a graphical interface that after entering and plotting the data allows visualize the obtained statistical values in the data set plot. The indices to be drawn are: minimum value, maximum value, average, median, mode, standard deviation, and range (the maximum minus minimum values). The steps for using the Data Statistics tool are listed below. Use for this the following laboratorian data for the face wear of the 50 piston rings:

```
37.8, 40.5, 46.6, 41.6, 41.3, 37.8, 40.9, 42.5, 43.1, 39.7, 39.4, 46.9, 35.5,
45.6, 46.0, 44.6, 36.2, 38.1, 39.0, 43.2, 36.6, 43.7, 36.3, 42.8, 40.9, 44.3,
43.6, 45.8, 45.7, 39.0, 43.4, 37.4, 35.4, 43.9, 41.0, 40.8, 45.9, 42.3, 42.4,
45.3, 44.7, 41.9, 37.2, 37.9, 45.6, 35.3, 40.9, 37.0, and 46.7 µm
```

Step 1. The data should be entered and plotted with the following commands

```
>>w=[37.8 40.5 46.6 41.6 41.3 37.8 40.9 ...
 42.5 43.1 39.7 39.4 46.9 35.5 45.6 ...
 46.0 44.6 36.2 38.1 39.0 43.2 36.6 ...
 43.7 36.3 42.8 40.9 44.3 43.6 45.8 ...
 45.7 39.0 43.4 37.4 35.4 43.9 41.0 ...
 40.8 45.9 42.3 42.4 45.3 44.7 41.9 ...
 37.2 37.9 45.6 35.3 40.9 37.0 46.7];
 >>plot(w,'o-')
```

Step 2. In the appeared Figure window, select the Data Statistics line in the popup menu of the Tools menu button – (Figure 1, a); the small-scale "Data Statistics -1" dialog box appears with statistics for the ordinate (x) and abscissa (y) values.

Figure 1. The data and popup menu of the Figure window "Tools" button (a); the Data Statistics dialog box (b) after launching

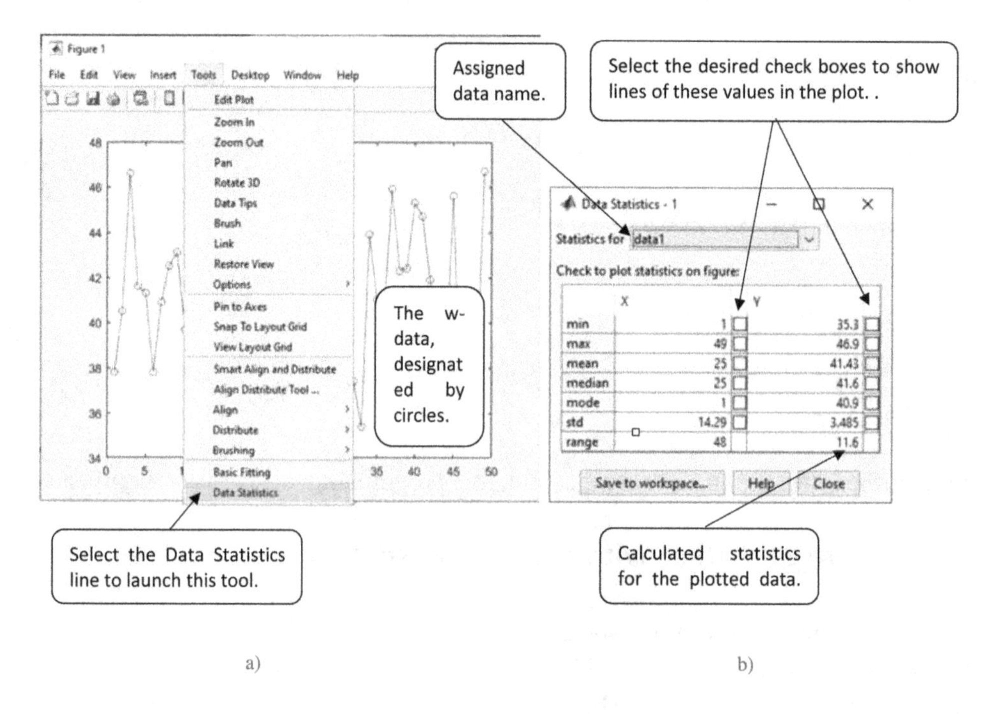

Step 3. The checkboxes with desired statistical indices should be marked in the appeared dialog box. In the plot, the lines appear showing the selected indices – Figure 2. All possible checkboxes are selected for displaying on this plot.

Figure 2. Resulting Data Statistics plot with the lines of selected statistical indices of the processed dataset

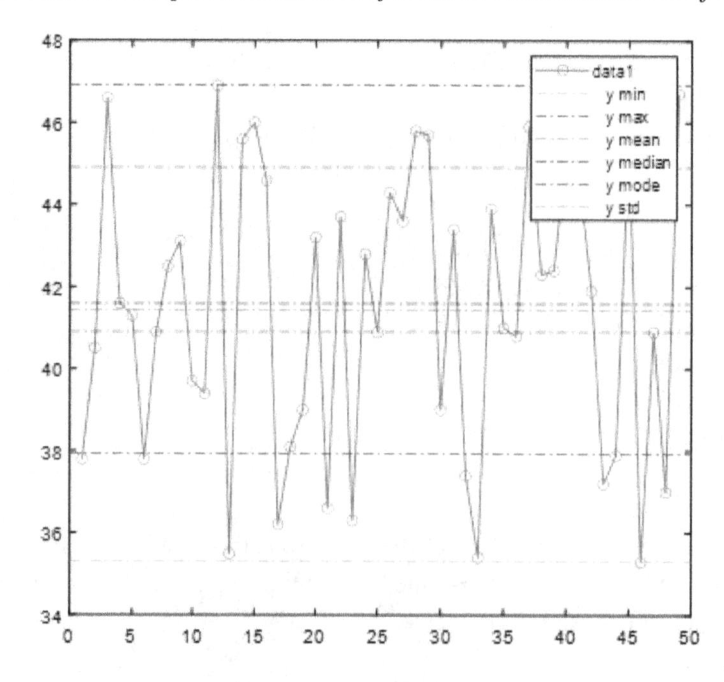

Note for Figure 2:
- in the discussed case, the x axis represents the serial numbers of the y values (namely, the piston ring serial numbers) and it is not reasonable to calculate statistical indices for this axis. In the case this is rational, vertical lines of the obtained values appear on the graph;
- the range is not the selectable for graphical representation on the plot;
- the resulting Data Statistics plot can be formatted with the same format command as an usual 2D plot.

Step 4 (optional). Obtained statistics can be transferred to the MATLAB® workspace; click for this the "Save to workspace …" button (see Figure 1, b) and mark in the appeared dialog box the desirable (x and/or y) statistics for saving in the workspace.

11.2. STATISTICAL GRAPHS

To visualize statistical indices MATLAB® has numerous set of the commands that allow represent experimental or theoretical data of a broad range of the M&T problems, e.g. in processing and drawing data of tests, visual representation of the mechanical and tribological properties of the industrial materials, or in creating control charts of technological process, etc. Most statistical means are available in the Statistics and Machine Learning Toolbox™ which is beyond the scope of the book. Here we describe some commands and Data Statistics tool that available in the basic MATLAB®.

11.2.1. The hist Command

One of the most common graphs representing the distribution of analyzing data is a histogram. To design a histogram, the range of the data values divided into several sub-intervals, called bins, and then the number of values falling into each sub-interval is counted. Obtained quantities, termed frequencies, are graphically represented as heights of the bars; the sub-intervals are represented as bar widths. To design a histogram the hist command can be used. The following is the simplest form of this command:

```
hist(y)
```

Problem: Plot a histogram for the experimental data on the width of 100 metallic rectangular prisms (parts of some optical devices) are: 3.52, 3.54, 3.49, 3.54, 3.41, 3.58, 3.43, 3.37, 3.60, 3.50, 3.53, 3.47, 3.43, 3.28, 3.54, 3.48, 3.49, 3.53, 3.40, 3.45, 3.58, 3.54, 3.41, 3.35, 3.46, 3.51, 3.57, 3.51, 3.42, 3.52, 3.42, 3.43, 3.45, 3.41, 3.48, 3.44, 3.48, 3.47, 3.47, 3.46, 3.59, 3.54, 3.50, 3.40, 3.39, 3.36, 3.68, 3.57, 3.52, 3.43, 3.42, 3.54, 3.61, 3.38, 3.47, 3.49, 3.36, 3.56, 3.47, 3.63, 3.47, 3.53, 3.42, 3.46, 3.42, 3.53, 3.59, 3.46, 3.53, 3.46, 3.51, 3.58, 3.51, 3.43, 3.44, 3.49, 3.49, 3.44, 3.51, 3.59, 3.48, 3.49, 3.35, 3.45, 3.40, 3.47, 3.54, 3.61, 3.63, 3.49, 3.48, 3.56, 3.53, 3.55, 3.55, 3.40, 3.54, 3.49, 3.45, and 3.45 mm. Present this data as a histogram with ten equally-spaced bins and axis headers.

To generate a histogram based on this data, the following commands should be entered:

```
>>width=[3.52 3.54 3.49 3.54 3.41 3.58 3.43 3.37 3.60 3.50 ...
 3.53 3.47 3.43 3.28 3.54 3.48 3.49 3.53 3.40 3.45 ...
 3.58 3.54 3.41 3.35 3.46 3.51 3.57 3.51 3.42 3.52 ...
 3.42 3.43 3.45 3.41 3.48 3.44 3.48 3.47 3.47 3.46 ...
 3.59 3.54 3.50 3.40 3.39 3.36 3.68 3.57 3.52 3.43 ...
 3.42 3.54 3.61 3.38 3.47 3.49 3.36 3.56 3.47 3.63 ...
 3.47 3.53 3.42 3.46 3.42 3.53 3.59 3.46 3.53 3.46 ...
 3.51 3.58 3.51 3.43 3.44 3.49 3.49 3.44 3.51 3.59 ...
 3.48 3.49 3.35 3.45 3.40 3.47 3.54 3.61 3.63 3.49 ...
 3.48 3.56 3.53 3.55 3.55 3.40 3.54 3.49 3.45 3.45];
>>hist(width)
>>xlabel('Width, mm')
>>ylabel('Number of widths per one bin')
```

The resulting plot is shown in Figure 3.

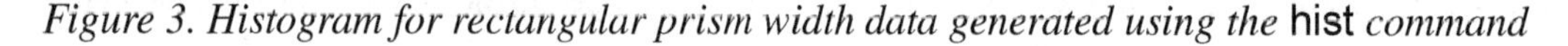

Figure 3. Histogram for rectangular prism width data generated using the hist *command*

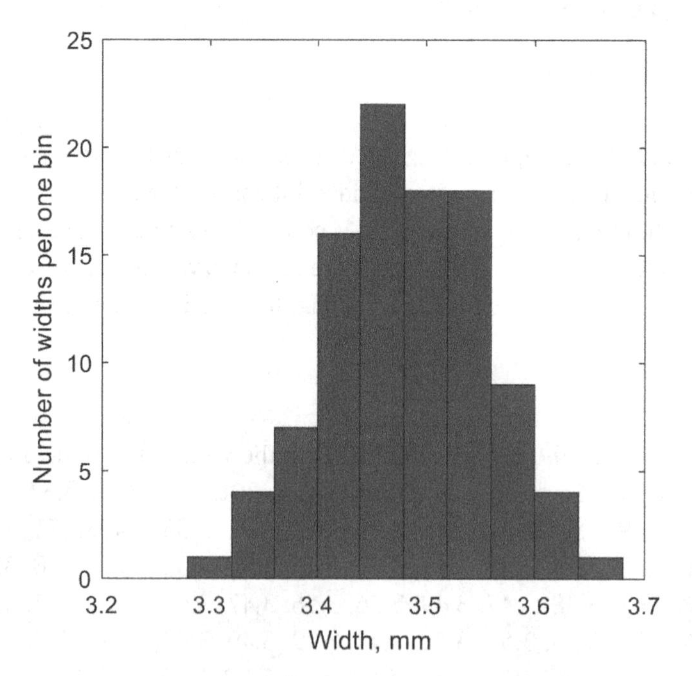

Another available form of the hist command is

```
[n,xc]=hist(y)
```

Note: in this form the hist command does not plot a histogram, but outputs a numerical result.

For the above problem, the command outputs the following frequencies and bin centers:

```
>> [n,xc]=hist(width)
n =
   1    4    7   16   22   18   18    9    4    1
xc =
 Columns 1 through 7
  3.3000   3.3400   3.3800   3.4200   3.4600   3.5000   3.5400
 Columns 8 through 10
  3.5800   3.6200   3.6600
```

The **hist** command also has some additional forms that allow you to enter desired *x* values for the each bin center; more detailed information can be obtained by entering the **doc hist** command.

In the last MATALAB® versions, it is recommended to replace the **hist** command with the **histogram** command that has more possibilities but is more complicate. Details are available by entering the **doc histogram** command.

11.2.2. The bar Command

This command draws the bar plot (see Table 5 in chapter 3) and can be used to plot the histogram when the bin centers and the frequencies are known. The command displays the values of a vector or matrix columns as vertical bars or groups of vertical bars, respectively. The form, additional to those in Table 5 in chapter 3, are

```
bar(x,y,color)
```

Problem: Represent the *x*-centers of the prism widths and width frequencies obtained in the previous subsection in two plots generated in one Figure window: the first plot with the **bar(x,y)** command and the second with the **bar (x,y,color)** command. Label the axes on the both graphs.

The following program named **ExBar** solves this problem:

```
xc=[3.30 3.34 3.38 3.42 3.46 3.50 3.54 3.58 3.62 3.66];            % centers
n=[1 4 7 16 22 18 18 9 4 1];                                    % frequencies
subplot(1,2,1)
bar(xc,n)                            %the [0 0.45 0.74] (default) bar color
xlabel('Width, mm')
ylabel('Frequences, units')
subplot(1,2,2)
bar(xc,n,'y')                                  % sets the yellow bar color
xlabel('Width, mm')
ylabel('Frequences, units')
```

Plots generated with the bar command are shown in the Figure 4.

Figure 4. Graphs for the metallic prism widths generated using the bar *command without (a) and with (b) the* color *specifier*

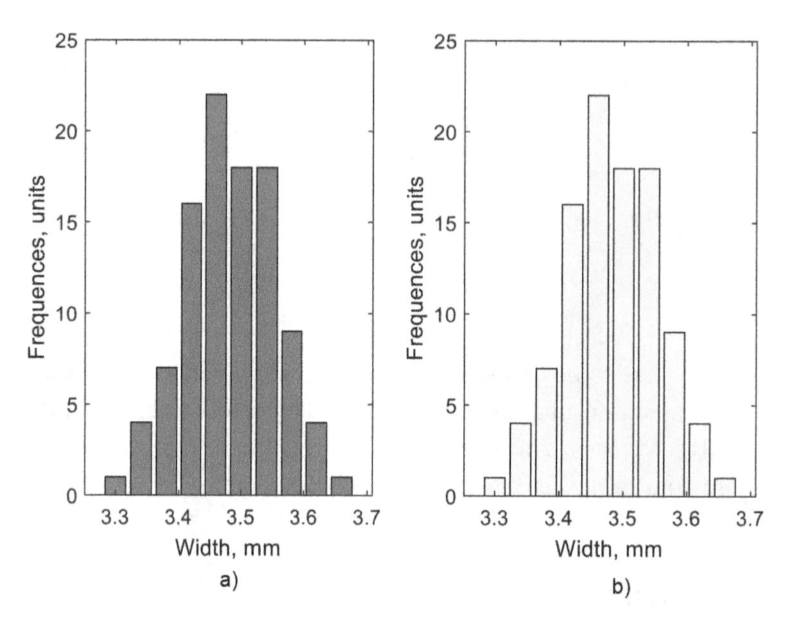

The bar command has more graphical capabilities than the hist command for plotting statistics. This command can be used not only for statistical representations but also to display trends or compare different groups of data or to track changes over time. Detailed information about the available command forms can be obtained by inputting the doc bar command in the Command Window.

11.2.3. Specialized Graphs From the Statistics and Machine Learning Toolbox

In addition to the above graphical commands of the basic MATLAB, the Statistics and Machine Learning Toolbox™ has supplementary commands for generating statistical graphs. Below we get some of commands that can be useful for M&T practice. The following are some of the commands that may be useful for practicing in the M&T fields.

11.2.3.1. Box-and-Whiskers Graph

The large-size technical data can be summarized into the so-termed box-and –whisker plot that shows data distribution in very compact form. The interpretation data in the form of box-and-whiskers can be performed using the boxplot command that in its simplest form reads

```
boxplot(x)
```

The command draws a box with the central line marking the data median and the bottom and top box edges marking the 25th and 75th percentiles, and the whiskers extended to the most extreme data points not considered outliers that are represented individually by the '+' symbol.

Generate for example a boxplot for the prism widths data used in the Subsection 11.2.1. To perform this, the following commands can be entered in the Command Window:

```
>>width=[3.53 3.47 3.43 3.28 3.54 3.48 3.49 3.53 3.40 3.45 ...
 3.58 3.54 3.41 3.35 3.46 3.51 3.57 3.51 3.42 3.52 ...
 3.42 3.43 3.45 3.41 3.48 3.44 3.48 3.47 3.47 3.46 ...
 3.59 3.54 3.50 3.40 3.39 3.36 3.68 3.57 3.52 3.43 ...
 3.42 3.54 3.61 3.38 3.47 3.49 3.36 3.56 3.47 3.63 ...
 3.47 3.53 3.42 3.46 3.42 3.53 3.59 3.46 3.53 3.46 ...
 3.51 3.58 3.51 3.43 3.44 3.49 3.49 3.44 3.51 3.59 ...
 3.48 3.49 3.35 3.45 3.40 3.47 3.54 3.61 3.63 3.49 ...
 3.48 3.56 3.53 3.55 3.55 3.40 3.54 3.49 3.45 3.45];
>>boxplot(width)
>>xlabel('Set number')
>>ylabel('Prism height data, micron')
```

Resulting graph is shown in Figure 5

Figure 5. Box-plot with one set of the prism height data

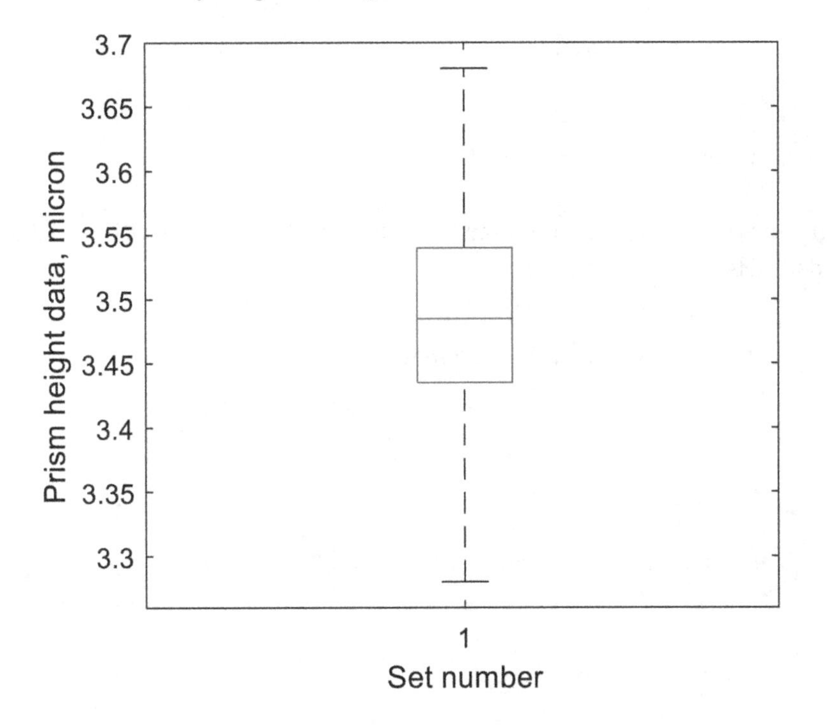

If there are two or more equally-sized groups of data, they should be presented as columns of the x matrix which will be shown as two or more boxes with whiskers on the same plot.

For example, in addition to measured values obtained on the one site, the heights of the another group of the 100 rectangular prisms were measured and the following results were obtained: 3.48, 3.38, 3.36, 3.32, 3.31, 3.40, 3.48, 3.50, 3.44, 3.38, 3.37, 3.43, 3.47, 3.48, 3.23, 3.40, 3.32, 3.36, 3.54, 3.34, 3.27, 3.41, 3.39, 3.37, 3.47, 3.45, 3.43, 3.38, 3.43, 3.38, 3.41, 3.45, 3.43, 3.35, 3.36, 3.32, 3.48, 3.40, 3.35, 3.45, 3.36, 3.44, 3.52, 3.48, 3.39, 3.39, 3.55, 3.39, 3.42, 3.43, 3.47, 3.42, 3.43, 3.43, 3.62, 3.47, 3.33, 3.30, 3.27, 3.55, 3.33, 3.48, 3.44, 3.28, 3.56, 3.44, 3.55, 3.43, 3.41, 3.32, 3.40, 3.41, 3.28, 3.40, 3.41, 3.34, 3.43, 3.30, 3.35, 3.41, 3.35, 3.43, 3.37, 3.33, 3.47, 3.45, 3.45, 3.45, 3.48, 3.48, 3.40, 3.36, 3.39, 3.50, 3.30, 3.35, 3.48, 3.47, 3.36, and 3.38 μm. Present these and preceded data in the two-columns matrix and generate boxplot with the following commands:

```
>>width=[width;...
 3.48 3.38 3.36 3.32 3.31 3.40 3.48 3.50 3.44 3.38 ...
 3.37 3.43 3.47 3.48 3.23 3.40 3.32 3.36 3.54 3.34 ...
 3.27 3.41 3.39 3.37 3.47 3.45 3.43 3.38 3.43 3.38 ...
 3.41 3.45 3.43 3.35 3.36 3.32 3.48 3.40 3.35 3.45 ...
 3.36 3.44 3.52 3.48 3.39 3.39 3.55 3.39 3.42 3.43 ...
 3.47 3.42 3.43 3.43 3.62 3.47 3.33 3.30 3.27 3.55 ...
 3.33 3.48 3.44 3.28 3.56 3.44 3.55 3.43 3.41 3.32 ...
 3.40 3.41 3.28 3.40 3.41 3.34 3.43 3.30 3.35 3.41 ...
 3.35 3.43 3.37 3.33 3.47 3.45 3.45 3.45 3.48 3.48 ...
 3.40 3.36 3.39 3.50 3.30 3.35 3.48 3.47 3.36 3.38]';
>>boxplot(width)
>>xlabel('Set number')
>>ylabel('Prism widht data, micron')
```

After entering, these commands produce the following graph (Figure 6) with two boxes-and-whiskers for each of the data sets:

Figure 6. Box-plot with two sets of the prism height data

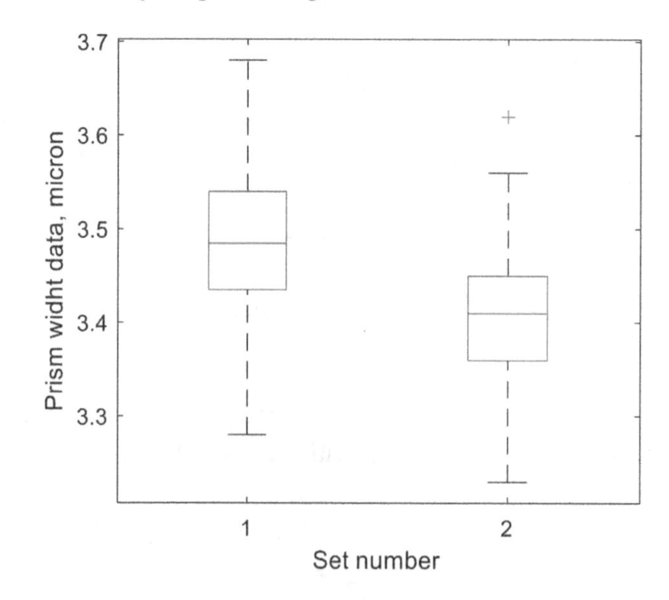

The point/s lying outside the tips of the whiskers, namely q_3+1.5($q_3 - q_1$) and below q_1-1.5($q_3 - q_1$), is/are outlier/s; in these expressions q_1 and q_3 designates the 25th and 75th percentiles respectively. In the generated boxplot the + character indicates the outlier that occurred in the second data set.

Using the boxplot command, you can generate various block shapes, color types, and chart orientation. For example,

- The notched box can be generated with the 'notch','on' property name-value pair as boxplot(width,'notch','on');
- The line color of the box can be changed to, e.g. the red color can be produced with the boxplot(width,'color','r') command;
- The orientation of the box can be selected as horizontal by addition of the 'orientation','horizontal' property-name pair as boxplot(width,'orientation','horizontal')

More detailed information about available properties and parameters for formatting the boxes and whiskers can be obtained with the help boxplot command.

11.2.3.2. Generating a Control Chart

A control chart is a graph that displays changes in a process characteristic over time. Data is plotted in time order. A special command allows construct a process control chart. The simplest form of this command is

```
controlchart(X)
```

The command constructs the chart containing: averaged by rows data points; central line of the data, in green; the lower and upper control limit lines, in red; and legend. The outlier points are automatically marked by 'o' and denote as 'o – Violation' in the legend.

Use for example the volume of a mechanical part that was measured 16 times, each time 4 parts were in the sample, the data are

```
649.8 651.6 650.5 647.6
649.6 652.4 651.2 646.5
646.3 650.2 650.2 648.9
649.4 648.1 650.1 652.5
650.5 650.4 650.6 650.4
653.7 652.1 654.0 649.8
647.8 649.9 650.5 649.8
649.0 650.6 650.2 651.6
649.4 648.6 652.4 650.5
648.6 652.3 650.1 650.7
649.0 650.6 649.0 646.9
647.1 649.4 651.4 649.7
649.8 647.5 649.5 650.0
649.2 649.0 650.3 648.7
650.8 648.5 651.3 650.0
650.1 649.3 651.2 649.1
```

Each of 16 rows contains four values that correspond to a sample formed at the same time. The commands that generate the control chart (Figure 7) are:

```
>> X=[649.8 651.6 650.5 647.6
649.6 652.4 651.2 646.5
646.3 650.2 650.2 648.9
649.4 648.1 650.1 652.5
650.5 650.4 650.6 650.4
653.7 652.1 654.0 649.8
647.8 649.9 650.5 649.8
649.0 650.6 650.2 651.6
649.4 648.6 652.4 650.5
648.6 652.3 650.1 650.7
649.0 650.6 649.0 646.9
647.1 649.4 651.4 649.7
649.8 647.5 649.5 650.0
649.2 649.0 650.3 648.7
650.8 648.5 651.3 650.0
650.1 649.3 651.2 649.1];
>> controlchart(X)
```

Figure 7. Control chart of the volume of a mechanical part

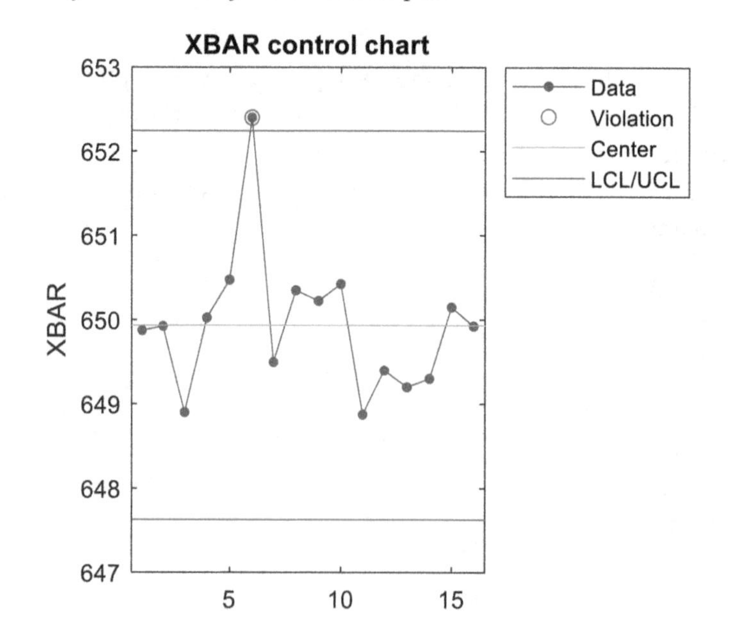

In the plot legend, the LCL and UCL denote the lower and upper control limits, respectively, represented by lines $\pm 3\sigma$ (σ -standard deviation); the Center represents the midline of the data; the outlier point is circled.

More complicate forms of this command permit to generate control chart of different view; in addition, the command can output some statistical data presented as a structure, e.g. >>stats=controlchart(X). The latter command assigns the following values to the **stats** structure: the average value of each sample, its standard deviation and the number of tested parts, the range of each sample, the total average of the data and their sigma. These values are in the structure fields that are named correspondingly: mean, std, n, range, mu, and sigma, e.g. if we want to display the each sample mean values for each sample, type stats.mean. More detailed explanations are beyond the scope of the book.

11.3. COMMANDS FOR PROBABILITY DISTRIBUTIONS AND RANDOM NUMBERS

Statistical methods are used the probability of occurrences of some parameter in the certain range, which is a probability distribution. As can be seen from the above, basic MATLAB® has commands for descriptive statistics, random numbers, and visualization, but a most wide set of commands from this area are available in the Statistics and Machine Learning Toolbox™; this tool extends MATLAB® possibilities to wide range engineering challenges including the M&T. In the two subsections below, we describe some of available probability distribution functions and random number generators. This subsections assume a somewhat familiarity with probability and statistics.

11.3.1. Density, Cumulative, and Invers Cumulative Functions: Normal Distribution

Commands for probability density function PDF, cumulative distribution function CDF, and inverse of cumulative distribution function InvCDF have the following simplest forms:

```
Y=normpdf(X,mu,sigma)
P=normcdf(X,mu,sigma)
X=norminv(P,mu,sigma)
```

We demonstrate the usage of these commands by the solution of the following problem.

Problem: A caliper indicates the mean length value 25.00 mm with standard deviation σ is equal 0.05 mm; assume that the values are distributed normally in the $2.5 \pm 3\sigma$ range. Write a live script namely ExNormDistr that calculates the PDF, CDF, and InvCDF functions and generates their graphs.

The ExNormDistr live script solving this problem with generated graphs are presented in Figure 8

Figure 8. The live script that uses the normpdf, normcdf, *and* norminv *commands to represent graphs with the PDF, CDF, and InvCDF functions.*

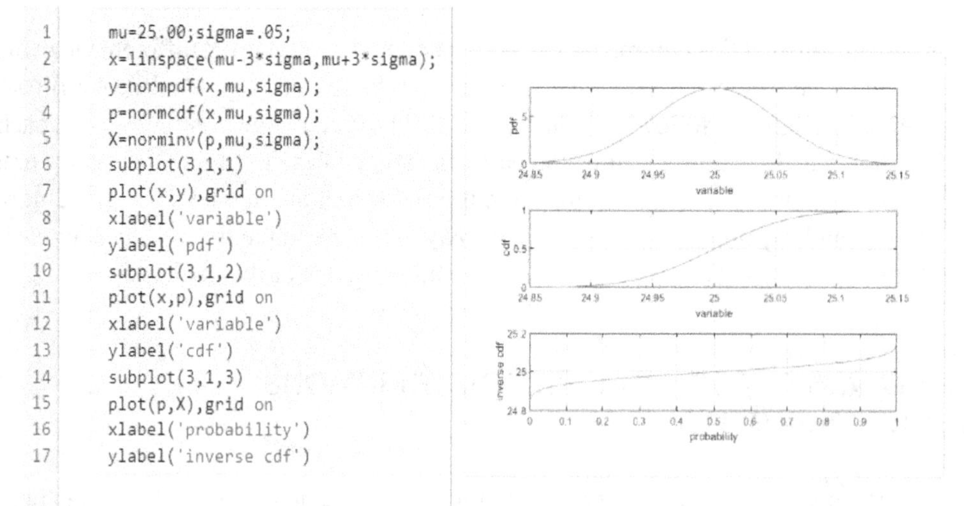

11.3.2. Universal Commands for Specified Distribution

Below are the simplest forms of three universal commands that calculate the PDF, CDF, and inverse CDF values for normal and about 30 other distributions (some of them see in Table 1):

```
Y = pdf('Distr_Name',X,A,…)
P = cdf('Distr_Name',X,A,…)
X = icdf('Distr_Name',P,A,…)
```

- 'Distr_Name' is a string with the actual name of the distribution;
- X is a random variable;
- Y is the calculated relative probability (density);
- P is the probability value of cumulative distribution function;
- A and the ellipsis (…) designate one, two or three parameter values in accordance to the chosen distribution (for example μ and σ in lognormal distribution), the ellipsis here marks unfinished list of parameters and should not be written in an actual command.

The distributions to use in the above commands, their names and input parameters are presented in the Table 1.

In the equations of this table, x is the random variable and Γ is the gamma function. In case the discrete distribution (finite outcome value) the probability density function called sometimes as probability mass function. Among represented in table, the binomial (regular and negative), hypergeometric, F and Poisson's distributions are discrete. The table includes commands to only the most used distributions; a complete list can be obtained in MATLAB® documentation by entering the doc pdf command in the Command Window.

Table 1. Distributions (alphabet order), names and parameters to input in the pdf, cdf, and icdf commands

Distribution Name and the Probability Density/Mass Function Expression	'Distr_Name' (Full and Short)	First Parameter	Second Parameter	Third Parameter
Beta distribution: $\dfrac{x^{a-1}\left(1-x\right)^{b-1}}{\Gamma\left(a\right)\Gamma\left(b\right)/\Gamma\left(a+b\right)}$	'Beta' or 'beta'	a is the first shape parameter	b is the second shape parameter	–
Binomial distribution: $\begin{pmatrix} n \\ x \end{pmatrix} p^x \left(1-p\right)^{n-x}$	'Binomial' or 'bino'	n - number of trials	p – success probability for each trial	–
Chi-square distribution: $\dfrac{x^{\frac{v-2}{2}} e^{-\frac{x}{2}}}{2^{\frac{v}{2}}\Gamma\left(\frac{v}{2}\right)}$	'Chi-Square' or 'chi2'	ν- degrees of freedom	–	–
Exponential distribution: $\dfrac{e^{-\frac{x}{\mu}}}{\mu}$	'Exponential' or 'exp'	µ- mean	–	–
F-distribution (Fisher–Snedecor distribution): $\dfrac{\Gamma\left(\frac{v_1+v_2}{2}\right)\left(\frac{v_1}{v_2}\right)^{\frac{v_1}{2}} x^{\frac{v_1-2}{2}}}{\Gamma\left(\frac{v_1}{2}\right)\Gamma\left(\frac{v_2}{2}\right)\left[1+\left(\frac{v_1}{v_2}\right)x\right]^{\frac{v_1+v_2}{2}}}$	'F' or 'f'	ν1 is the numerator degrees of freedom	ν2 is the denominator degrees of freedom	–
Gamma-distribution: $\dfrac{x^{a-1}e^{-\frac{x}{b}}}{b^a\Gamma\left(a\right)}$	'Gamma' or 'gam'	a is the first shape parameters	b is the second shape parameters	–
Hypergeometric distribution: $\dfrac{\begin{pmatrix} K \\ x \end{pmatrix}\begin{pmatrix} M-K \\ n-x \end{pmatrix}}{\begin{pmatrix} M \\ n \end{pmatrix}}$	'Hypergeometric' or 'hyge'	M is the size of population;	K is the number of items with desired population characteristic	n – number of defected samples

continued on following page

Table 1. Continued

Distribution Name and the Probability Density/Mass Function Expression	'Distr_Name' (Full and Short)	First Parameter	Second Parameter	Third Parameter
Lognormal distribution: $\frac{e^{\frac{-(\log x-\mu)^2}{2\sigma^2}}}{x\sigma\sqrt{2\pi}}$	'Lognormal' or 'logn'	μ is the natural logarithm of the mean	σ is the standard deviation	–
Negative binomial distribution: $\binom{nr+x-1}{x}p^r(1-p)^x$	'Negative Binomial' or 'nbin'	r is the number of successes	p is the probability of successes in a single trial	–
Normal distribution: $\frac{e^{\frac{-(x-\mu)^2}{2\sigma^2}}}{\sigma\sqrt{2\pi}}$	'Normal' or 'norm'	μ is the mean	σ is the standard deviation	–
Poisson distribution: $\frac{\lambda^x e^{-\lambda}}{x!}$, x=1,2, ...; λ>0	'Poisson' or 'poiss'	λ - rate parameter, indicates mean times of events in a given interval	–	–
Student's distribution: $\frac{\Gamma\left(\frac{v+1}{2}\right)}{\Gamma\left(\frac{v}{2}\right)\sqrt{v\pi}\left(1+\frac{x^2}{v}\right)^{\frac{v+1}{2}}}$	'T' or 't'	v – degrees of freedom	–	–
Uniform distribution: $\begin{cases}\frac{1}{b-a}, a \le x \le b\\ 0, otherwise\end{cases}$	'Uniform' or 'unif'	a is the minimal x value	a is the maximal x value	–
Weibull distribution: $\frac{b}{a}\left(\frac{x}{a}\right)^{b-1}e^{-(x/a)^b}, x>0$	'Weibull' or 'wbl'	a is the scale parameter	b is the shape parameter	–

Consider as an example the following problem to demonstrate usage of one of the tabular distributions.

Problem: An inspector tests of the surfaces of metallic parts. According to the control plan, the manufacturing process should be interrupted if more than five contaminated sections will be detected on a part. Find answers to the following questions: (a) what is the probability to stop the process after the first inspection if surfaces have two contaminated sections in average, and (b) what is the probability that a surface has less than six contaminated sections? Use the cumulative function for Poisson distribution (discrete distribution) for calculations; the cumulative function gives the probability that the random variable X is less or equal to the amount x. Write life script named **ExPoissCFD** solving this problem and displaying probability values with explanatory headers.

The designed live script **ExPoissCFD** and the obtained numbers are presented in Figure 9

Figure 9. The live script that uses the cdf *commands for the Poisson distribution.*

```
1  x=5;                        % stated value by the control plan
2  lamb1=3;                                        % a-question
3  lamb2=5;                                        % b-question
4  prob2stop = 1-cdf('pois',x,lamb1);   % answer to a-question
5  prob_less6 = cdf('pois',x,lamb2);    % answer to b-question
6  fprintf([' a) Probability  to stop process by\n' ...
7  'after the first inspection are:%5.2f \n'],prob2stop)
8  fprintf([' b) Probability that a surface has less\n' ...
9          'than six contaminated sections are:%5.2f\n'],prob_less6)
```

```
a) Probability  to stop process by
after the first inspection are: 0.08

b) Probability that a surface has less
than six contaminated sections are: 0.62
```

These results reveal that in case (a) there is only 8% probability that the process will be stopped after the first inspection, or, in other words, 8% probability that more than five contaminated sections will be identified during the first inspection; in case (b) there is a 62% probability that a part surface has less than six contaminated sections ($x \leq 5$) and the manufacturing process can be continued.

11.3.3. Generating Random Numbers Using a Universal Command

Many mathematical models and engineering simulations in M&T areas require using numbers that have random values. Some of these commands were presented in Chapter 2, i.e. the **rand**, **randn**, and **randi** commands that generate random numbers for normal and uniform distributions. In addition to these commands, the Statistics and Machine Learning Toolbox™ provide a universal command that generates random numbers using many other distributions. The general transcription for one of the available forms of this command is:

```
Y=random('Distr_Name',A,…,m,n)
```

- 'Distr_Name', A and the ellipsis (…) designate the same input parameters as in the previous subsection;
- m and n are the number of rows and columns of the desired matrix with random numbers;
- Y is the resulted matrix containing the generated random numbers.

Available distributions, their names and input parameters of this command are the same as in Table 1, except for the Student's distribution, which cannot be used with this command.

Figure 10. The ExRandom *live script producing five random numbers using negative binomial distribution.*

```
1  rng default %restores generator settings to the start value
2  r=3;    % number of 'successes' (finding 3 defective units)
3  p=0.05;            % probability to successed in single trial
4  m=1;                                % desired number of rows
5  n=5;                             % desired number of columns
6  units = random('nbin',r,p,m,n)+3
7  % 3 provides minimal number to the genarated units

units = 1x5
    47    57    38    39    21
```

We demonstrate this command use by the following example (MATLAB® documentation, the nbinrnd command):

Problem: A production process should be designed with 0.05 defect probability. Write a live script named ExRandom that simulate five random numbers representing the amount of the manufactured units that might be inspected before finding three defective units. Use the negative binomial distribution (discrete distribution).

The live script that solves this problem is (Figure 10):

11.4. SUPPLEMENTARY SPECIALIZED COMMANDS FOR STATISTICAL CALCULATIONS AND VISUALIZATIONS

In addition to the described commands, there are many others that may be necessary for calculating and visualizing methods used in the M&T fields. Some of them are listed in the tables below.

11.4.1. Some Additional Commands for Probability Distributions and Random Numbers

The commands for calculating the probability distributions characteristics and generating the random numbers are summarized in Table 2. Note, the actual four-letter distribution name (see Table 1) must be printed instead of the xxxx characters in each of commands represented in this table.

The input and output arguments X, A,.., Y, and P of table commands can be vectors or matrices of the same size; when a part or one of the input arguments is vector or matrix other input arguments are expanded respectively to a constant vector or array with the same size. Commands for normal distribution - normpdf and normcdf - can be written without the μ and σ arguments that by default are assumed to be 0 and 1 respectively.

11.4.2. Additional Commands for Visualizing Statistics

Table 3 lists some additional visualization commands that are intended to process data and generate various plots.

Table 3 represents one of the possible command forms; use the Search option in the Help window to get full information.

Table 2. Supplementary commands for probability distributions[1] and random numbers

Name and Command Form	Command Parameters	Example (Input and Output)
Probability density function: Y=xxxxpdf(X,A,…)	dist is the name of the desired distribution that should be selected (from Table 1*) and written instead of these characters; X is the value at which the probability value should be calculated; A and ellipsis (…) designate parameters of the selected distribution; Y is the output vector with the obtained probability densities. Note, the ellipsis marks unfinished list of arguments and should not be written in an actual command.	>> % F-distribution >>X=1:3;num=2;denom=2; >>y = fpdf(X,num,denom) y = 0.2500 0.1111 0.0625 >> % binomial distribution >>X=10;tests=200;prob=0.02; >>y=binopdf(X,tests,prob) y = 0.0049
Cumulative distribution function: P=xxxxcdf(X,A,…)	The dist, X, A, and ellipsis (…) denote the same parameters as for previous command; P is the cumulative probability value.	>> % F-distribution >>X = 2:5; >>nu1 = 4;nu2 = 6; >>P = fcdf(X,nu1,nu2) P = 0.7863 0.8889 0.9355 0.9594 >> % Uniform distribution >> X=0.7;a=-1;b=1; >>P = unifcdf(X,a,b) P = 0.8500
Inverse cumulative function: X=xxxxinv(P,A,…)	The dist, P, A, ellipsis (…), and X are the same as for previous command; The command calculates value of the variable X for given probability P of the selected distribution.	>> % Chi-square distribution >>P=0.95,nu=8 >>X = chi2inv(P,nu) x = 15.5073 >> P=0.5;mu=700; >> % exponential distribution >> X=expinv(P,mu) X = 346.5736
Random number generation: xxxxrand(A, …,m,n)	m and n are the rows and columns of the matrix in which the calculated numbers will be presented; The dist and A are the same as for previous command.	>> % binomial distribution >>rng default >> N=100;m=2;n=3; >>binornd(N,1./N,2,3) ans = 0 1 0 1 1 0 >> % normal distribution >>rng default >> mu=0;sigm=1;m=3;n=2; >> normrnd(mu,sigm,m,n) ans = 0.5377 0.8622 1.8339 0.3188 -2.2588 -1.3077
Calculates mean M and variance V: [M,V]=xxxxstat(A,…)	The dist, P, A, ellipsis (…), and X are the same as for previous command.	>> % binomial distribution >>n=10;p=0.5; >> [m,v] = binostat(n,p) m = 5 v = 2.5000

Table 3. Some additional graphical commands of the Statistics and Machine Learning Toolbox™

Command	Purpose and designations	Example with generated plots

capaplot(data,specification_limits)	demonstrates accordance of the normally distributed data to the specification limits.
	data is the vector with data;
	specification_limits is the two-element vector with the lower and upper specification limits
		`>>rng default;`
		`>>data = normrnd(4.1,0.005,100,1);`
		`>>spec_lim=[4.09, 4.11];`
		`>>capaplot(data, spec_lim)`
		`>>grid on` 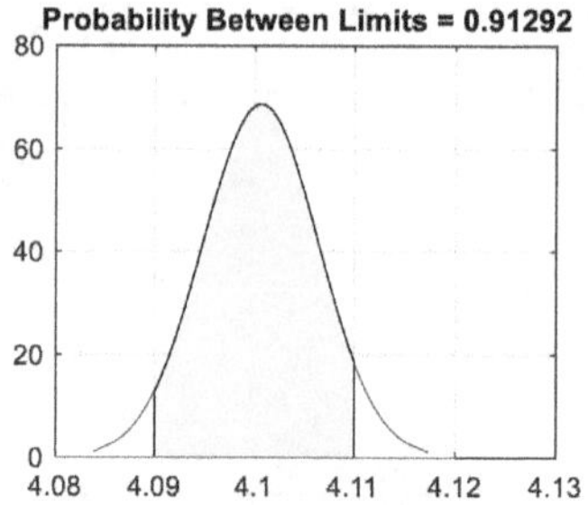
		Probability Between Limits = 0.91292

scatter(x,y,s,'filled')	generates a plot with circled points.
	x and y are the point coordinates;
	s - circle size in square points, one point = 1/72 inch;
	'filled' - is the string that is added to fill the circles with color.
		`>>x = linspace(0,pi);`
		`>>y=sin(x)+rand(1,100);`
		`>>scatter(x,y,80,'filled')`
		`grid on` 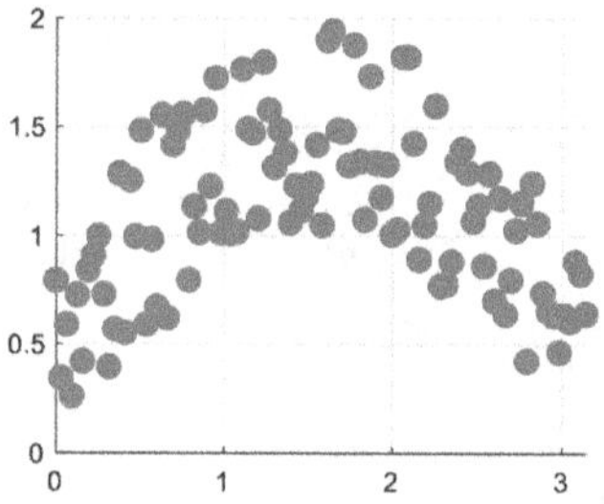

continued on following page

Table 3. Continued

pareto(y,y_names)	ranks causes from the most to the least significant and demonstrates its as a bar plot with a line of the cumulative percentage.
	y – is a vector of the ranked data;
	y_names – string vector with name of each of the y values.
		`>>% observed defects of a mechanical part`
		`>>y=[30 400 20 40 5];`
		`>>% defects due to ...`
		`>>y_names=char('traces','solders','misalignments','metal','others');`
		`>>pareto(y,y_names)` 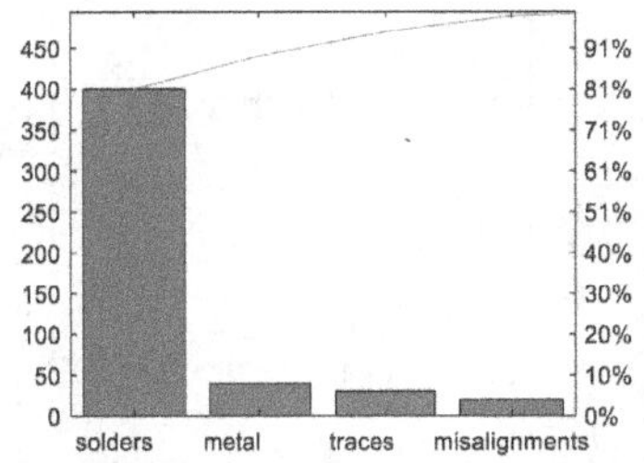

disttool	opens the interactive "Probability Distribution Function Tool" window allowing the user to visualize the pdf and cdf functions for various distributions.	`>>disttool` 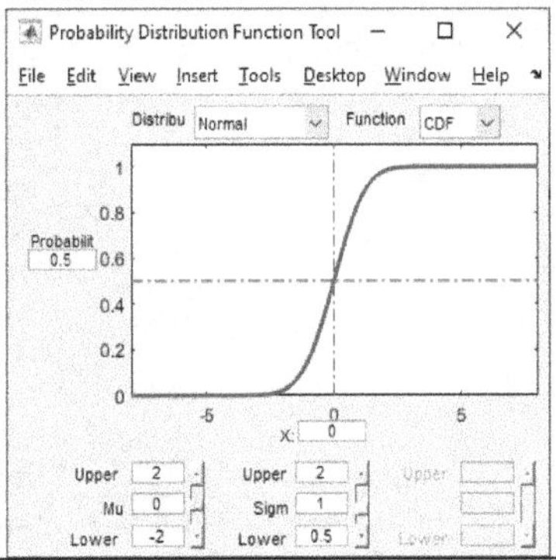
distributionFitter	opens the "Distribution Fitting" window allowing interactively fits probability distributions to data selected from the workspace.	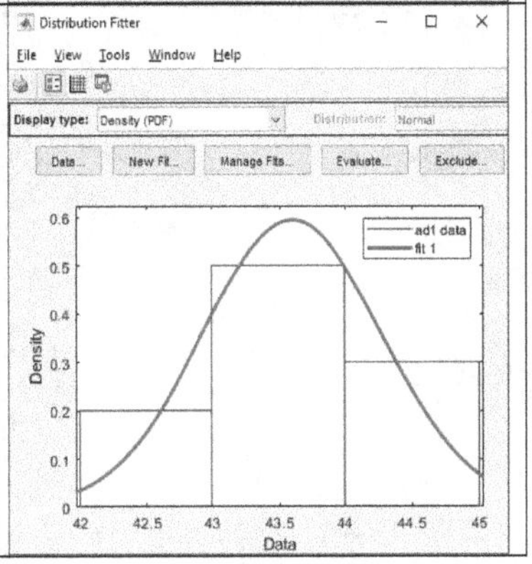

continued on following page

Table 3. Continued

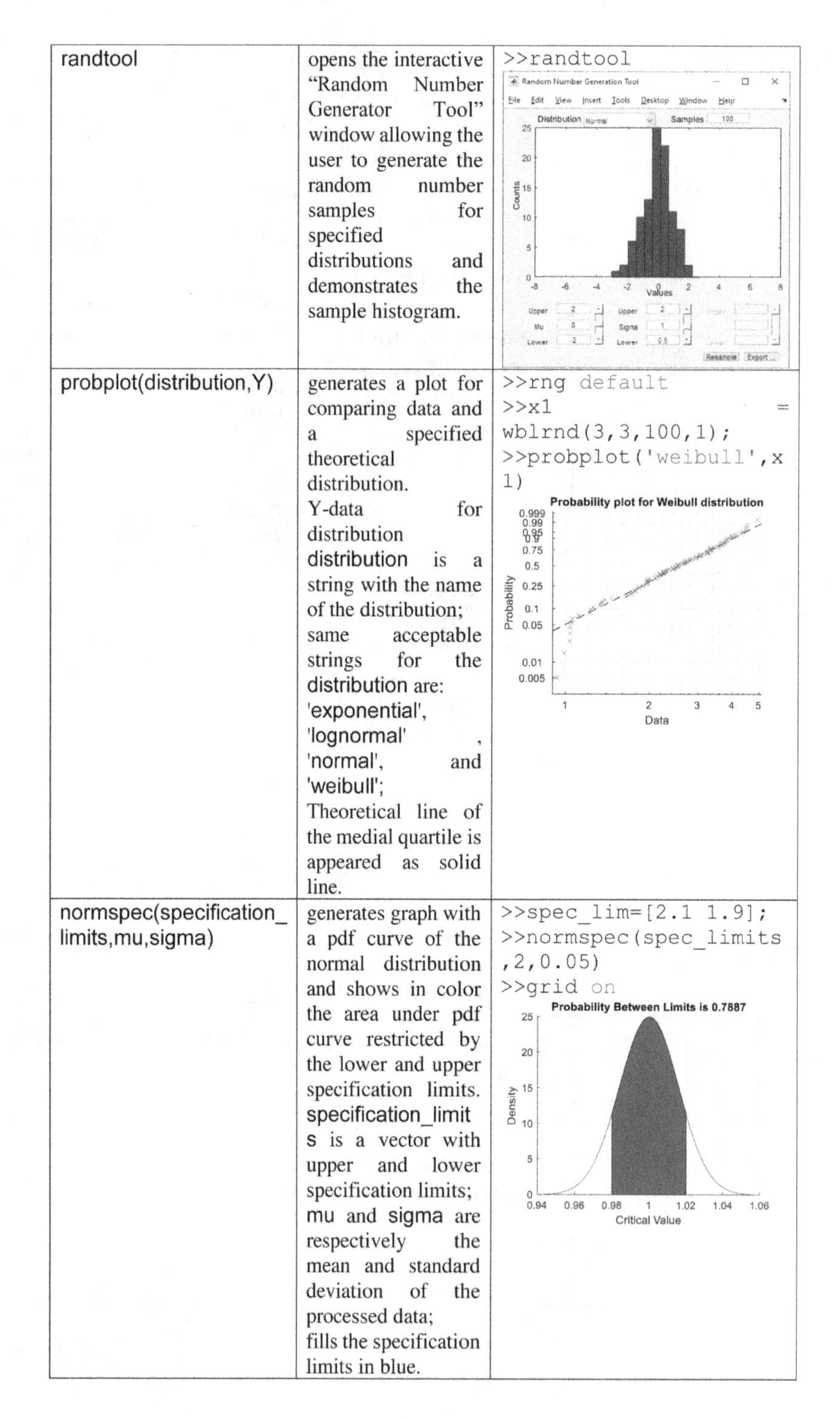

randtool	opens the interactive "Random Number Generator Tool" window allowing the user to generate the random number samples for specified distributions and demonstrates the sample histogram.	>>randtool
probplot(distribution,Y)	generates a plot for comparing data and a specified theoretical distribution. Y-data for distribution distribution is a string with the name of the distribution; same acceptable strings for the distribution are: 'exponential', 'lognormal', 'normal', and 'weibull'; Theoretical line of the medial quartile is appeared as solid line.	>>rng default >>x1 = wblrnd(3,3,100,1); >>probplot('weibull',x1)
normspec(specification_limits,mu,sigma)	generates graph with a pdf curve of the normal distribution and shows in color the area under pdf curve restricted by the lower and upper specification limits. specification_limits is a vector with upper and lower specification limits; mu and sigma are respectively the mean and standard deviation of the processed data; fills the specification limits in blue.	>>spec_lim=[2.1 1.9]; >>normspec(spec_limits,2,0.05) >>grid on

Note:

- The disttool, distributionFilter, and randtool commands open interactive windows, after which the desired distribution functions and their parameters should be inputted in appropriate fields to generate respectively a graph, fitting curve, and random numbers; the obtained results can also be exported to the MATLAB® workspace.
- In the normspec command, the specification_limits vector can be written with only one of the specification limits; in this case the inf (no upper limit) or -inf (no lower limit) should be written instead of the limit value, for example specification_limits =[0.98 inf] should be written when the lower specification level is 0.98 and the upper is not specified.

11.5. STATISTICAL EVALUATION OF ENGINEERING DATA USING HYPOTHESIS TESTS

Hypothesis testing has grown out of the necessity to decide some practical problems in science and technology in general and in M&T in particular. For example, is it possible to consider a technical parameter determined from a small number of measured parts (termed a sample) valid for the entire population of parts; or the values of a parameter obtained for two samples of different sizes are identical or the differences are significant; or are the results of two different laboratories identical? To solve such problems, the hypothesis testing of the probability statistics are used.

In the below subsections, the available test commands of the Hypothesis tests section of the Statistics and Machine Learning Toolbox™ are described by examples from the M&T area. Basic familiarity with statistics and probability theory is assumed.

11.5.1. Hypothesis Test Outlines

The variety of possible tests is carried out using the same steps: a tested so-termed null hypothesis is postulated, then an analysis plan is formulated, the sample data is analyzed, and finally the decision about the acceptation or rejection the null hypothesis is made.

Step 1. Hypothesis:

Two mutually exclusive hypotheses are state – a null and an alternative. They should be formulated so that if one is true, then the other is false.

Step 2. Analysis plan:

Now the following items should be specified:

- The significance level α, the recommended value of which is 0.05 or 0.01;
- Method and type of test, which implies the choice of certain statistics and distribution for the data;
- The acceptable value of the sample probability to accept/reject the null hypothesis.

Step 3. Sample data analysis:

On this stage the calculations of the sample statistics should be performed, and, as a result, the statistical probability value for the test sample, called the p-value, must be determined that comfirmed (or did not) the null hyphothesis.

Step 4. Decision:

In conclusion, the decision must be made based on comparison of the p-value with the significance level α: if the p is less than α then the null hypothesis is rejected, otherwise the null hypothesis can be accepted or in equivalent but more cautious formulation – it cannot be rejected. The latter formulation is due to the probabilistic nature of the statistical tests. It is considered that there are two types errors in such decision. Apparently, the significance level α is the probability of rejecting the null hypothesis when it is actually true - a type I error. However, even if the null hypothesis is not rejected, it may still be false - a type II error. Thus, it is reasonable to check the alternative hypothesis. The distribution of the test statistic under the alternative hypothesis determines the probability β of a type II error (often causes due to small sample sizes). The probability to correctly reject a false null hypothesis called power p of the test and is $1-\beta$. Summarizing, the probability of making a type I error is α and the probability to make a type II error is β.

The Statistics and Machine Learning Toolbox™ has different test commands allowing perform a various types of tests. These commands analyze the sample data and determine the true or false null hypothesis. Therefore, the sample data analysis, step 2, are perform by the test commands, that should be selecting before testing and suitable for this problem.

Available tests are dividing in two categories: parametric and nonparametric. A parametric statistical test is a test that makes assumptions about the distribution and parameters of the observed data, while a nonparametric test does not make such assumptions. Accordingly, some of the available test commands of the Statistics and Machine Learning Toolbox™ implements parametric tests, for example, such as *t*-tests, Jarque-Bera test, Lilienfors test, and Kolmogorov-Smirnov test, and another part implements nonparametric tests, for example, Wilcoxon signed-rank test and Wilcoxon rank-sum test. Each of these two groups of commands can be subdivided in three subgroups – paired, unpaired, and multiple (more than two samples) data tests. Below we briefly describe some test commands and their usage for analyzing sample data.

11.5.2. *T*-test Using the ttest Command

When the sample size is small, and variance or standard deviation of the underlying normal distribution does not known, then the *t*-test is recommended. The test belongs to the parametric group of the hypothesis tests, since it assumes some statistical preconditions about the sample parameter/s, such as the normality and mean. The test can be applied to the one-sample data or to the paired-sample data (latter test is often termed pre/post-test); two data sets called paired when each of them has the same number of data points, and each point in one data set is related with an appropriate point in the other. In the first case, the mean of the sample data is compared with the same mean of the population presenting by the normal distribution. In the second case, the two means of the equally sized samples are compared.

The two most commonly used forms of this command that actualize the *t*-test are:

```
[h,p] = ttest(x,m,'alpha',alpha_val)
```

and

```
[h,p] = ttest(x,y,'alpha',alpha_val)
```

- x and y are vector of the same size, each of which must contain data from one of two samples; in case one-sample test, the first form of this command is used, and for two-sample – the second form. For the first form, the data is transformed to the test statistic form $t = \frac{\overline{x}_1 - m}{s / \sqrt{n}}$ where $\overline{x}$ and s are the sample mean and standard deviation respectively, n is the sample size; For the second form, the data transform as $= \frac{\overline{x}_1 - \overline{x}_2}{s / \sqrt{n}}$;
- m is the given or hypothesized population mean;

- In the property pair 'alpha',alpha_val the 'alpha' is a string containing the word alpha and alpha_val indicates the value of the significance level α;
- h is the test decision, if h=1 then the null hypothesis is rejected and if h=0 then the null hypothesis is accepted; the null hypothesis is "there is not difference between the sample and the population means";
- p is the obtained *p*-value, on basis of which it is concluded that the null hypothesis is true or false.

The first of the ttest commands performs a *t*-test for the null hypothesis that the *x*-data normally distributed with mean *m* and unknown variance, against the alternative that the mean is not *m*. The second ttest command performs a paired-sample t-test for the null hypothesis that the difference *x*-*y* normally distributed with mean 0 and unknown variance, against the alternative that the mean is not zero; *x* and *y* must be equivalent vectors. The command can be written without the alpha value, in this case α=0.05 by default. The output argument p can be omitted in the both presented ttest commands.

Demonstrate the usage of the command on the following example.

Problem: Lifetime data for some mechanical devices: 9.5, 11.1, 4.7, 5.8, 12.0, 11.1, 6.3, 5.2, 13.1, and 9.6, in years. Devices of this model should have 9.5 years mean lifetime. Can we say that the lifetime data came from normal distribution with mean 9.5 years? Write the program as live script named Ex_ttest that used the first of the above ttest commands for solving the problem. The program should print the defined h and p values and a conclusion on the conformity or inconformity of the average values of the lifetime.

First we postulate the null hypothesis for this problem - the lifetime data are normally distributed with the mean *m*=9.5. Alternative hypothesis: the mean *m* is not equal to 9.5. We further assume that the significance level for this hypothesis is 0.01 (or 1%).

The following program, written as the Ex_ttest live script, solves the problem:

```
x=[9.5 11.1 4.7 5.8 12.0 11.1 6.3 5.2 13.1 9.6];                % lifetime data
alpha=0.01;                                                  % significance level
m=9.5;                                                        % mean to comparing
[h,p_value] = ttest(x,m,'alpha',alpha)       % the first of the t-test com-ds
h = 0
p_value = 0.5152
if h==0                                                  % null-hypothesis is true
  disp({'The average values of the lifetimes for'; ...
     'the sample and population are not different'})
elseif h==1                                             % null-hypothesis is false
  disp({'The average values of the lifetimes for'; ...
     'the sample and population are different'})
end
  'The average values of the lifetimes for'
  'the sample and population are not different'
```

The final h=0 (as the obtained *p*-value is greater than α) shows that the null hypothesis is true, therefore the sample and population averages are equal for the given α =0.01. The program provides a conclusion text after obtaining h with the ttest command.

In case we have two equally sized samples with lifetimes of devices of the same model, then the second form of the ttest command should be used for the *t*-test.

To perform *t*-test for two non-paired samples, the **ttest2** command should be used – see Table 4.

Table 4. Available hypothesis test commands

Command and Test Name	Null and Alternative Hypotheses	Description	Example (Obtaining Data, Test Command, Conclusion)
[h,p]=chi2gof(x) Chi-square goodness-of-fit test.	The null hypothesis: the data are from a normal distribution with mean or variance estimated from x. The alternative hypothesis: the data are not normally distributed.	x is the vector with sample data values; h –is the number 0 or 1: if h=1 then the null hypothesis can be rejected; otherwise, h=0, it cannot. The significance level is 5% (default). p is the *p*-value determined in the test	>>rng default >>% data >>x=random('norm',45,4,90,1); >>[h,p]=chi2gof(x) h = 0 p = 0.2065 Conclusion: The null hypothesis cannot be rejected
[h,P]=jbtest(x) Jarque-Bera test.	The null hypothesis: the data are from a normal distribution with unknown mean and variance. The alternative hypothesis: the data are not from normal distribution.	x is the vector with sample data values; h –is the number 0 or 1: if h=1, then the null hypothesis can be rejected; otherwise, h=0, it cannot. The significance level is 5% (default). p is the *p*-value determined in the test	>>% mileage data, >>% MGP -miles per >>% gallon >>load carbig >>[h,p] = jbtest(MPG) h = logical 1 p = 0.0022 Conclusion: The null hypothesis should be rejected
[h,p] = kstest(x) One-sample Kolmogorov-Smirnov test.	The null hypothesis: the data are from a standard normal distribution. The alternative hypothesis: the data are not from that distribution.	x –is the vector with sample data values; h –is the number 0 or 1: if h=1, then then the null hypothesis can be rejected; otherwise, h=0, it cannot. The significance level is 5% (default). p is the *p*-value determined in the test	>>x=-2:1:4;% data >>[h,p]=kstest(x) h = logical 0 p = 0.1359 Conclusion: The null hypothesis cannot be rejected.
[h,p]=kstest2(x1,x2) Two-sample Kolmogorov-Smirnov test.	The null hypothesis: the data in two samples are from the same continuous distribution. The alternative hypothesis: the data are from different continuous distributions.	x1 and x2 are the vectors each with its own sample data; h –is the number 0 or 1: if h=1, then the null hypothesis can be rejected; otherwise, h=0, it cannot. The significance level is 5% (default). p is the *p*-value determined in the test.	>>x=-1:1:6;%sample 1 >>rng default >>y= randn(19,1);% sample 2 >>[h,p]=kstest2(x,y) h= logical 0 p = 0.1200 Conclusion: The null hypothesis cannot be rejected.

continued on following page

Table 4. Continued

Command and Test Name	Null and Alternative Hypotheses	Description	Example (Obtaining Data, Test Command, Conclusion)
[p,h]=signrank(x) Two-sided Wilcoxon signed rank test.	The null hypothesis: The sample data is from a distribution with zero median (data is symmetric about its median). The alternative hypothesis: the sample data is not from a distribution with zero median.	x is the vector with sample data values; h –is the number 0 or 1: if h=1, then the null hypothesis can be rejected, otherwise, h=0, it cannot. The significance level is 5% (default). p is the *p*-value determined in the test.	>>rng default >>% data >>x=randn(1,22)+1.2; >>[p,h]=signrank(x) p = 2.0135e-04 h = logical 1 Conclusion: The null hypothesis should be rejected.
[h,p]=lillietest(x) Lilliefors test.	The null hypothesis: the data is from a normal distribution. The alternative hypothesis: The data is not from a normal distribution.	x is the vector with sample data values; h –is the number 0 or 1: if h=1, then the null hypothesis can be rejected; otherwise, h=0, it cannot. The significance level is 5% (default). p is the *p*-value determined in the test.	>>rng default >>x=random('weibull',45,4,400,1); %data >>[h,p]=lillietest(x) h = logical 0 p = 0.0949 Conclusion: The null hypothesis cannot be rejected.
[h,p] = ttest2(x1,x2) Two-sample *t*-test.	The null hypothesis: the data in two samples are from the normal distributions with equal means and equal but unknown variances. The alternative hypothesis: the means are unequal.	x1 and x2 are the vectors each with its own sample data; h –is the number 0 or 1: if h=1, then the null hypothesis can be rejected; otherwise, h=0, it cannot. The significance level is 5% (default). p is the *p*-value determined in the test.	>>rng default >>mu1=0;mu2=0.1; >>s1=1;s2=1; >> x1=random('norm',mu1,s1,1,1000);%sample 1 >> x2=random('norm',mu2,s2,1,800) ;%sample 2 >>[h,p]=ttest2(x1,x2) h = logical 1 p = 3.3751e-04 Conclusion: The null hypothesis should be rejected.
[h,p] = ztest(x,m,sigma) One-sample *z*-test.	The null hypothesis: the data are from a normal distribution with known mean and standard deviation values. The alternative hypothesis: the mean is not m.	x is the vector with sample data values; m is the sample mean; sigma is the sample standard deviation; h –is the number 0 or 1: if h=1, then the null hypothesis can be rejected; otherwise, h = 0, it cannot. The significance level is 5% (default). p is the *p*-value determined in the test.	>>rng default >>s_m=0.1;% mean >>s_std=1;% std >>x = random('norm',s_m,s_ std,1,100);%sample >> [h,p] = ztest(x,s_m,s_std) h = logical 0 p = 0.2184 Conclusion: The null hypothesis cannot be rejected.

11.5.3. Wilcoxon Rank Sum Test

This test is used when the two independent unequal-sized samples should be compared to verify identity of their medians. It belongs to the group of nonparametric hypothesis tests as it does not require any assumptions about the distribution function. One form of the commands that provides this test reads

```
[p,h]= ranksum(x,y,'alpha',alfa_val)
```

- x and y are the vectors each with its own sample data; in general case the length of the x and y vectors can be different;
- 'alpha' is the name of the significance level, α, and alfa_val is the values of α; the 'alpha',alfa_val property pair is optional and can be omitted, in this case α=0.05 by default;
- p and h is the *p*-value and test decision respectively; h = 1 (when p<α) indicates that the null hypothesis should be rejected, and h = 0 (when p>α) indicates that the null hypothesis can be accepted.

Demonstrate the usage of the command on the following example.

Problem: Operating times (time before first failure) of an actual and a new model of computer were tested in a laboratory, the obtained data are respectively 8.6 10.2 3.8 4.9 19.0 10.0 5.4 4.3 12.2 8.6 and 10.1, 9.2 7.8 14.5 16.1 3.2 4.9 8.8 11.4 20.2, in days. Can we say that the median difference of the pairs is zero? Write the program as live script named Ex_ranksum that used the ranksum command for solving the problem. The program should print the defined h and p values and a conclusion on the conformity or inconformity of the median values of the operating times.

The null hypothesis for this: the median difference in the paired lifetime values is insignificant, h=0, and the alternative – significant, h=1. Assume the significance level α is 0.05.

The following commands written as the Ex_ranksum live script program solve the problem:

```
x=[3.1 8.4 7.2 3.4 2.7 11.7 2.3 2.5 13.9 8.7];                      % actual model
y=[13.1 5.7 9.8 12.6 11.8 21.1 7.3 10.0];                         % enhanced model
[p,h]=ranksum(x,y);                      % Wilcoxon test with alpha=0.05 by default
disp(['p=',num2str(p);]),disp(['h=',num2str(h)])                  % display results
p=0.034279
h=1
if h==0                                                  % null-hypothesis is true
  disp({'Medians of operating times for'; ...
      'two samples do not differ'})
elseif h==1                                             % null-hypothesis is false
  disp({'Medians of operating times for'; ...
      'two samples are different'})
end
  'Medians of operating times for'
  'two samples are different'
```

The p-value of the test is greater than the significance level and null-hypothesis cannot be rejected, thus decision of the test is h=0 which means that there is no real lifetime difference. Note, if we assume that the significance level is 0.01, the test reveals opposite results – the medians do not different for the two samples.

The ranksum command is used in this program without the 'alpha',alfa_val pare, in this case the value α=0.05 is used. The output parameters of this command are displayed with the disp commands, in each of which square brackets are used to assemble a string and a number in a row vector (for their simultaneous output) and the num2str command is used to convert a number to a string (because all elements of the vector must be data of one type).

11.5.4. Sample Size and Statistical Test Power

The statistical power of a hypothesis test is the probability that test correctly rejects the null hypothesis, or, in opposite, the probability of accepting the alternative hypothesis if it is true.High power value (generally accepted in range 0.8… 1) indicates the correctness of the rejection of the null hypothesis when the test shown that null hypothesis is false. In MATLAB® there is the sampsizepwr command that can calculate test power or size, the two of available forms are:

```
power = sampsizepwr(testtype,p0,p1,[],n)
n = sampsizepwr(testtype,p0,p1,power)
```

- testtype is a string with the test name; some of the available names is 'z', 't' and 't2' – for *z*-, *t*-test, and two-sample pooled tests respectively;
- p0 is a two-element vector written in form [mu0 sigma0] where mu0 is the mean and sigma0 is the standard deviation for the null hypothesis, in case of the t_2 –test these values should be taken for the first sample;
- p1 is the true mean value tested under alternative hypothesis, in case the t_2-test p1 is that for the second sample;
- [] is the empty square brackets; no information is required here;
- n is the size of the sample, should be given as positive integer value;
- power is the value of the power achieved for the performed (given) hypothesis test; the power of test depends on sample size, the difference of the variances, the significance level, and the difference between the means of two populations.

We illustrate the use sampsizepwr with the following problem.

Problem: In nine tests of a certain industrial machine, it was determined that the machine operates on average 105 min to complete some operation with the standard deviation 15 min. The guaranteed time for this operation, declared by the manufacturer, is 90 min (taken as true value). Find the probability β of incorrectly concluding that the mean machine operating time value does not exceed 90 minutes specified by the manufacturer. If obtained probability is not enough small (for example β>0.4), define *n* as required number of the machine tests to provide the β = 0.05. Write the program as live script named Ex_sample_p_n that used two above forms of the sampsizepwr commands: the first form to defining the power of the *t*-test and the second form to determine the number of the machine tests that are required so that the probability of an incorrect output is no more than 0.4 . The program should display the power, the β- probability, and confirm or suggest the required number of machine trials (sample size) to provide a 5% uncertainty in the conclusion.

In accordance with the named data, the mean and standard deviation under the null-hypothesis are 105 and 15 respectively, and 90 is the mean value under alternative hypothesis.

The live script program Ex_sample_p_n for solving the problem is

```
n=9;                                                    % sample size
testtype='t';                                                % t-test
mu0=105;
sigma0=15;
p0=[mu0,sigma0];                                   % null hypothesis
p1=90;                                      % alternative hypothesis
power=sampsizepwr(testtype,p0,p1,[],n)               % power of test
power = 0.7480
beta=1-power          % probability that sample mean is not greater p1
beta = 0.2520
        % if we want probability not more 0.05 to reach incorrect conclusion
if beta<0.4
p1=95;                                  % new alternative hypothesis
required_n=sampsizepwr(testtype,p0,p1,power);   % required sample size
disp(['Required sample size is ' num2str(required_n)])
else
disp('The given sample size is OK')
end
Required sample size is 18
```

Thus in the case of the *t*-test, the power value 0.748 shows that *n*=9 for the machine operating time data is not sufficient to conclude that the mean operating time is not greater than 90 min, there is a high probability of rejecting the null hypothesis when the null hypothesis is maybe true. However, with n = 18, a safe result can be obtained.

In the list of input arguments of the sampsizepwr command the significance level and tail type can be added, for example n = sampsizepwr(testtype,p0,p1,power,[],'tail','right'); in some more details see Subsection 11.4.3.

11.5.5. Supplementary Commands for the Hypothesis Tests

In addition to the commands described above, the Statistics and Machine Learning Toolbox™ provides many other commands for hypothesis tests. Some of them are presented in Table 4.

In some examples of this table the carbig file is used; this file is one of the available set of the files with various data providing by the Statistics and Machine Learning Toolbox™ and listed in the Sample Data Sets section of the toolbox documentation. This and other data files can be loaded by typing the load file_name command; the file_name is a name of one of the files, e.g. carbig. The carbig file contains 13 differently named vectors/matrices with various parameters of some car models from 1970–1982.

The commands in this table can be written with single or without output parameters; in the latter case, only the h value is displayed. The non-default significance level or/and standard deviation can be inputted in the chi2gof, jbtest, kstest, kstest2, lilietest, and ztest commands, as a rule just after the input arguments presented in the table, e.g. ztest(x,m,sigma,alpha); related details see by typing the help with appropriate test name in the Command Window. The tabled commands written for two-tiled tests by default, the ztest and ttest2 commands can be used for the one-tailed tests, in this case they has more complicate form and is out of scope this book, the details can be obtaining with the help command.

11.6. APPLICATION EXAMPLES

11.6.1. Surface Roughness Indices Using the Descriptive Statistics Commands

The surface texture is characterized by roughness as one of its important component. Surface roughness indices can be defined from measurements using the following expressions:

$R_a = \frac{\sum_{i=1}^{n} y_{p_i}}{n}$ – average of the highest profile peaks y_{p_i} from the mean (central) line, recorded within the evaluation length, µm;

$R_z = \frac{\sum_{i=1}^{n} y_{p_i} + \sum_{i=1}^{n} y_{v_i}}{n}$ – average of the five peaks y_{p_i} and five valleys y_{v_i} located within the sampling length, µm;

$R_p = \max\left(y_p\right)$ – highest peak within the sampling length, µm;

$R_v = \max\left(y_v\right)$ – lowest valley within the sampling length, µm;

$R_{max} = R_p + R_v$ – maximum peak-to-valley roughnes within the sampling length, µm;

$R_q = \sqrt{\frac{\sum_{i=1}^{n} y_{p_i}^{2}}{n}}$ – root-mean-square roughness within the evaluation length, µm;

$R_{ku} = \frac{\sum_{i=1}^{n} y_{p_i}^{4}}{nR_q^4}$ – kurtosis parameter (dimensionless) using for evaluation of the randomness of heights and closeness of roughnesses to the normal distribution, $R_{ku} = 3$ for normal distribution.

i in these expressions is the serial number of the peak or valley.

The term “evaluating length” refers to the surface profile length selected for measurements; and the term “sampling length” means a segment of the evaluation length. The evaluation length should contain at least one sampling length.

Problem: The peaks and valleys of the surface profile, measured by a profilometer within a one of the sampling lengths are 13, 16, 15, 14, 12 and 15 14 16 13 16 µm respectively. The rest peak data for the entire evaluating length are 11, 13, 12, 10, 14, 15, 9, 12, 15, and 16 µm. Write a live script named

Ch_11_ApExample_1 that calculates the above roughness indices and outputs them with the appropriate header.

The Ch_11_ApExample_1 program solving the problem are (Figure 11):

Figure 11. The Ch_11_ApExample_1 *live script calculating the roughness indices*

```
1   y_p=[13 16 15 14 15];              %micron, sampling length peaks
2   y_v=[15 14 16 15 16];            %micron, sampling length valleys
3   y_p_e=[11, 13, 12, 10, 14, 15, 9, 12, 15, 16];   %micron, peaks
4   Ra=mean([y_p y_p_e]);        % or rpm - mean peak profile height
5   Rz=Ra+mean(y_v); %ten-point height(5 highest peaks + 5 valleys)
6   Rp=max(y_p);                                      % maximal peak
7   Rv=max(y_v);                                     % lowest valley
8   Rmax=max(y_p+y_v);                       % maximal peak-to-valley
9   Rq=sqrt(mean(y_p.^2));                %root-mean-square rouhgness
10  Rku=mean(y_p.^4)./Rq^4;                                %kurtosis
11  out=[Ra Rz Rp Rv Rmax Rq Rku]';
12  fprintf('  Ra    Rz    Rp    Rv    Rmax  Rq    Rku\n')
13  fprintf('%6.2f%6.2f%6.2f%6.2f%6.2f%6.2f%6.2f\n',out)
```

```
   Ra    Rz    Rp    Rv  Rmax    Rq   Rku
13.33 28.53 16.00 16.00 31.00 14.64  1.02
```

This live script is arranged as follows:

- The first three lines contain the peak and valley data for sampling length and peak data for the evaluation length;
- The next seven lines represent equations with the descriptive statistics command that calculates required roughness indices;
- The formed column vector out contains all calculated surface roughness parameters for further output with the fprintf command;
- Two final fprint commands output a header line and a line with calculated roughness parameters (to the right of the commands).

Comparing these values and the values required by the appropriate standard, the conclusion can be made about their coincidence or non-coincidence for the tested surface.

11.6.2. Runout of a Rotating Mechanical Part: Histogram and Theoretical PDF

An inaccuracy of the rotating mechanical parts is characterized by the runout parameter that should be provided by the manufacturer and checked periodically during operation of machine.

Problem: Runout data of 83 rotating mechanical parts are -0.1280, -0.3206, -0.0898, -0.1783, 0.2705, 0.1002, -0.4956, -0.0141, -0.3156, -0.1282, -0.0595, 0.0362, -0.1312, 0.1720, -0.1652, -0.0327, 0.0225, -0.0724, 0.0580, -0.0334, -0.2149, -0.3823, 0.2182, -0.1855, -0.0695, 0.0059, -0.0678, -0.2347, -0.1631, -0.1069, 0.1560, -0.2275, 0.0422, -0.0358, -0.1247, -0.2596, -0.1330, -0.1006, -0.3272, -0.0549, -0.2212, -0.1018, -0.0313, -0.2347, -0.1336, -0.0290, -0.3873, 0.0834, 0.3110, 0.0708, -0.1382, -0.0161, -0.2562,

0.2214, 0.0678, 0.0426, -0.2300, -0.0340, -0.1779, -0.1374, -0.1798, -0.2545, -0.2353, -0.1208, -0.3648, 0.0132, -0.1057, 0.0282, -0.0422, -0.0054, -0.3295, -0.2536, -0.1597, -0.0684, 0.0986, -0.1339, 0.1204, -0.0085, 0.1060, -0.0656, -0.1941, -0.3292, and -0.0609 mm. Write a script named Ch_11_ApExample_2 that compares graphically distribution this data with normal distribution by the probplot command (Table 11.3) and generates a histogram (with the bar command) together with the probability distribution function of normal distribution. Normalize the histogram bins (runout frequencies) n as $\frac{n}{\int_{\min(x)}^{\max(x)} n(x)dx}$ where x is the runout data. Place two plots in one Figure window and format them.

The appropriate Ch_11_ApExample_2 program is:

```
data=[-0.1280 -0.3206 -0.0898 -0.1783 0.2705 0.1002 -0.4956...
   -0.0141 -0.3156 -0.1282 -0.0595 0.0362 -0.1312 0.1720...
   -0.1652 -0.0327 0.0225 -0.0724 0.0580 -0.0334 -0.2149...
   -0.3823 0.2182 -0.1855 -0.0695 0.0059 -0.0678 -0.2347...
   -0.1631 -0.1069 0.1560 -0.2275 0.0422 -0.0358 -0.1247...
   -0.2596 -0.1330 -0.1006 -0.3272 -0.0549 -0.2212 -0.1018...
   -0.0313 -0.2347 -0.1336 -0.0290 -0.3873 0.0834 0.3110...
    0.0708 -0.1382 -0.0161 -0.2562 0.2214 0.0678 0.0426...
   -0.2300 -0.0340 -0.1779 -0.1374 -0.1798 -0.2545 -0.2353...
   -0.1208 -0.3648 0.0132 -0.1057 0.0282 -0.0422 -0.0054...
   -0.3295 -0.2536 -0.1597 -0.0684 0.0986 -0.1339 0.1204...
   -0.0085 0.1060 -0.0656 -0.1941 -0.3292 -0.0609];
st=std(data);
av=mean(data);
[n,x]=hist(data);            % outputs runout numbers in bins and x-centers
subplot 121
probplot('normal',data)              % compares data to the normal distrib.
legend('Theoretical','Data','location',...
   'northwest')                                   % top left legend location
grid on
axis square
n_norm=n/trapz(x,n);                            % normalizes the frequencies
subplot 122
bar(x,n_norm)                                  % bar plot with defined x and n
n_pdf=pdf('norm',x,av,st);               % calculates theoretical values of n
hold on
plot(x,n_pdf,'-o')                            % adds pdf curve to the bar plot
legend('pdf by data','theoretical pdf',...
   'location','northoutside')                     % places legend out of plot
xlabel('Runout, mm')
ylabel('Probability density, nondim')
axis tight square
hold off
```

After running, a Figure window appears with the following two plots

```
>> Ch_11_ApExample_2
```

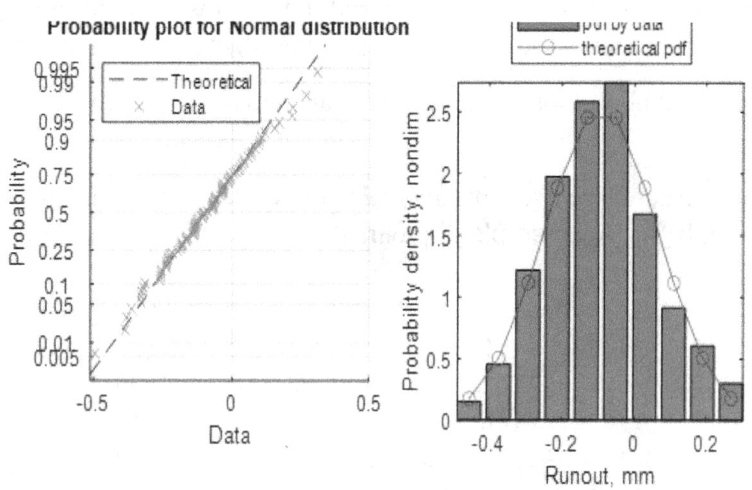

This program is designed as follows:

- The three first commands assign the runout data to the data variable and calculates its average and standard deviation;
- The **hist** command in the used form does not generate plot but outputs the number of the runouts **n** in each of ten (default) bins and bin coordinates **x**, the **n**-values should be further normalized and used together with the **x** in the **bar** and **pdf** commands;
- The **subplot** command divides the Figure window in two pains for further plotting;
- The **probplot** and three following commands generate and format the graph that compares the data to the normal distribution; if most of the data points appear along the reference line, thus they are normally distributed;
- The **n/trapz(x,n)** command uses the trapezoidal rule (subsection 5.4.2.) for integrating the $n(x)$ function and calculates normalized n-values representing the actual pdf values;
- The **bar** command generate graph with the n bars along x and the **hold on** command save this graph for further manipulations;
- The **pdf** command calculates theoretical pdf values of the normal distribution at defined early **x**, data mean **av** and standard deviation **st**;
- The **plot** command adds a theoretical pdf curve to the bar graph that allows you to compare theoretical pdf and pdf calculated by the data; in order to best see the correspondence or inconsistency between the bars and the theoretical curve, the curve points on which it was built are marked by "o";
- The final commands format the graph with bars and the theoretical probability density functions.

The generated plots apparently show that the data can be represented by a normal distribution.

11.6.3. Piston Ring Gaps: Process Capability by Simulation

The ability of a manufacturing process to produce some part parameter within the lower and upper specification limits *LSP* and *USL* is measured by the process capability ratio C_p. For large sample the process capability of the normally distributed sample data is defined as

$$C_p = \frac{USL - LSL}{6\sigma}$$

The measure if the sample mean lays in the center of the *USL* … *LSL* interval is the capability ratio C_{pk}, which can be calculated with the expression

$$C_{pk} = \min\left[\frac{USL - m}{3\sigma}, \frac{m - LSL}{3\sigma}\right]$$

where m and σ is the sample mean and standard deviation respectively.

If $C_p<1$, then it is likely that a large number of nonconforming units are produced in the process; if C_{pk} <0 then the process falls outside the specification limits.

Problem: To simulate the piston ring process capability to meet specifications for ring gap, the random command generates three samples with 60 values of the ring gaps each; for data simulation take the sample mean μ=0.42 mm and standard deviations σ equal 0.05, 0.12, and 0.15 mm for the 1st, 2nd, and 3rd samples respectively. Specified values for the ring gap are: m=0.4, LSL=0.1 and USL=0.7 mm. Write a live script named Ch_11_ApExample_3 that calculate capabilities C_p and C_{pk}, and generate plot for each of the capabilities with the capaplot command; place the three generated plots with the subplot commands and show the defined Cp and Cpk on each of the plot.

The following live script, Ch_11_ApExample_3, solves the problem and outputs the required values of the process capabilities (Figure 12).

This program is constructed as follows:

- The three first lines introduce the values to simulate sample data – mean mu, size N, and specification limits LSL and USL – and input the sigma vector with three values of the standard deviation;
- Next five lines contain commands intended to generate the data matrix with three columns containing 60 values each; these values are simulated by the random command in which the sigma value is changed within every pass of the for … end loop;
- Six further lines contain commands calculating saple mean and standard deviation and the C_p and C_{pk} capabilities by the expressions above.
- The next command assigns specification limits to the two-element vector spec that is used as one of the arguments of the capaplot command.
- The command of the final lines generate three plots using the subplot command for each of the plots. In each of the capaplot commands one column of data is used with a column number equal to the plot number. On each of the plots, the calculated values of the C_p and C_{pk} are placed using the text commands. The string parameter of each of these commands is formed as two element

Figure 12. The Ch_11_ApExample_3 live script for the process capability calculations of the ring gap

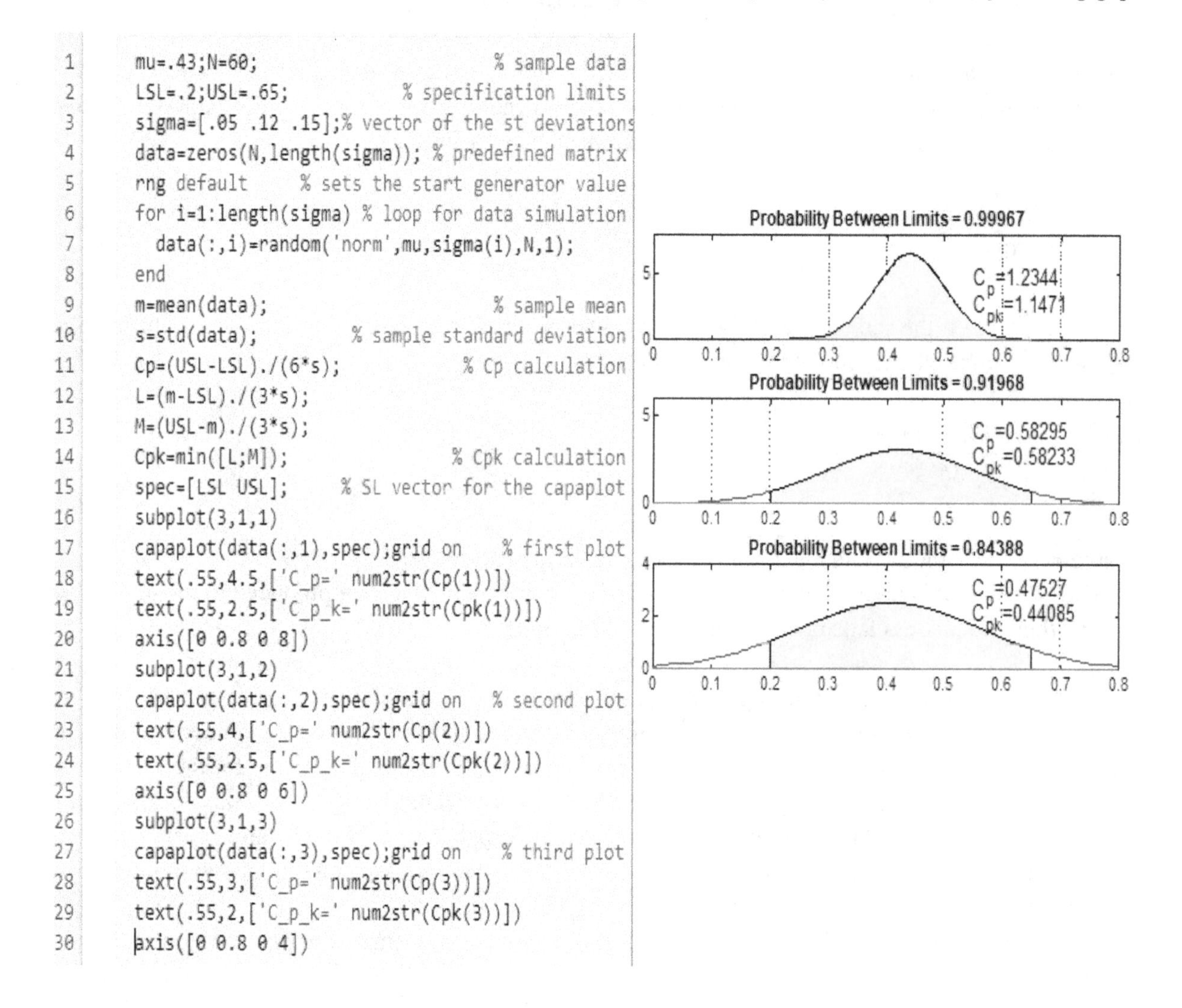

vector of the characters 'C_p=' or 'C_pk=' and the value Cp or Cpk converted to string with the num2str command.

As can be seen, the dispersion of the given sample mean with the standard deviation σ=0.05 keeps ring gaps within the specified limits with probability of 0.9997; in other considered cases, the calculated probabilities are less than those are usually recommended for manufacturing.

11.6.4. Comparison of the Friction Torque Means for Two Oil Additives

The friction torque of two different engine oil additives were tested in a tribological laboratory. The results for the first additive are 43.2, 43.7, 42.9, 44.1, 42.4, 44.2, 43.9, 43.4, 43.5, 44.7 N·m, and for the second additive - 38.5, 39.9, 39.7, 40.5, 41.7, 42.4, 41.1, 43.5 N·m. Which of the additives is likely preferable?

Problem: Is a statistical difference between the friction torque average of these additives at the significance level 5%? Write a live script program named Ch_11_ApExample_4 hat generates a normal probability plot using the probplot command (Table 3), tests the sample means with the ttest2 command, and display the result of the *t*2-test with the text conclusion.

We postulate the null hypothesis (h=0) for this problem as follows - the mean friction torque values of the both samples are statistically equal. Alternative hypothesis (h=1): the means are not equal. We assume additionally that the significance level for this hypothesis is 0.05 (or 5%).

The following live script, Ch_11_ApExample_4, solves the problem and outputs the normal probability plot (Figure 13) and the *t*2 test conclusion

Figure 13. The Ch_11_ApExample_4 *live script with normal probability plot for the friction torque average of two oil additives*

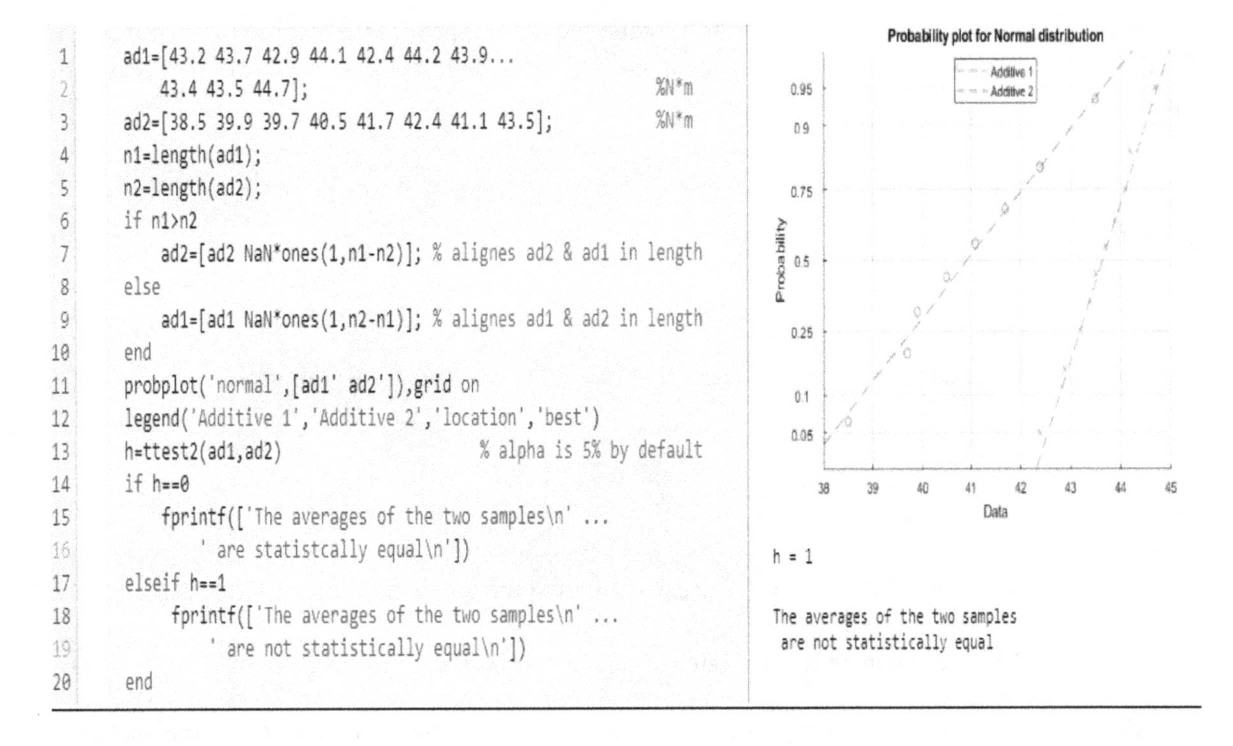

This program is arranged as follows:

- The first four commands are entered the sample data in the variables ad1 and ad2 and obtain the lengths of the each sample;
- The if … else statement define the lengthiest sample and adds to the shorter the NaN (Not-a-Number) values to equalize the sample lengths for further usage in the probplot command;
- The probplot command generates graph to match each sample with normal distribution; the following legend command adds legend to the plot axes;

- The ttest2 command executes two-sample *t*-test in which the null hypothesis about the sample means is checked and the result is assigned to variable h; if h=0, then the null hypothesis ("means are equal") can be accepted, otherwise (h=1) the null hypothesis should be rejected;
- The final if … else statement displays the concluding sentence about equality, h=0, or inequality (h=1) of the averages of the two samples.

The generated plot reveals that the data of each of the both samples are closed to the normal distribution, as they follow the solid lines of the medial quartiles of the normal distribution.

The *t*-test indicates that the null hypothesis should be rejected, therefore, the difference between the two additives is significant, and the first additive creates a greater friction torque than the second.

11.6.5. Sample Size Evaluation for Desired Precision

To evaluate the needed size *n* of a sample to provide the desired precision (specified by the max and min limits) of the testing parameter *x* of some a unit, the following expressions can be used (based on Bonett, 2016):

$$n = 4\tilde{\sigma}^2 \left(\frac{z_{\frac{\alpha}{2}}}{w} \right)^2 + \frac{z_{\frac{\alpha}{2}}^2}{2}$$

where α is the significance level, $\tilde{\sigma}$ is a crude planning value of standard deviation

$$\tilde{\sigma} = \frac{\max(x) - \min(x)}{4}$$

$z_{\frac{\alpha}{2}}$ is a two-sided critical z-score value, *w* is the width of the confidence interval, that can be calculated roughly as $2\tilde{\sigma}z_{\frac{\alpha}{2}}$, when there is the n tested data then the confidence interval width is $2\tilde{\sigma}t_{\frac{\alpha}{2},df} / \sqrt{n}$; max(x) and min(x) can be taken as upper USL and lower LSL specification limits $t_{\frac{\alpha}{2},p}$ is the *t*-score, *df* =*n*-1 is the so-called degrees of freedom.

Problem: Write a live script named Ch_11_ApExample_5 that evaluates the sample number needs to check diameter of a circular mechanical part when USL and LSL are 25.2 and 24.8 mm. Assume α=0.05 (5% significance level) and use the inverse cumulative distribution function norminv command for defining the *z*-score value. After determining *n*, use the following tested diameters 24.7, 25.1, and 24.9 mm to recalculate the sample size and confirm the previous *n* value or obtain a more reliable *n* value. In this case, the *z*-score must be changed using the *t*-score in the *w* and *n* equations; in the latter equation, such a replacement can be made only in the first term.

The live script program Ch_11_ApExample_5, which solves this problem are:

```
alpha=0.05;                                   % significance level, two-sided
USL=25.4;LSL=24.6;                                           % specified limits
sigma_s=(USL-LSL)/4;                                        %planed sample std
z_s=norminv(1-alpha/2);               % z-score at 1-alpha/2 confidence interval
w=2*z_s*sigma_s;                               % width of the confidence interval
n_p=4*sigma_s^2*(z_s/w)^2+z_s^2/2;                        % sample size evaluation
n_p=round(n_p);                                                % nearest integer
                                                         % post-hoc calculations
data=[24.7 25.2 24.9];                                           % measured data
n=length(data);                                              % data sample size
sigma_s=std(data);                                            % data sample std
df=n-1;                                                      % degree of freedom
t_s=icdf('T',1-alpha/2,df);          %t-score: confid. Interv.=1-alpha/2, df=n-1
w=2*sigma_s*t_s/sqrt(n);                       % width of the confidence interval
n=4*sigma_s^2*(t_s/w)^2+z_s^2/2;
n=round(n);                                                    % recalculated n
fprintf('Sample size: evaluation n=%3.0f; and post-hoc n=%3.0f\n',n_p,n)
Sample size: evaluation n= 3; and post-hoc n= 5
```

This live script is arranged as follows:

- The first seven commands enter the significance level and specified limits for the unit diameter, and then, calculate standard deviation, width of the confidence interval, and z-score for the $1\text{-}\alpha/2$ confidence interval (the norminv command), then the sample size is evaluated and rounded to the nearest integer;
- In subsequent commands the sample size are calculated retrospectively; for this the measured diameter values are entered, then their length, standard deviation, and degrees of freedom are defined; then t-score for the $1\text{-}\alpha/2$ confidence interval an n-1 degrees of freedom are calculated with the *icdf* command; the width of the confidence interval and sample size are calculate with the t-score instead of the previously used z-score (in the latter case, the t-score is used only in the first term of the n expression);
- The final *fprintf* command outputs first estimated and post-hoc sample size values.

As can be seen, the estimated sample size value are not accurate enough, the post-hoc value is more than 1.5 times larger; therefore, the adjusted sample size should be used when testing the diameter of the part.

11.7. CONCLUSION

The commands for statistical calculations and their graphical presentation have been examined on various experiments/tests data, mainly from the M&T field. It was shown how to process data with the Data

Statistics tool and how to generate histogram, probability distribution, and boxplot using the appropriate MATLAB® commands. Basic and supplementary commands for hypothesis testing are applied to evaluate the lifetime of an engineering device, the operating time of a some new computer model, and the required sample size to the reliable action time estimation of an industrial machine. Statistic calculations were used to create programs that process M&T test data to estimate surface roughness, evaluate the runout of a rotating mechanical part, determine manufacturing process capability of the piston rings, compare the friction torque for two oil additives, and calculate the required sample size to provide the desired diameter accuracy of a disk. All this demonstrates the good ability and the rationality of statistical processing of M&T data using the described commands of the Statistics and Machine Learning Toolbox™.

REFERENCES

Bonett, D. G. (2016). *Sample size planning for behavioral science research.* Retrieved from https://people.ucsc.edu/~dgbonett/sample.html

ENDNOTE

[1] Full list of possible distributions can be obtained in the Help window by finding the Function Reference Section of the Statistics and Machine Learning Toolbox™ documentation.

APPENDIX

Table 5. List of examples, problems, and applications discussed in the chapter

No	Example, Problem, or Application	Subsection
1	Descriptive statistics for bore data of a worn cylinder.	11.1.1.
2	Data Statistics tool for presenting statistics on wear of piston ring ends.	11.1.2
3	Histogram for the width data of metallic rectangular prisms.	11.2.1
4	Two bar graphs for the width data of metallic rectangular prisms.	11.2.2
5	Box plots for one and two groups of the width data of metallic rectangular prisms.	11.2.3.1.
6	Control chart by the volume data of a mechanical part.	11.2.3.2
7	PDF, CDF, and InvCDF curves by caliper meassurements.	11.3.1
8	Evaluation of the surface quality of a metallic part using the Poisson CDF.	11.3.1.2
9	Simulation amount of the manufactured units that should be inspected using the random command.	11.4
10	T-test for a lifetime data.	11.6.2
11	Wilcoxon rank sum test for comparison of the operating time of the actual and new computer models.	11.6.3
12	Required number of machine tests using the sampsizepwr hypothesis test command.	11.6.4
13	Surface roughness indices using the descriptive statistics commands.	11.7.1.
14	Runout of a rotating mechanical part: data histogram and theoretical PDF.	11.7.2
15	Piston ring gaps: process capability by simulation.	11.7.3
16	Comparisons of the friction torque means for two oil additives.	11.7.4
17	Sample size evaluation for desired precision.	11.7.5

Note, some small examples, mostly related to non-M&T issues, are not included in the list.

Appendix

A1.1. PREDEFINED VARIABLES

The variables that are provided automatically and are permanently stored in memory are termed predefined. They are presented in Table 1. In this and other tables, the letter t next to the page number means that the desired value, variable, command or example is in the table located on this page.

Table 1. Predefined variables

No	Variable	Value	Definition
1	ans	Last calculated value	Variable name, contains the most recent number
2	eps	2.220446049250313·10-16	Smallest difference between two numbers
3	i	$\sqrt{-1}$	Imaginary unit
4	inf	∞	Infinity
5	j	$\sqrt{-1}$	the same as i
6	NaN	fff8000000000000 (IEEE code)	Not a number
7	pi	3.141592653589793	Number π

A1.2. SPECIAL CHARACTERS, OPERATORS, ALTERNATIVE COMMANDS, AND PUNCTUATIONS

The symbols, special characters, alternative commands (if exist) with references to places in the book are presented in Table 2.

Table 2. Punctuation and special characters/operators used for mathematical operations.

No	Symbol	Definition	Name or Alternative Command (After Horizontal Double Line)
1	=	assignes a value/set of values to a variable	Assignment
2	%	introduces comments and specifies an output number format	Percent
3	()	used for input arguments, sets the priority of the operation, accesses elements in a vector / matrix / array	Parentheses
4	[ ]	is used for input of the vector/matrix/array elements, or for output aguments the of usre-defined functions	Brackets
5	(space)	separates elements of vectors/arrays/matrices	Space
6	.	used in element-by-element array operations, or referes to a structure field; separates whole and decimal parts of a number	Dot
7	:	creates vectors, used also for loops iterations and for selecting all array elements	Colon
8	,	separates commands on the same line and elements into arrays	Comma
9	;	prevents outputting when wrote after the command; separates commands on the same line and marix rows	Semicolon
10	...	denotes that a long statement to be continued on the next line; also the same as % when is written after command	Ellipsis
11	'	transposes vector, array or matrix, and used for a text string generation	Apostrophe
12	~	uses to ignore function argument or output; is used also as logical not	Tilde
13	>>	command prompt	Right shift
14	+	Addition	plus
15	-	Subtraction	minus
16	*	Scalar and matrix multiplication	mtimes
17	.*	Element-wise multiplication	times
18	/	Right matrix division	mrdivide
19	./	Element-wise right division	rdivide
20	\	Left matrix division	mldivide
21	.\	Element-wise left division	ldivide
22	^	Exponentiation or matrix power	mpower
23	.^	Element-wise power	power

A1.3. OPERATORS FOR RELATIONAL AND LOGICAL STATEMENTS

The operators that was used along the book in rational and logical commands are described in Table 3 together with alternative commands that can be used instead of them. When these commands are used for array or matrices they are element-wise.

Table 3. Relational and logical operators.

No	Symbol/s	Definition	Alternative Command
1	==	Equal (double)	eq
2	>	Greater than	gt
3	>=	Greater than or equal to	qe
4	<	Less than	lt
5	&	AND (logical)	and
6	~	NOT (logical)	not
7	|	OR (logical)	or
8	<=	Less than or equal to	le
9	~=	Not equal	ne

A1.4. COMMANDS REPRESENTED IN THE BOOK

The commands implemented in this book are conditionaly arranged below in three groups:

- commands used to manage the program and perform numerical calculations;
- graphic, drawing, and color commands;
- commands for symbolic calculations and plotting by symbolic expressions.

A1.4.1. Commands for Program Management and Numerical Calculations

Commands for input / output, flow control, array/matrix manipulation, and mathematical calculations are called "non-graphical" here and are presented in Table 4.

Table 4. List of non-graphical commands.

No	Command (Alphabetically)	Definition
1	abs	Absolute value
2	acos	Inverse cosine for angle in radians
3	acosd	Inverse cosine for angle in degrees
4	acot	Inverse cotangent for angle in radians
5	acotd	Inverse cotangent for angle in degrees

continued on following page

Table 4. Continued

No	Command (Alphabetically)	Definition
6	asin	Inverse sine for angle in radians
7	asind	Inverse sine for angle in degrees
8	atan	Inverse tangent for angle in radians
9	atand	Inverse tangent for angle in degrees
10	besselj	Bessel function of first kind
11	bvp4c	Solves boundary value problem of ODE
12	bvp5c	Solves boundary value problem of ODE
13	cdf	The CDF for a specified distribution
14	ceil	Rounds towards infinity
15	char	Creates a row matrix with string elements
16	ch2gof	Chi-square goodness-of-fit test
17	clc	Clear the command window
18	clear	Remove variables from the Workspace
19	convpress	Convert pressure units
20	cos	Cosine for angle in radians
21	cosd	Cosine for angle in degrees
22	cosh	Hyperbolic cosine
23	cot	Cotangent for angle in radians
24	cotd	Cotangent for angle in degrees
25	coth	Hyperbolic tangent
26	cross	Calculates cross product of a 3D vector
27	det	Calculates a determinant
28	diag	Creates a diagonal matrix from a vector
29	diff	Calculates a difference, approximates a derivative
30	disp	Display output
31	doc	Displays HTML documentation in the Help window
32	dot	Calculates scalar product of two vectors
33	edit	Opens Editor window
34	else,elseif	Is used with if; conditionally executes if statement condition
35	end	Terminates scope of for, while, if statements, or serves as last index
36	erf	Error function
37	exp	Exponential
38	eye	Creates a unit matrix
39	factorial	Factorial function
40	find	Finds indices of certain elements of array
41	fix	Rounds towards zero
42	floor	Rounds off toward minus infinity

continued on following page

Table 4. Continued

No	Command (Alphabetically)	Definition
43	fminbnd	Finds a function minimum within the interval
44	fminsearch	Multidimensional optimization, fitting
45	for	Is used to repeat execution of command/s
46	format	Sets current output format
47	fprintf	Displays formatted output and saves it to file
48	function	Defines an user-defined function
49	fzero	Solves a single-variable equation
50	gamma	Gamma function
51	global	Declares a global variable
52	help	Displays explanations in the Command Window
53	icdf	Invers CDF
54	if	Conditionally execute
55	input	Prompts to user input
56	integral	Calculates integral
57	interp1	One-dimensional interpolation
58	inv	Calculates the inverse matrix
59	jbtest	Jarque-Berra hypothesis test
60	kstest	Kolmogorov-Smirnov goodness-of-fit one-sample hypothesis test
61	kstest2	Kolmogorov-Smirnov goodness-of-fit two-sample hypothesis test
62	length	Number of elements in vector
63	lillietest	Lilliefors' goodness-of-fit hypothesis test
64	linspace	Generates a linearly spaced vector
65	log	Natural logarithm
66	log10	Decimal logarithm
67	lookfor	Search for the word in all help entries
68	max	Returns maximal value
69	mean	Calculates mean value
70	median	Calculates median value
71	min	Returns minimal value
72	mode	most frequent value in a sample
73	normcdf	Normal cimmulative distribution function
74	norminv	Inverse of the normal CDF
75	normpdf	Normal probability density function
76	num2str	Converts numbers to a string.
77	ode113	Solves nonstiff ODEs
78	ode15s	Solves stiff ODEs
79	ode23	Solves nonstiff ODEs

continued on following page

Table 4. Continued

No	Command (Alphabetically)	Definition
80	ode23s	Solves stiff ODEs
81	ode23t	Solves stiff ODEs
82	ode23tb	Solves stiff ODEs
83	ode45	Solves nonstiff ODEs
84	odeset	Sets ODE options
85	ones	Creates an array with ones
86	pchip	Interpolates by the Piecewise Cubic method
87	pdepe	Solves 1D (spatially) PDE
88	pdf	The PDF for a specified distribution
89	polyder	Calculates the derivative for a polynimial
90	polyfit	Fits the points with a polynomial
91	polyval	Evaluates the polynomial value
92	quad	Numerical integration with the Simpson's rule
93	quadl	Calculates integral, Lobatto quadrature
94	quadgk	Calculates integral, Gauss-Kronrod quadrature
95	quad2d	Calculates double integral
96	rand	Generates uniformly distributed random numbers
97	randi	Generates integer random numbers from uniform discrete distribution
98	random	Generates random numbers for a specified distribution
99	randn	Generates an array with normally distributed numbers
100	ranksum	Wilcoxon rank sum test, equal medians
101	repmat	Duplicates a matrix
102	reshape	Changes size of array/matrix
103	rng	Controls random number generator
104	roots	Defines the roots of a polynomial
105	round	Rounds off toward nearest decimal or integer
106	sampsizepwr	Calculates sample size and power for hypothesis test
107	sec	Secand for angle in radians
108	secd	Secand for angle in degrees
109	sign	Sign function
110	sin	Sine for angle in radianns
111	sind	Sine for angle in degrees
112	sinh	Hyperbolic sine
113	signrank	Two-sided Wilcoxon rank sum test
113	size	Size of vector/array/matrix
114	sort	Arranges elements in ascending or descending order
115	spline	Interpolates with the cubic spline

continued on following page

Table 4. Continued

No	Command (Alphabetically)	Definition
116	sqrt	Square root
117	std	Calculates standard deviation
118	strvcat	Concatenates strings vertically
119	sum	Calculates sum of elements together
120	tan	Tangent for angle in radians
121	tand	Tangent for angle in degrees
123	tahh	Hyperbolic tangent
124	tri2grid	Converts solution from triangular to rectangular grid
125	ttest	One and paired-sample hypothesis test
126	ttest2	Two-sample hypothesis test
127	trapz	Numerical integration with the trapezoidal rule
128	var	Sample variance
129	ver	Displays versions of the MATLAB and toolboxes
130	while	Repeats execution of command/s
131	who	Displays variables stored in the Workspace
132	whos	Displays variables with additional information about them stored in the Workspace
133	xxxxcdf	Calculates CDF for a given distribution
134	xxxxinv	Inverse CDF for a given distribution
135	xxxxpdf	Calculates PDF for a given distribution
136	xxxxrand	Generates pseudorandom numbers for a given distribution
137	xxxxstat	Calculates mean and variance for a given distribution
138	zeros	Creates an array with zero
139	ztest	One-sample Z-test

A1.4.2. Commands for Plotting, Drawing, and Interactive Tools

The commands for drawing, plotting, formatting graphs, and opening interactive tools are presented in Table 5. The table includes the commands of both the basic Matlab toolbox and the Statistics and Machine Learning Toolbox™.

Table 5. List of commands for plotting, drawing, formatting graphs, and interactive tools.

No	Command (Alphabetically)	Definition
1	axis	Controls axis scaling and appearance
2	bar	Draws vertical bars on the plot
3	barh	Draws horizontal bars on the plot
4	bar3	Generates 3D vertical bars on the plot

continued on following page

Table 5. Continued

No	Command (Alphabetically)	Definition
5	boxplot	Generates box-whiskers plot
6	capaplot	Generates capability plot
7	clabel	Labels iso-level lines
8	close	Closes one or more Figure Windows
9	colormap	Sets colors
10	controlchart	Generates control (Shewhart) chart
11	contour	Creates a 2D-contour plot
12	contour3	Creates a 3D-contour plot
13	cylinder	Generates a 3D plot with cylinder
14	distributionFitter	Opens tool for fitting distributions to data
15	disttool	Opens window for interactive distribution plotting
16	errorbar	Creates a plot with error bounded points
17	figure	Creates the Figure window.
18	fplot	Creates a 2D plot of a function
19	grid on/off	Adds/removes grid lines
20	gtext	Adds a text with the help of the mouse
21	help graph2d	Displays list of 2D graph commands
22	help graph3d	Displays list of 3D graph commands
23	help specgraph	Displays list of specialized graph commands
24	hist	Plots a histogram
25	histogram	Plots a histogram
26	hold on/off	Keeps current graph open/close
27	legend	Adds a legend to the plot
28	loglog	Generates a 2D plot with log axes
29	mesh	Creates a 3D plot with meshed surface
30	meshgrid	Creates X,Y matrices for further plotting
31	normspec	Draws normal density between specified limits
32	pareto	Generates Pareto chart
33	pdeModeler	Opens an interface for solving PDEs
34	pderect	Draws a rectangle or square
35	pdeellip	Draws an ellipse or circle
36	pdecirc	Draws a circle
37	pdesmech	Calculates tensor functions in Structural mechanics PDE application
37	pdepoly	Draws a plygon
38	pie	Creates a 2D pie plot
39	pie3	Creates a 3D pie plot
40	plot	Creates a 2D line plot
41	plot3	Creates a 3D plot with points/lines

continued on following page

Table 5. Continued

No	Command (Alphabetically)	Definition
42	plottools	Opens the plot editing tools
43	polar	Generates a plot in polar coordinates
44	probplot	Generates a comparative normal normal distribution plot
45	randtool	Opens an interface for random number generation and plotting
46	rotate3d on/off	Interactively rotates/stops rotation of a 3D plot
47	scatter	Generates a 2D scatter plot
48	semilogx	Creates a 2D plot with log-scaled x-axis
49	semilogy	Creates a 2D plot with log-scaled y-axis
50	sphere	Generates a sphere plot
51	stairs	Creates star-like plot
52	stem	Creates a 2D stem plot
53	stem3	Creates a 3D stem plot
54	subplot	Places multiple plots on the same page
55	surf	Creates a 3D surface plot
56	surfc	Generates surface and counter plots together
57	text	Adds a text to the plot
58	title	Adds a caption to the plot
59	view	Specifies a viewpoint for 3D graph
60	waterfall	Generates mesh plot without column lines
61	xlabel	Adds a label to x-axis
62	ylabel	Adds a label to y-axis
63	yyaxis left/right	Activates left/right y-axes of the plot
64	zlabel	Adds a label to z-axis

A1.4.3. Commands for Symbolic Calculations and Plotting

The commands for symbolic calculations and plotting graphs by symbolic expressions that were used in the book are presented in Table 6. The table includes the commands of both the basic Matlab toolbox and the Statistics and Machine Learning Toolbox™.

Table 6. Commands for symbolic calculations and plotting.

No	Command (Alphabetically)	Definition
1	collect	Assembles terms in an expression
2	diff	Differentiates symbolic expression
3	double	Converts symbolic numbers to their numerical form
4	dsolve	Solves symbolic ODEs
5	expand	Expands symbolic espression
6	ezplot	Generates 2D graph by symbolic expression
7	ezsurf	Generates 3D graph by symbolic expression
8	factor	Returns vector containing product of polynomials of lower degrees
9	ilaplace	Inverse Laplace transform
10	int	Integrates symbolic expression
11	laplace	Laplace transform
12	limit	Defines limit of an expression
13	piecewise	sets a math function defined by several sub-functions
14	pretty	Displays an expression in a format closing to mathematical
15	rewrite	Rewrites an expression in terms of another
16	simplify	Simplifies an expression
17	solve	Solves single/set of equation/s
18	subs	Substitutes numericvalues in an expression
19	sym	Creats symbolic object
20	syms	Creats symbolic variables or matrices
21	symvar	Searches symbolic variables in an expression
22	taylor	Taylor's polynomial approximation
23	vpa	Converts symbolic numbers to a numeric format with up to 32 digits
24	vpasolve	Solves symbolic equation/s numerically

About the Author

Leonid Burstein served as associate staff member at ORT Braude College of Engineering, Program Engineering Department. Before that he taught at the Technion -IIT, at Kinneret Academic College, Engineering School, at Haifa University, and at several other academic institutions. Following an MA in thermophysics at Lomonosov Technological Institute at Odessa, Ukraine, and a doctorate at the National Research Institute for Physical and Radio Engineering Measurements at Moscow, he obtained his PhD in physical properties of materials from the Heat/Mass Transfer Institute of the Belarus Academy of Science, Minsk, in 1974. After a short period of work in Russia and Belarus, Dr Burstein started his carrier at the Piston Ring Institute in Odessa, where he served as Head of Projects and Head of the CAD/CAM group. In 1991-2002 he worked at the Technion – IIT, Israel, at the Faculty of ME, in the QAR Program at the Faculty of IEM. Scientific works of Dr Burstein has been reported in more than 60 publications in leading scientific journals. He is author and contributor of published textbooks, monographs, and an Editorial Board member and reviewer for a number of international scientific journals.

Index

D

E

F

M

N

O

P

Q

R

S

T

U

V

W

X

Y

Z

CPSIA information can be obtained
at www.ICGtesting.com
Printed in the USA
BVHW012320110421
604581BV00003B/24

9 781799 870784